Miomir Vukobratović

Introduction to Robotics

In collaboration with
Milan Djurović, Dragan Hristić, Branko Karan,
Manja Kirćanski, Nenad Kirćanski,
Dragan Stokić, Dragoljub Vujić, Vesna Živković

With 228 Figures

Springer-Verlag Berlin Heidelberg NewYork
London Paris Tokyo 1989

Miomir Vukobratović, Ph. D., D. Sc.
Milan Djurović   B . Sc.
Dragan Hristić   Ph. D.
Branko Karan   B . Sc.
Manja Kirćanski Ph. D.
Nenad Kirćanski Ph. D.
Dragan Stokić   Ph. D.
Dragoljub Vujić Ph. D.
Vesna Živković   Ph. D.

Institute Mihajlo Pupin
YU-11000 Beograd

Based on the original Uvod u Robotiku published by
Institute Mihajlo Pupin, Beograd, Yugoslavia, 1986.

ISBN-13:978-3-642-82999-4   e-ISBN-13:978-3-642-82997-0
DOI: 10.1007/978-3-642-82997-0

Library of Congress Cataloging-in-Publication Data
Vukobratović, Miomir. Introduction to robotics. Translation of: Uvod u robotiku.
Includes index. 1. Robotics. I. Title.
TJ211.V8613 1988 629.8'92 88-10307
ISBN-13:978-3-642-82999-4 (U.S.)

2161/3020-543210 – Printed on acid-free paper

Preface

The publication of *Introduction to Robotics* is one among the results of many years' work by the authors of this book in the study of robotics. During the first stage of this work, devoted to locomotion robots and anthropomorphic mechanisms in particular, a number of papers were published in the period 1974–76. The most prominent of them is M. Vukobratović's *Legged Locomotion Systems and Anthropomorphic Mechanisms*, published in English, Russian, Japanese and Chinese.

The contributions by the associates of the Mihajlo Pupin Institute's Robotics Department in the second stage of their work in the field of robotics are included in *Scientific Fundamentals of Robotics* – a series of monographs covering the study of manipulation robots. Six volumes of this series were published in English by Springer-Verlag during 1982–85. Some of these volumes were translated into other languages. Volume 1 was translated into Japanese and published by Springer-Verlag, Volume 2 was published by Nauka in Russian (the translation of Volume 5 into Russian is finished and will be published by Mir in 1988), and all six volumes of the series were translated into Chinese. The work on writing the series for Springer-Verlag continues.

The topics treated in the books that have been published by Springer-Verlag are as follows. The first in the series provides a study of dynamics and the application of dynamic models to the dynamic calculations of robot mechanisms, and the second book studies the problems of synthesizing control laws for manipulation robots on the basis of their complete dynamic models. Volume 3 describes efficient methods for forming the kinematic models of manipulation robots and the synthesis of robot trajectories in unobstacled and obstacled operating environments. Volume 4 deals with the algorithms for efficient construction of the mathematical models of robot dynamics, that are suitable for use in nonadaptive and adaptive control of robot mechanisms as well as the analysis of the numerical complexity of various control laws intended to permit their microcomputer implementation. Volume 5 treats the synthesis of nonadaptive and adaptive control and the analysis of the numerical complexity of various control laws. Volume 6 outlines the convenience of using the mathematical models of manipulation robot dynamics in choosing different types of robot actuators, and the results concerning some specific manipulation tasks in which dynamic reactions occur on a robot gripper.

The books described above and work published by an increasing number of other authors represent adequate introduction for postgraduate robotics courses. However, it has also been necessary to create textbook material suitable for the basic robotics courses of undergraduate studies at engineering faculties. Such literature on robotics is lacking both in quantity and appropriate presentation. In addition to textbooks, books on robotics that introduce readers in a more easily understandable way to this important and propulsive field of technical science appear to be lacking as well.

In writing a book intended to introduce a technically qualified reader to the field of robotics, the authors are always faced with the risk of either over-simplifying the subject matter or providing just a condensed version of the text of a previously published monograph.

The authors of this book have attempted to present the relatively wide scope of robotics subject matter in a different way. Their intention has been to discuss all the details of robotics in such a manner as to arouse interest to this research area without facing the reader with serious difficulties in following the text, from the standpoints of classical mechanics, systems theory and computer engineering. We think that we have attained this aim which makes the whole text easy to follow by final-year students in mechanical and electrical engineering faculties as well as by students of science and mathematics specializing in mechanics, mathematics and physics, with some basic knowledge of automatic control theory.

The authors of this text, and of the chapter on robot dynamics in particular, were in a dilemma as to how to present this subject because of its exceptional importance in robotics and its role in the control and modern dynamic analysis of robot mechanisms. It is well known that more than fifteen years ago, within the Belgrade school of robotics some of the authors of this book were among the first in the world to lay the foundations of computer-oriented methods for constructing the dynamic equations of active spatial mechanisms. But, because of the character of this book and space limitations, there was some fear, that the automated algorithm for forming dynamic robot models could be presented in a way understandable to readers who are making their first steps in robotic studies. However, consistent with our opinion that the "manual" construction of robot models is now an anachronism and accepting a certain amount of risk, we decided to take this contemporary approach to the study of robot dynamics.

This book is organized into nine chapters.

Chapter 1 presents general robot characteristics concerning the classification of robot systems, the general specification of robot mechanisms and the specification of manipulation robots.

Chapter 2 is devoted to manipulation robot kinematics. It treats the kinematic structure of manipulation robots, the types of kinematic configurations, the kinematic model of manipulation robots, based on the Denavit-Hartenberg (homogeneous) coordinates. The inverse problem of kinematics is formulated and the basic principles of robot trajectory synthesis are given.

Chapter 3 contains the fundamentals of studying robot dynamics. It presents the automated construction of the dynamic equations of motion, the modelling of actuator dynamics and the effect of fundamental vibration on the robot dynamics. The class of tasks involving constrained gripper motion is also presented.

The problems of manipulation robot control are discussed in Chapter 4. Particular attention is given to the synthesis of programmed control based on the complete or partial dynamic models of robots and to the synthesis of local controllers. Some results relating to load feedback are presented and the control problem of automated mechanical assembly is considered.

Chapter 5 is devoted to the microcomputer implementation of control algorithms. It presents the basic elements of today's robot control systems as well as software modules among which communication-command module, kinematic module and dynamic (servosystem) module deserve to be especially mentioned.

Manipulation robot programming is treated in Chapter 6. Apart from manipulator motion programming, special attention is devoted to robot communication with the environment and to a short survey of programming methods.

Chapter 7 deals with the sensors used in robotics: position sensors (potentiometers, encoders, resolvers), environment sensors (force sensors, tactile sensors, proximity sensors) and vision sensors (scene illumination, special vision sensors, lasers).

The elements of industrial robot design and application are discussed in Chapter 8. Some specific features of industrial robots as mechanical constructions are presented, design solutions for the basic modules are described and numerous examples illustrating the practical application of industrial robots are given (spot and arc welding, die-casting, forging, painting, machine tool serving, etc.).

Considerable research and development work has been performed on flexible automation systems. In Chapter 9 we have tried to unify in a systematic way major problems (and possible solutions) encountered with factory automation by integrating one or more robots and several machine tools into work cell, flexible manufacturing line and assembly systems.

The authors of this book have tried to present the text in such a way as will arouse the interest of all those who wish to have a solid background for systematic robotic studies.

Beograd, Yugoslavia   M. Vukobratović
September 1987

Contents

Chapter 1

General Introduction to Robots

1.1 Dedication and Classification of Robotic Systems

Robotic systems represent in principle new technical means of complex automation production process. Handwork can be totally eliminated by their use, both in basic and auxiliary technological operations.

In contemporary production, high automation of the basic technological processes is characteristic, but the auxiliary operations are performed manually. These operations are monotonous, primitive and often strenuous, damaging, and even hazardous; with a large part of work effort still spent on these operations.

Practice has demonstrated that many auxiliary manual work operations cannot be automated by traditional means. Hence essential needs in the realization and broad application of industrial robots arose – of multi-segment mechanisms with all degrees of freedom active, i.e. powered. By action of the robot automatic control system, its manipulators often perform motions in its work activity which are similar to human arm motion.

The robot control systems are easily switched to various kinds of operations. In that way, industrial robots represent multi-purpose machines, satisfying the contemporary demands of flexible production automation and the realization of economical technology and dignified work in production complexes, mines, underwater and other environments.

The general classification of robotic systems, can be listed as:

1. **Manipulation robotic systems:**
2. **Mobile robotic systems:**
3. **Information and control robotic systems.**

Greatest development and practical application in industry has been attained by manipulation robotic systems of various types.

Mobile robotic systems are, in fact, mobile platforms, the motion of which is controlled by an automatic system. Beside the motion trajectory programme, they also possess automatic loading to the aim point and they can be loaded and unloaded automatically. In production plants they are used for automatic transport of parts and tools to the working machines and from these to the magazines. Manipulation mechanisms can be placed onto such mobile systems.

In agricultural production mobile robotic systems are automatically driven devices for field and plantation works, as are autonomous tractors. Mobile systems are also needed for work on the ocean bottom (Fig. 1.1) for serving oil and gas installations, etc.

In the case of mobile robots various motion principles are used. Motion is realized by means of wheels, mechanical legs, tracks; they can be flying (pilotless aircraft and similar) or floating (crewless submarines).

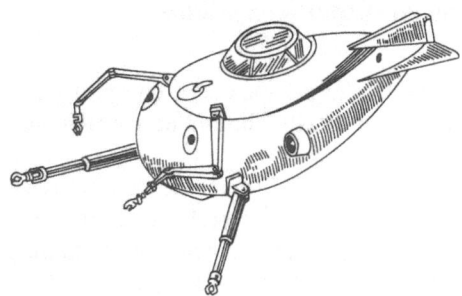

Fig. 1.1. Underwater robot

Information and control robotic systems represent complex measuring – information and controlling devices, for acquiring, processing and transferring data, as well as for use in forming various control signals.

In production plants, these are automatic control systems in unmanned production processes, complexly mechanized with industrial robots used in groups (Fig. 1.2).

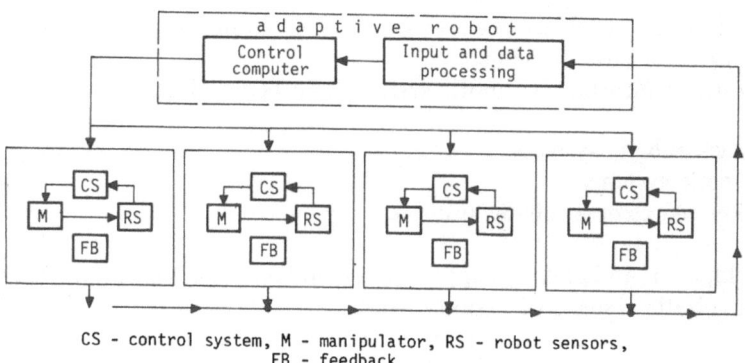

Fig. 1.2. Scheme of group robot control. CS control system; M manipulator; RS robot sensors; FB feedback

In underwater conditions, these are devices equipped by measuring-information and control systems and automatic cinematic apparatus for determining the properties of water and the sea bottom (Fig. 1.3), and for recognition of objects with automatic display of information, etc.

Fig. 1.3. Information underwater robot

Manipulation robot systems can be divided into three sorts (Fig. 1.4):

1. Automatic robots, automatic manipulators and robotized technological complexes:
2. Remote control robots, manipulators and technological complexes:
3. Manually controlled, directly linked to the motion of arms, sometimes of legs, of the human operator.

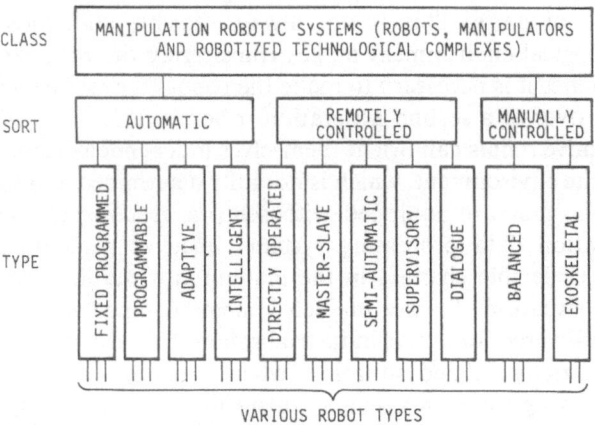

Fig. 1.4. Manipulation robots classification

Automatic robots are mainly used in industrial production (industrial robots and robotized complexes); remote control robots in hostile conditions (e.g. nuclear radiation, air contamination, explosive danger, high or low temperatures and high pressure); manually controlled robots are used for loading and unloading and heavy work.

Automatic manipulation devices are divided into four types: fixed programmes, programmable, adaptive and "intelligent". Instead of the term "type", "generation" is often used. As manipulators with fixed programmes cannot represent robots, they can be considered as the "zero" ("pre-robotic") generation. Thus the programmable robots are considered as the first, the adaptive as the second and the intelligent as the third generation. However, as opposed to computer techniques, these generations do not supersede one another, but coexist. Hence fourth generation robots do not exist and the artificial intelligence of the third generation can develop alongside the advance of science together with the possibility of using newer microcomputer generations.

Let us characterize each of these generations of automatic robotic systems.

Manipulators with fixed programmes (Fig. 1.4) do not possess a programmable control system. These are mechanical arms. They are firmly linked to the technological equipment, maintaining a certain programme of the technological process as a whole. Their application is notably characteristic for substituting manual work in mass production, e.g. on assembly lines of watch mechanisms and similar.

Programmable robots (first robot generation) have controlled drives in all joints and their control system is easily adapted to various manual operations. However, after each adjustment these robots repeat the same fixed programme in strictly defined conditions, with a pre-determined arrangement of objects. The majority of contemporary industrial robots are of such type and are used for performing auxiliary operations in pressing, welding, machine tools, casting machines etc. Such robots demand technological arrangement of the work environment and position of parts. This is not always feasible and, most important, "fixed" technological environment makes transferring the robot to new operations difficult. Hence it is necessary to make the robot control system more complex, i.e. to pass over to a second generation robot.

Second generation–adaptive robots can orient themselves independently, to a higher or lesser degree, in the environment, which is not fully determined and to which they adapt. For this, they are equipped with sensors, reacting to the situation, and an information data processing system aimed at generating adaptive control signals, i.e. flexible changes in the manipulator motion programme according to the situation. To-day in such systems compact microprocessor systems are broadly used. Adaptive industrial robots are needed when it is difficult to ensure a strictly defined situation, when avoiding obstacles, working with parts on a moving track, in assembly operations, in arc welding, painting, applying protective layers and other operations. Second generation adaptive robots are continually being developed and used in production processes.

Third generation–intelligent robots possess more varied sensors with microcomputer processing of information, recognition of situations and automatic generation of the solutions for further actions by the robot itself, aimed at performing the necessary technological operations in an undetermined environment. These are robots with elements of artificial intelligence.

As already stated, robot generations do not supersede one another. Each of them is applied, where suitable. Naturally, with development of the components the robots will become better, safer and faster.

The general scheme of an automatic robotic system is presented in Fig. 1.5. The system can possess several manipulators and units of technological equipment, as well as transfer devices. Hence the manipulators can be at various points of the technological complex and possess their individual control units, as well as a common control system.

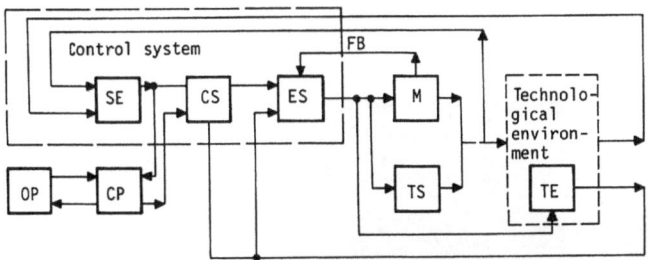

Fig. 1.5. General scheme of automatic robot system. SE sensor element; OP operator; ES executive system; CS computer system; CP command panel; FB feedback; M manipulators; TS transfer system

Figure 1.5 shows a basic organization of an automatic robotic system, containing the typical elements necessary for its functioning.

In the scheme of second and third generation robots sensors are used. With second generation robots force, tactile, location (ultra-sound) and similar sensors can be applied. With third generation robots the presence of technical vision systems is characteristic, which together with the advanced microcomputer data processing forms the artificial intelligence itself, i.e. the behaviour of the robot in this case is more complete and corresponds to a certain degree to rational human behaviour in working activity. Also, into the sensory device complex means for production quality control, as well as properties of the outer world can be included, in case this is required by the automatic control of the work regime.

The vertical lines in Fig. 1.4 show the types into which each type of robot is divided. They differ in the principles and techniques of control system design, the actuators in the manipulator joints, the number of manipulator links, load mass, sensor type, and mathematical programme support, etc. Three different types of Yugoslav industrial robots are shown in Fig. 1.6 (electromechanical), Fig. 1.7 (electropneumatical), and Fig. 1.8 (electrohydraulic).

Fig. 1.6. Electromechanical robot UMS-2

Fig. 1.7. Electrohydraulic robot UMS-3

In production processes the robot, performing some determined task, is coupled into a unique system of the corresponding technological equipment. Hence it should be regarded as an element of a complex automated technological line – the task usually not being solved by one robot but by a group of them in a unique technological system (flexible production cell or line).

It should be pointed out that, contrary to man, the robot can be placed more arbitrarily with respect to the other production equipment (on the floor, ceiling-hanging, on the wall, etc).

Analogous classification can be applied to robotized technological plants, too. They can be firmly programmed or adapted to changes in external conditions. This represents the basis for the realization of flexible automated complexes, of factories practically without personnel.

Fig. 1.8. Electropneumatical robot RPD

Remotely controlled robots and manipulators are divided into five types according to the classification scheme (Fig. 1.4): direct control manipulators, master–slave manipulators, semi-automatic manipulators, robots with supervisory control, robots with dialogue (interactive) control.

Only the last two types were denominated robots because they possess, beside remote control, fully automatic work regimes.

Direct control manipulators are defined as the human operator remotely switching on the actuator of each manipulator joint by pressing the corresponding button. With such a regime teaching of industrial robots is often performed from the command panel. The so-called teleoperators for hazardous work areas are operated analogously.

Master–slave manipulators, being in a hazardous area, are controlled remotely by the human operator from a distant safe position by means of a device which is kinematically similar to the manipulator itself (Fig. 1.9b). Therefore motion of each command manipulator joint is transferred to the corresponding joint of the working (executive) mechanism (manipulator) using the tracking system principle. Such manipulators are used in nuclear environment, contaminated atmosphere and other aggressive conditions.

Semi-automatic manipulators, compared with the master–slave, on the command panel of the operator as a command mechanism, the kinematic of which can be arbitrary to suit small movements of a human hand, use a multifunctional joy-stick. The electric signals from the joy-stick are transformed by means of a special-purpose computer (Fig. 1.9c) into the input command signals of the manipulator actuators. Different control algorithms are possible [3].

Robots with supervisory control (Fig. 1.4) are defined as having all the elements of the operations they perform being pre-programmed and generated automatically. The human operator supervising from a distance the robot work being in a hazardous zone sends only individual command signals which according to the individual programmes of the automatic robot work are then

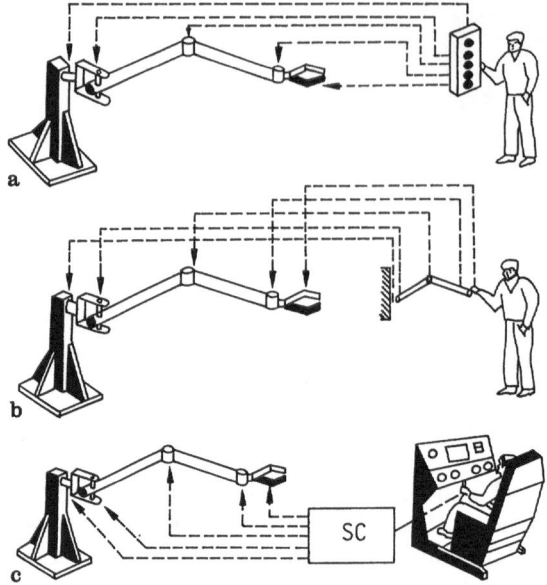

Fig. 1.9 a–c. Remotely controlled manipulator. **a** direct control, **b** master–slave control, **c** semi-automatic control. SC special-purpose computer

switched on. To the human operator only environment (scene) recognition and decision making remains. After receiving the goal command, the robot performs any further particular programme. If the robot is adaptive, the human operator can give more global commands.

It should be mentioned that robots with such control mode are used today as a real means, as special vehicles operating under specific conditions, when it is necessary to safeguard the operator.

Robots with combined control are those which the automatic regimes (similar to robots with supervisory control) are combined with manual-control regimes (as with semi-automatic or master–slave manipulators). They are applied in the case of underwater pilotless devices, in an explosive environment, in mines without human workers, nuclear power plants, etc. Such combined control is used in the case of various teleoperator types.

Robots with dialogue (interactive) control generally (but not obligatorily) are intelligent and differ from the supervisory ones by the fact that they receive not only human commands, but intervene actively themselves in situation recognition and decision making, helping the human operator [3].

Function schemes of the supervisory and dialogue (interactive) robotic systems is presented in Fig. 1.10. If necessary, a joy-stick is used on the command panel to enable the operator to take over the control of manipulator motion in a semi-automatic or master–slave regime. Such safety methods by means of combinations of various remote control principles are necessary, because man himself cannot enter the hazardous zone in which the robot is situated. Such robot functioning scheme represents the most complex control system and offers

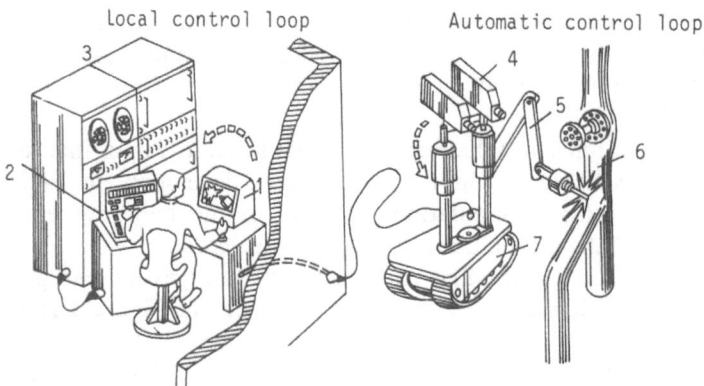

Local control loop   Automatic control loop

Fig. 1.10. Functional scheme of interactive control systems of robots (supervisory, dialogue, combined). *1* display; *2* operator's panel; *3* control computer; *4* sensors; *5* manipulators; *6* workpiece; *7* robotic system computer

the broadest possibilities for introducing adaptive mobile systems in the most diversified working conditions.

Finally, the third sort of (manual) robotic manipulation systems divides into hinged-balancing and exoskeletal (power amplifiers of human extremities).

The hinged-balancing manipulator (Fig. 1.11) represents a multi-segment mechanism with drives in some of the joints, which is in an equilibrium state with arbitrary load and configuration (within its possibilities). Hence the operator can move a big load easily. By moving the control stick, the operator generates control signals, whereby all load transferring work is done by the drives in the manipulator joints. Such manipulation systems are suitable for loading and unloading operations with heavier loads [3].

Exoskeletons are multi-segment mechanisms, the segments of which are directly connected with the segments of the human arm (Fig. 1.12) or human

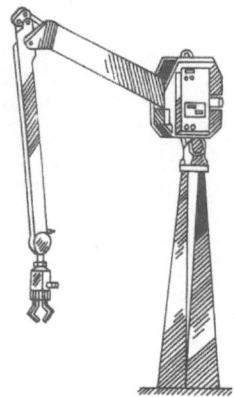

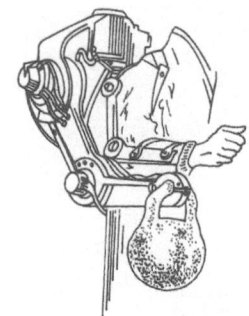

Fig. 1.11. Hinged-balancing manipulator   **Fig. 1.12.** Arm exoskeleton

legs. In the exoskeleton mechanism joints are also controlled drives, doing all the work. Motions of the human extremity only form the control signals. Such systems can amplify the power of human extremities (and of his body), both healthy and handicapped.

As a result of research into rehabilitation in the Mihailo Pupin Institute in Beograd, the first exoskeletons in the world were developed, to establish locomotor and later also manipulation activities of the handicapped. The following are a few of the possibilities of active exoskeletons.

Figure 1.13 shows one of the first exoskeleton prototypes for paraplegics, Fig. 1.14 shows a complete electrically driven exoskeleton. Figure 1.15 dem-

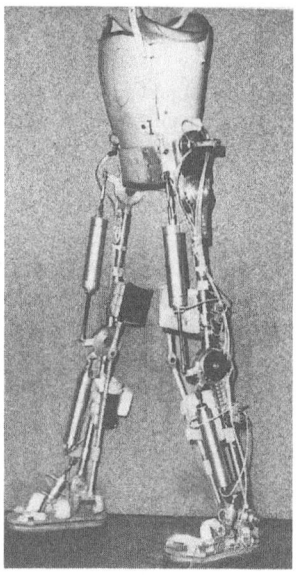

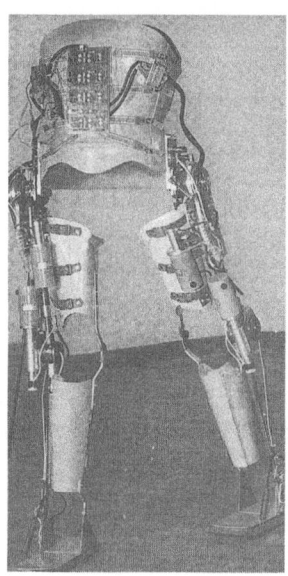

Fig. 1.13. Pneumatic exoskeleton for paraplegics (1971)

Fig. 1.14. Electrical exoskeleton for paraplegics (1974)

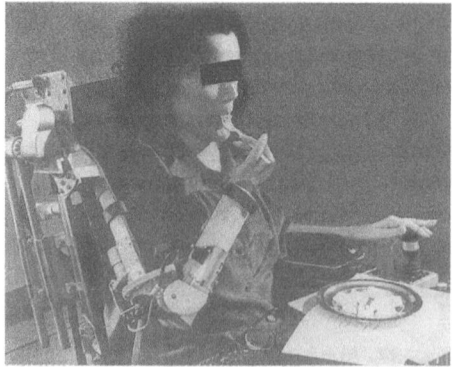

Fig. 1.15. Electronic arm for dystrophics (1981)

onstrates the exoskeleton type electronic arm for dystrophics in an advanced phase of the illness.

1.2   General Features of Robotic Mechanisms and Their Classification

From the point of view of mechanism theory, active mechanisms in robotics are complex kinematic chains of variable structure, having a great number of members, some of which can be of variable length, with controlled degrees of freedom.

Regarding control theory, they are complex, non-linear, multi-variable dynamic systems. Active mechanisms can be divided, according to the number of kinematic chains, into:

– simple (consisting of a single kinematic chain);
– complex (comprising a number of simple chains).

According to their form, simple kinematic chains may be open or closed. Complex chains may be classified as:

– branched (comprising only simple open chains);
– combined (comprising both open and closed chains).

Depending on the kinematic constraints imposed on their end members, active mechanisms may be divided into:

– free or open;
– connected or closed (connected by kinematic pairs to the fixed base).

Members of active mechanisms are interconnected by means of kinematic pairs. There is no difference between kinematic pairs of active and "classical" spatial mechanisms. Execution of kinematic pairs of both mechanisms classes are practically identical, except for the differences created by actuators mounted in the mechanism joints.

The class of a kinematic pair is determined by the number of constraining conditions on the connections concerning the free relative motion of its members. In the table shown in Fig. 1.16 kinematic pairs are arranged into five classes according to the number of the members of the number relative motion, d.o.f. Kinematic pairs in the fifth class have one d.o.f. and the pairs in the first class have five d.o.f. of relative motion. Besides being partitioned into classes, kinematic pairs are divided into types, depending on the number of relative rotations within the scope of the total number of d.o.f. in the joint. Pairs of the first type allow the maximal number (three) of relative rotations, pairs of the

class of task	nr. of links	nr. of d.o.f.	TYPES OF PAIRS					
			I			II		
I	1	5	nr. of movem.	rot. 3	lin. 2			
			allowed	3	2			
			restricted	0	1			

class of task	nr. of links	nr. of d.o.f.	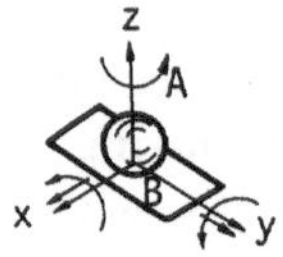 I			II			III		
II	2	4	nr. of movem.	rot.	lin.	nr. of movem.	rot.	lin.			
			allowed	3	1	allowed	2	2			
			restricted 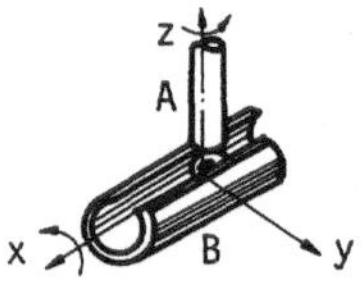	0	2	restricted 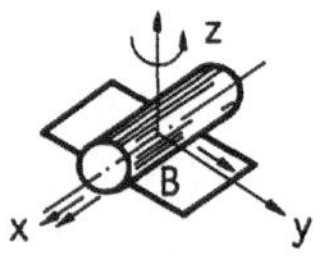	1	1			
III	3	3	nr. of movem.	rot.	lin.	nr. of movem.	rot.	lin.	nr. of movem.	rot.	lin.
			allowed	3	0	allowed	2	1	allowed	1	2
			restricted 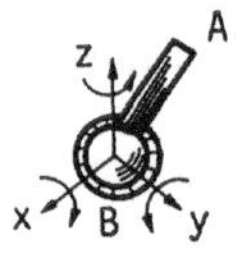	0	3	restricted 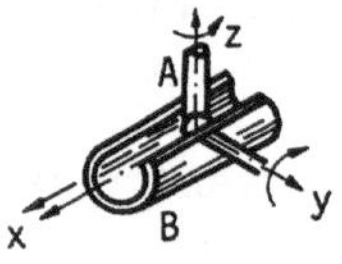	1	2	restricted 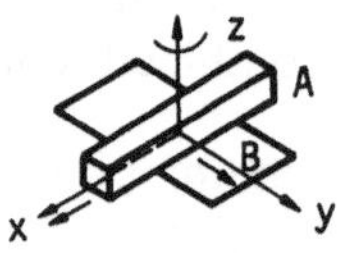	2	1
IV	4	2	nr. of movem.	rot.	lin.	nr. of movem.	rot.	lin.			
			allowed	2	0	allowed	1	1			
			restricted 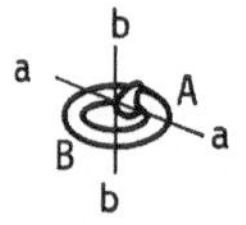	1	3	restricted 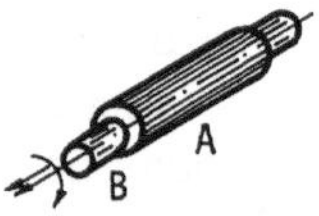	2	2			
V	5	1	nr. of movem.	rot.	lin.	nr. of movem.	rot.	lin.			
			allowed	1	0	allowed	0	1			
			restricted 	2	3	restricted 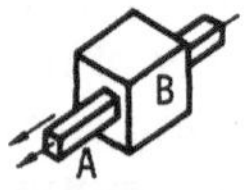	3	2			

Fig. 1.16. Table of kinematic pairs

second type two rotations, and pairs of the third type only one relative rotational motion. As well as the pairs, in which relative motions of members are mutually independent, there are pairs with interconnected motion. The simplest example is the screw-nut kinematic pair, in which the linear and rotational motions are linearly dependent [4].

In kinematic schemes, symbolic presentation for various kinematic pairs is adopted (Fig. 1.18). Kinematic pairs form kinematic chains. One example of spatial mechanism of the steering wheel is presented in Fig. 1.17. The mechanism consists of four kinematic pairs, two each entering into two kinematic pairs only. One part of mechanism (in the figure, member 1) is fixed. This is usually designated the basis. The kinematic chain presented in Fig. 1.17 represents a simple closed chain. In a closed chain, each member enters in two kinematic pairs [4].

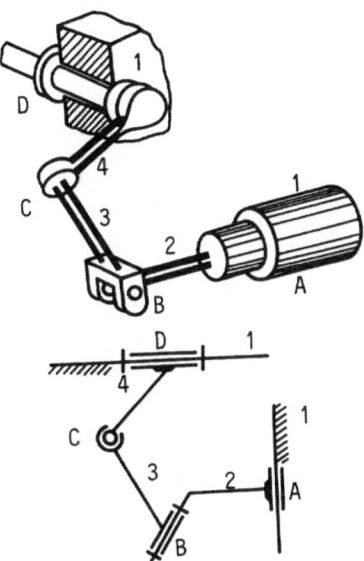

Fig. 1.17. Steering wheel mechanism

Another example of spatial mechanism is presented in Fig. 1.19. This is a mechanism of a five degrees-of-freedom manipulator. The chain consists of six kinematic pairs; only the last member, differing from the previous example, enters into one kinematic pair. This is an example of a simple kinematic chain.

In the examples considered, each mechanism member enters into one kinematic pair only. However, mechanisms are known in which one member can belong to several pairs. In Fig. 1.20, the scheme of an anthropomorphic robot (exoskeleton) is presented. The model contains twelve moving members connected into III class pairs of the I type. With such a mechanism members 4 and 8 form the robot body and enter into each of the kinematic pairs [4].

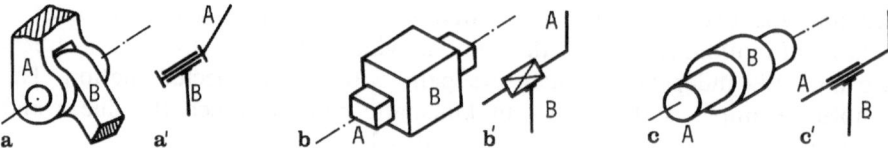

Fig. 1.18 a–c. Presentation of kinematic pairs

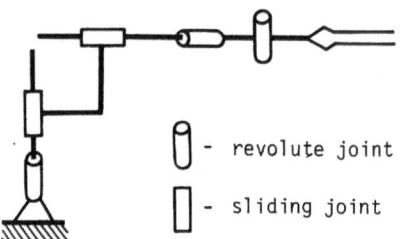

$\bigcap$ – revolute joint

$\bigsqcup$ – sliding joint

Fig. 1.19. One mechanism of the "telescopic" manipulator

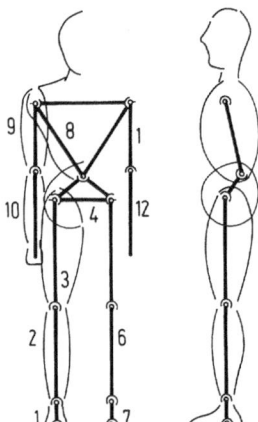

Fig. 1.20. Scheme of the anthropomorphic mechanism

In the theory of machines and mechanisms, kinematic chains are classified as simple or complex, complex chains being formed by several simple ones. Simple kinematic chains can be open or closed. In a closed chain, each member enters into two kinematic pairs, while in an open chain, the last member enters into one kinematic pair only. With complex kinematic chains, the individual members enter into three or more kinematic pairs. Here the notion and properties of open and closed kinematic chains should be examined more closely.

In the literature on active mechanisms, neither the notion nor the conditions of the closed (open) state of the open and closed chain is discussed. The kinematic chain is closed when its terminal members are connected by means of kinematic pairs to one (or more) member(s), which can be fixed (support), a member of another kinematic chain or a member of the initial chain. It should also be emphasized that in the course of working the active mechanism chains (either of manipulators or locomotion machines) change their configuration once or several times from open to closed or vice versa. The kinematic chain of the manipulator during its motion through the working space is open but during execution of the operation itself (e.g. insertion or screwing in) it becomes closed. The mechanism of the locomotion biped is open during the swing phase of the step (when one foot is not on the ground) but in the double support phase becomes closed (Fig. 1.21). However, during the single support phase the anthropomorphic mechanism can also possess two configurations. The "foot" can rotate around its edges (Fig. 1.22a, b). The corresponding kinematic schemes are given in Fig. 1.22c, d. As can be seen, when the foot is supported alternately on one and then the other foot edge, the position of hinge "O" changes abruptly.

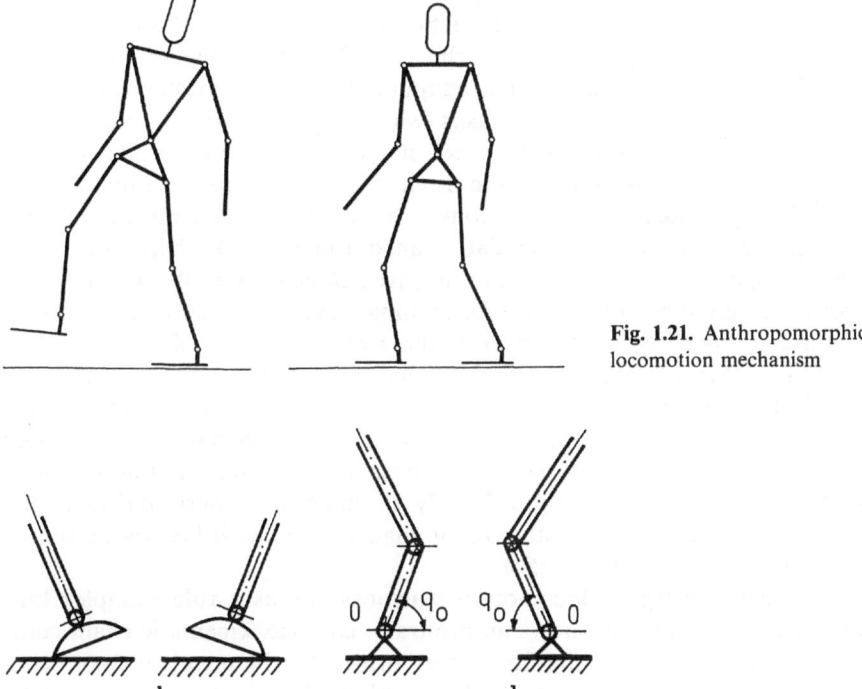

Fig. 1.21. Anthropomorphic locomotion mechanism

Fig. 1.22 a–d. Schematic of the uncontrollable d.o.f. of foot

In addition, joint O, because of constant changes in its position, cannot be equipped by a corresponding drive (actuator). On the other hand, the change of the coordinate q_0 is important because with greater values of q_0 the system becomes statically unstable (it overturns). This feature of the anthropomorphic mechanism creates a special control problem because the uncontrollable d.o.f. must be controlled by means of the other d.o.f. Such an anthropomorphic mechanism also has the corresponding kinematic constraints at each joint, thereby imitating the human state. In addition, the mechanism of the loco-motion biped (differing from the manipulator) is connected to the support surface by frictional force only. Thus, the mechanism of an exoskeleton dem-onstrates a variable structure, the presence of an uncontrollable d.o.f., kinematic constraints and an essential influence of the frictional force.

The variable structure of active mechanism chains presents an essential difference from the classical spatial mechanisms, the structure of which does not change during work. A second difference is in the number of d.o.f. With spatial mechanism with a driving member, the number of d.o.f. is rarely greater than two and by synchronizing the motion of the working members, the execution of the working operation is achieved in advance. The number of d.o.f. of active mechanisms is noticeably greater (up to ten and even more), so drives in the joints are indispensable. Only by the action of the torques and forces of these motors (during working operation) is the desired motion achieved. On the one hand this allows for an exceptional adaptability of these mechanisms to the working environment and various tasks (which in the case of automation type mechanisms is impossible); on the other, it imposes exceptional difficulties in realizing control because some of the mechanism d.o.f. appears redundant.

As already stated, robots demonstrate active mechanisms of variable structure. For instance, one manipulator can, during its work, change the group to which it would belong according to the given classification. We illustrate this change of structure by considering the example of an industrial manipulator in the course of inserting a cylindrical working object into a hole (Figs. 1.23 to 1.25). At first (Fig. 1.23a) the manipulator has an open kinematic scheme as in Fig. 1.23b (simple open chain). In the phase of transferring the working object (Fig. 1.24a) the kinematic chain does not change (Fig. 1.24b) but the last member (now the gripper and object together) changes its dimensions and mass, which cause the dynamics to change too. Finally, during object insertion (Fig. 1.25a), the kinematic scheme of the manipulator changes too and it becomes a simple closed kinematic chain (Fig. 1.25b).

Mechanisms of legged locomotion machines are, as a rule, complex kin-ematic chains. Figure 1.26 shows an arbitrary, complex kinematic chain com-prising four simple chains, the first three (formed by the members 1–6) being closed and the fourth (formed by the members 7 and 8) being open. The kinematic chains connected to the support are basic chains, while the chains connected to them, but not by means of the support, are satellite chains (satellites). With this the procedure for separating one complex chain into a number of simple ones is defined. The notion of an independent kinematic chain can be defined too. A kinematic chain is said to be independent if its motion with respect to the support is independent of the satellite chains, namely, if its last members are connected to the support. This means that a basic chain is

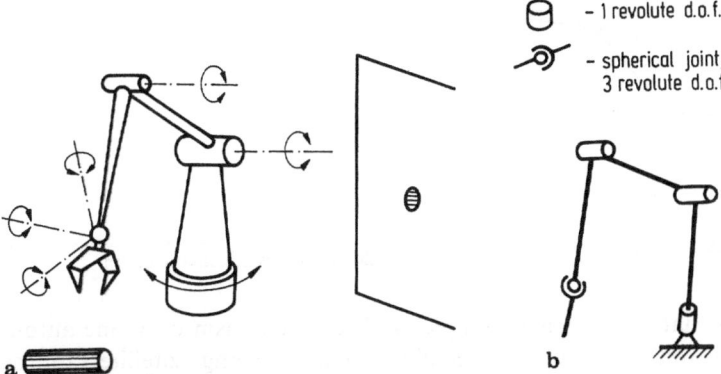

Fig. 1.23 a, b. Manipulator before grasping the object

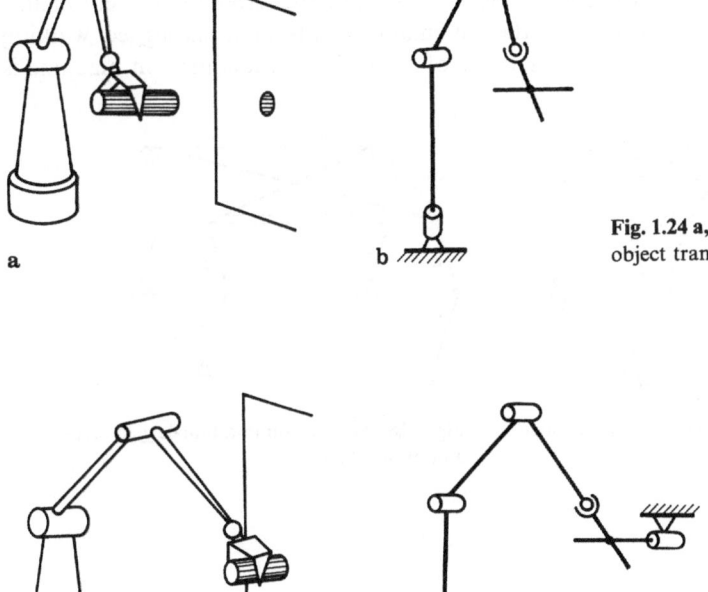

Fig. 1.24 a, b. Phase of working object transfer

Fig. 1.25 a, b. Phase of object insertion into the hole

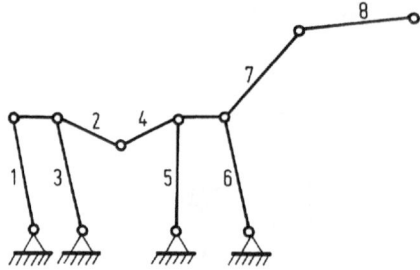

Fig. 1.26. Complex kinematic chain

independent. In that way in each complex linked mechanism only one autonomous chain is obtained, all the remaining chains being satellites to the autonomous one. Taking the model of a human presented in Fig. 1.20 as an example, it is possible to isolate the chains of "legs", "body" and "arms"; here, the chain consisting of "legs" is independent since the chains of "body" and "arms" impose no kinematic constraints on it. For the mechanism presented in Fig. 1.26, if the motion of member 1 is known, the chain I is independent, with all the remaining ones being satellites. The sequence to be followed in performing kinematic and dynamic analyses has thus been determined; namely, the basic independent chains should be analysed first, and the satellite (guided) chains second. Figure 1.27 illustrates the kinematic chains of a six-legged walking machine, while in Fig. 1.28 the scheme of a four-legged locomotion machine is presented.

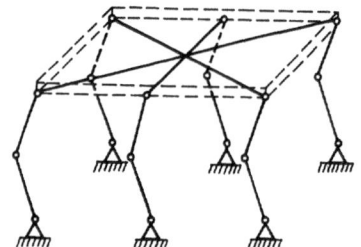

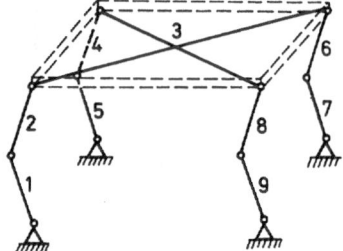

Fig. 1.27. Mechanism of a six-legged locomotion machine

Fig. 1.28. Mechanism of a four-legged locomotion machine

References

[1] Vukobratović, M., "Applied Dynamics of Manipulation Robots", Springer-Verlag, 1988.
[2] Vukobratović, M., Potkonjak V., "Applied Dynamic and CAD of Manipulation Robots", Monograph, Springer-Verlag, 1985.
[3] *Robototechnika* (in Russian), edited by E. P. Popov and E. I. Yureevitch, Mashinostroenie, Moscow, 1984.
[4] Stepanenko, Y., "Dynamics of Spatial Mechanisms", (in Russian), Mathematical Institute, Beograd, 1974.

Chapter 2
Manipulator Kinematic Model

2.1 Introduction

Kinematic modelling of manipulators plays an important role in contemporary robot control. It describes the relationship between robot end-effector position and orientation in space and manipulator joint angles. It also describes the correlation between linear and angular velocities of the end-effector and joint velocities. Since kinematic modelling is an inevitable step in modern robot control, in this chapter we will consider the main principles of manipulator kinematic model generation.

2.2 Definitions

In this section we will introduce some basic notations and definitions relevant to manipulator kinematics formulation. We will be concerned with the manipulator structure, link, kinematic pair, kinematic chain, joint coordinates, world coordinates, direct and inverse kinematic problems and redundancy.

Let us consider the manipulator shown in Fig. 2.1. It consists of n rigid bodies interconnected by joints. The joints are either revolute or prismatic

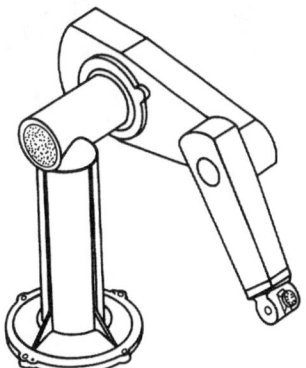

Fig. 2.1. Six degree-of-freedom manipulator

(sliding). Revolute joints enable rotational motion of one link with respect to the other (Figure 2.2). In prismatic joints the motion is translational (Figure 2.3).

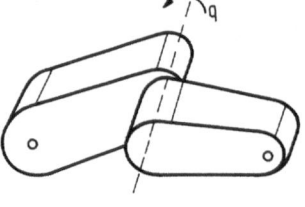

Fig. 2.2. Revolute joint

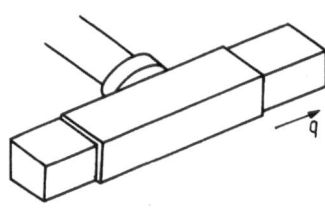

Fig. 2.3. Prismatic joint

2.2.1 Mechanical Structure

The mechanical structure of the mechanism depends on the type of the joints that are included in the given robot and the disposition of the joints. For example, the robot shown in Fig. 2.1 has six revolute degrees of freedom, with the second and the third always being in parallel (anthropomorphic manipulator).

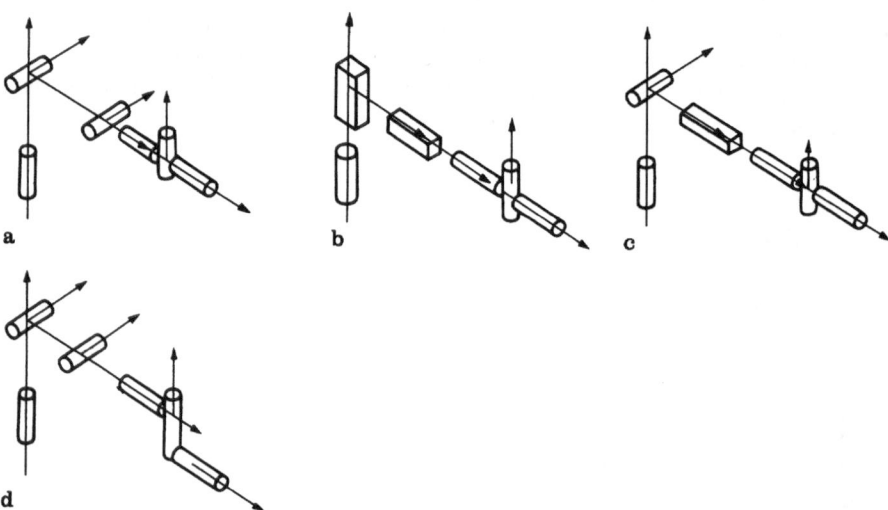

Fig. 2.4 a–d. Schematic representation of typical mechanical structures of robots

Different schematic representations have been introduced in order to describe manipulator configurations simply. Figure 2.4(a)–(d) shows several types of robots. Here, revolute joints are denoted by cylinders, while prismatic joints are represented by parallelepipeds.

2.2.2 Link

Manipulator link is a rigid body described by its kinematic and dynamic parameters. Figure 2.5 shows one typical manipulator link. Kinematic parameters describing the link may be defined in different ways. Generally speaking, these parameters are the length of the manipulator link, the angle between the axes lying at two ends of the link, etc. Precise definitions of link parameters for the Denavit–Hartenberg kinematic notation will be presented in Sect. 2.3.

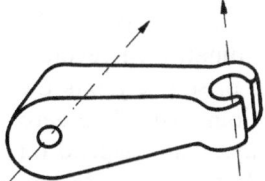

Fig. 2.5. Manipulator link

Dynamic parameters describing the link are its mass, the location of its centre of mass and the inertia tensor. These parameters are used in dynamic modelling of the mechanism.

2.2.3 Kinematic Pair

A kinematic pair represents a set of two adjacent links interconnected by a joint. We will consider only kinematic pairs that have single degree-of-freedom rotation or translation.

2.2.4 Kinematic Chain

A kinematic chain is a set of n interconnected kinematic pairs. According to the structure of connections, chains are classified into simple, complex, open and closed.

A chain in which no link enters more than two kinematic pairs is said to be a simple kinematic pair (Fig. 2.6). On the other hand, a complex kinematic chain contains at least one link that enters more than two kinematic pairs.

An open kinematic chain has at least one link that belongs to one kinematic pair only. If each link enters into at least two kinematic pairs, the chain is said to be closed.

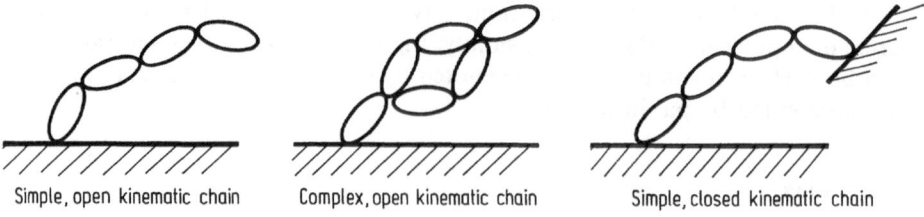

Simple, open kinematic chain   Complex, open kinematic chain   Simple, closed kinematic chain

Fig. 2.6. Various types of kinematic chains

Here we will consider only simple, open kinematic chains. The closed kinematic chains will be discussed in Chap. 3 which is concerned only with constrained end-effector motion.

2.2.5 Joint Coordinates

Scalar quantities that determine the relative disposition of the links of the kinematic pair are referred to as manipulator joint coordinates. In revolute joints, the joint coordinate is the rotation angle, while in prismatic joints the joint coordinate represents the displacement along the joint axis. The zero values of joint coordinates may be chosen in different ways. They depend on how the link coordinate frames are attached to the links. Joint coordinates are usually denoted by $q_1, q_2, \ldots, q_n$. They form the vector of joint coordinates $q = [q_1 \, q_2 \, \ldots \, q_n]^\mathsf{T}$. Each joint coordinate q_i, $1 \leqslant i \leqslant n$ varies in the range defined by mechanical constraints $q_{i\,\min} \leqslant q_i \leqslant q_{i\,\max}$, where $q_{i\,\min}$ and $q_{i\,\max}$ denote minimal and maximal value of joint coordinate q_i, respectively. Joint coordinate space $Q \subset R^n$, is the set of all joint coordinate vectors when they vary between their physical constraints, $Q = \{q : q_{i\,\min} \leqslant q_i \leqslant q_{i\,\max}\}$.

2.2.6 External Coordinates

External (operational, world) coordinates describe manipulator end-effector position and orientation with respect to some reference coordinate system. The reference system is chosen to suit a particular application. Most frequently, a fixed coordinate frame attached to manipulator base is considered as the reference system. The manipulator hand position is usually described by Cartesian coordinates x, y and z (Fig. 2.7), although sometimes cylindrical or spherical coordinates are used (if it is convenient to describe the manipulation task in these coordinates).

Precise execution of manipulation tasks requires the control not only of the manipulator hand position, but also of hand orientation with the objects located in the manipulator work space. The orientation is usually specified by Euler angles between the coordinate frame attached to the last (n-th) link and the reference system (Fig. 2.7). We will consider the yaw angle ψ, the pitch angle θ

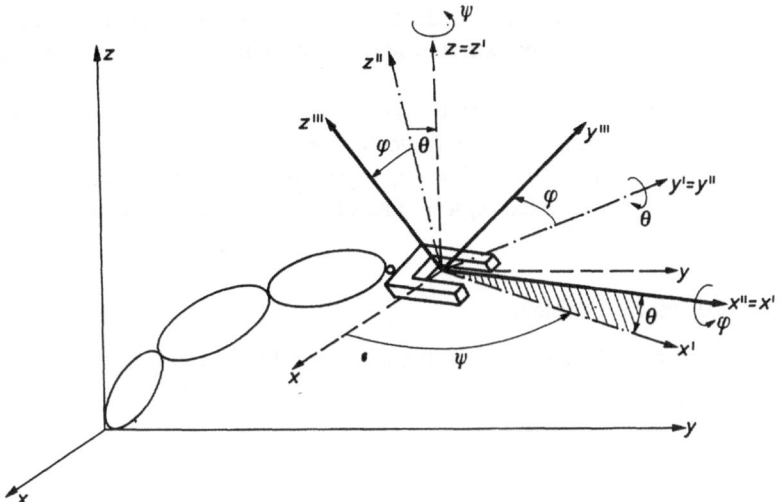

Fig. 2.7. World coordinates

and the roll angle φ. The yaw angle corresponds to a rotation ψ about the z axis, pitch corresponds to a rotation θ about the new y axis, and roll corresponds to a rotation φ about the new x axis.

In the general case, the vector of external coordinates x_e is an m-dimensional vector, where m is the number of external coordinates required to specify a certain class of manipulation tasks, $m \leqslant n$. However, most frequently m equals 6, i.e. the external coordinates vector describes the manipulator position and orientation completely:

$$x_e = [x \quad y \quad z \quad \psi \quad \theta \quad \varphi]^T \tag{2.1}$$

If a certain class of tasks can be specified by a fewer number of coordinates, the vector of external coordinates is obtained by simply rejecting some coordinates. For example, if only the positioning task is required the external coordinates vector is $x_e = [x \quad y \quad z]^T$.

2.2.7 Direct Kinematic Problem

The external coordinates x_{ei}, $i = 1, \ldots, m$ change their values when the joint coordinates q_i, $i = 1, \ldots, n$ are changed. This relationship may be described by the equation

$$x_e = f(q) \tag{2.2}$$

where $f: R^n \rightarrow R^m$ is a nonlinear, continuous, differentiable function that maps joint coordinates into external coordinates. The problem of evaluating the external coordinates, given the joint coordinates is known as the direct kinematic problem. The solution of the direct kinematic problem will be discussed in more detail in Sect. 2.3.

The external coordinates space is the set of external coordinates vectors corresponding to all the joint coordinate vectors when joint coordinates vary between their physical constraints, $X_e = \{x_e : x_e = f(q),\ q \in Q\} \subset R^m$. The space X_e is at the same time a generalized region of reachable manipulator work space.

2.2.8 Inverse Kinematic Problem

Evaluating joint coordinates, given a vector of external coordinates, i.e. solving the equation

$$q = f^{-1}(x_e) \tag{2.3}$$

is known as the inverse kinematic problem. This problem is far more complex than the direct kinematic problem, since it is equivalent to obtaining solutions to a set of nonlinear trigonometric equations. Solving the inverse kinematics is required when the manipulator hand path is specified in external coordinates, and the corresponding evolution of joint coordinates is needed. This problem will be discussed in more detail in Sect. 2.4.

2.2.9 Redundancy

A manipulator is regarded as non-redundant in a certain class of manipulation tasks specified in the external coordinate space $X_e \subset R^m$, if the dimension m of the external coordinates vector is equal to the number of degrees of freedom n, i.e. $m = \dim x_e = \dim q = n$. If $n > m$ is valid, i.e. the number of degrees of freedom is greater than the number of external coordinates, the manipulator is redundant with respect to the manipulation task. In that case there exists an unlimited number of joint coordinate vectors that correspond to the same position and orientation of the hand.

Solving the inverse kinematic problem for redundant manipulators requires special methods which are not used for non-redundant manipulators, since one solution of the inverse kinematics has to be chosen from the unlimited number of solutions. Thus, the motion synthesis for redundant manipulators is numerically much more complex than for non-redundant manipulators. On the other hand, redundant manipulators are required in certain tasks, such as obstacle avoidance. Practically, only non-redundant manipulators are used in industrial practice today. Therefore, we will consider only inverse kinematics solution and motion synthesis for nonredundant manipulators in this book (various methods for redundant manipulator motion synthesis are discussed in [1]).

2.3   Direct Kinematic Problem

Various objects, tools and obstacles are located in the manipulator work space. Their positions can easily be described in relation to a reference coordinate frame. In order to determine the relative position of the manipulator hand with respect to these objects, given the joint coordinates, the direct kinematic problem has to be solved.

As we have already mentioned, the relationship between joint and world coordinates is non-linear. This means that a linear change of joint angles results in a hand path which is not a straight line, neither is the orientation linear. This non-linearity directly follows from the fact that the function relating between the joint and external coordinates consists of the products and sums of trigonometric functions.

Let us now consider how the manipulator kinematic model can be formed.

2.3.1   Manipulator Hand Position

There exist several methods of forming the manipulator kinematic model. The most widely accepted is the Denavit–Hartenberg kinematic modelling technique [2, 3] which is described here.

Local coordinate systems have to be attached to each link in order to model a manipulator kinematically. Before considering how the link coordinate frames are attached to the link by this method, we will introduce homogeneous transformation matrices which will be used here.

2.3.1.1   Homogeneous Transformation Matrices

Homogeneous transformation matrices are 4×4 matrices containing information both on the rotation between two coordinate frames and on the distance between their origins. The purpose of introducing these matrices is to enable more compact writing of the position vectors expressed in different coordinate frames.

Let us consider two coordinate frames $0_{i-1} x_{i-1} y_{i-1} z_{i-1}$ and $0_i x_i y_i z_i$ shown in Fig. 2.8. The position vector $^{i-1}\vec{r}_A$ of the point A with respect to system

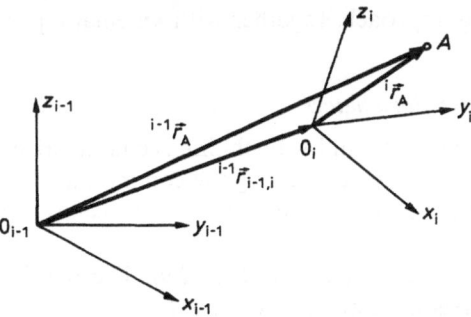

Fig. 2.8. Positon vectors

$(i-1)$ can be described by the following equation

$$^{i-1}\vec{r}_A = {}^{i-1}A_i \, {}^i\vec{r}_A + {}^{i-1}\vec{r}_{i-1,i}$$ (2.4)

where $^{i-1}A_i$ is the rotation matrix between systems i and $(i-1)$, $^i\vec{r}_A$ is the position vector of point A with respect to system i, $^{i-1}\vec{r}_{i-1,i}$ is the distance vector between the origins of systems i and $(i-1)$ expressed with respect to system $(i-1)$. The corresponding homogeneous transformation matrix between systems i and $(i-1)$ has the form

$$^{i-1}T_i = \left[\begin{array}{c|c} ^{i-1}A_i & ^{i-1}\vec{r}_{i-1,i} \\ \hline 0 \quad 0 \quad 0 & 1 \end{array} \right]$$ (2.5)

Equation (2.4), i.e. the equation

$$\begin{bmatrix} ^{i-1}r_{Ax} \\ ^{i-1}r_{Ay} \\ ^{i-1}r_{Az} \end{bmatrix} = {}^{i-1}A_i \begin{bmatrix} ^i r_{Ax} \\ ^i r_{Ay} \\ ^i r_{Az} \end{bmatrix} + \begin{bmatrix} ^{i-1}r_{i-1,ix} \\ ^{i-1}r_{i-1,iy} \\ ^{i-1}r_{i-1,iz} \end{bmatrix}$$

can be written in a simpler form using the homogeneous transformation (2.5)

$$^{i-1}\vec{r}_A = {}^{i-1}T_i \, {}^i\vec{r}_A$$ (2.6)

$$\begin{bmatrix} ^{i-1}r_{Ax} \\ ^{i-1}r_{Ay} \\ ^{i-1}r_{Az} \\ 1 \end{bmatrix} = \left[\begin{array}{c|c} ^{i-1}A_i & \begin{array}{c} ^{i-1}r_{i-1,ix} \\ ^{i-1}r_{i-1,iy} \\ ^{i-1}r_{i-1,iz} \end{array} \\ \hline 0 \quad 0 \quad 0 & 1 \end{array} \right] \begin{bmatrix} ^i r_{Ax} \\ ^i r_{Ay} \\ ^i r_{Az} \\ 1 \end{bmatrix}$$ (2.7)

Thus, if there exist several coordinate frames, the position vector of a point is expressed only by means of products of homogeneous transformations, instead of successive products and sums as in Eq. (2.4).

Let us now consider how the link coordinate frames are assigned to the links according to the Denavit–Hartenberg approach, together with kinematic parameters which are introduced here.

2.3.1.2   Denavit–Hartenberg Kinematic Parameters

Consider a simple open kinematic chain with n links. Each link is characterized by two dimensions: the common normal distance a_i (along the common normal between axes of joint i and $(i+1)$), and the twist angle α_i between these axes in the plane perpendicular to a_i.

Each joint axis has two normals to it: a_{i-1} and a_i (Fig. 2.9). The relative position of these normals along the axis of joint i is given by d_i.

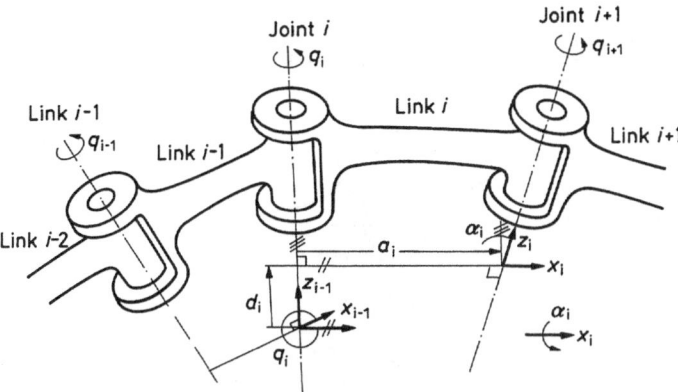

Fig. 2.9. Denavit–Hartenberg kinematic parameters for a revolute kinematic pair

Denote by $0_i x_i y_i z_i$ the local coordinate system assigned to link i. We will first consider revolute joints (Fig. 2.9). The origin of the coordinate frame of link i is set to be at the intersection of the common normal between the axis of joint $(i+1)$ and i, and the axis of joint i. In the case of intersecting joint axes, the origin is set to be at the point of intersection of the joint axes. If the axes are parallel, the origin is chosen to make the joint distance zero for the next link whose coordinate origin is defined. The axes of the link coordinate system $0_i x_i y_i z_i$ are to be selected in the following way. The $\vec{z}_i$ axis of system i should coincide with the axis of joint $(i+1)$, about which rotation q_{i+1} is performed. The $\vec{x}_i$ axis will be aligned with any common normal which exists (usually the normal between axes of joint i and $(i+1)$) and is directed from joint i to joint $(i+1)$. In the case of intersecting joint axes, the $\vec{x}_i$ axis is chosen to be parallel or antiparallel to the vector-cross product $\vec{z}_{i-1} \times \vec{z}_i$. The $\vec{y}_i$ axis satisfies $\vec{x}_i \times \vec{y}_i = \vec{z}_i$.

The joint coordinate q_i for a revolute joint is now defined as the angle between axes $\vec{x}_{i-1}$ and $\vec{x}_i$ (Fig. 2.9). It is zero when these axes are parallel and have the same direction. The twist angle α_i is measured from axis $\vec{z}_{i-1}$ to $\vec{z}_i$, i.e. as a rotation about $\vec{x}_i$ axis.

Let us now consider the prismatic joint (Fig. 2.10). Here, the distance d_i becomes joint variable q_i, while parameter a_i has no meaning and it is set to zero. The origin of the coordinate system corresponding to the sliding joint i is chosen to coincide with the next defined link origin. The $\vec{z}_i$ axis is aligned with the axis of joint $(i+1)$. The $\vec{x}_i$ axis is chosen to be parallel or antiparallel to vector $\vec{z}_{i-1} \times \vec{z}_i$. Zero joint coordinate $q_i = d_i$ is defined when the origin of systems $0_{i-1} x_{i-1} y_{i-1} z_{i-1}$ and $0_i x_i y_i z_i$ coincide. For a prismatic joint, angle θ_i between axis $\vec{x}_{i-1}$ and $\vec{x}_i$ is fixed and represents a kinematic parameter together with the twist angle α_i.

The origin of the reference coordinate system $0xyz$ is set to be coincident with the origin of the first link. Zero value of joint coordinate q_1 occurs when axes $\vec{x}$ and $\vec{x}_1$ are parallel and have the same directions.

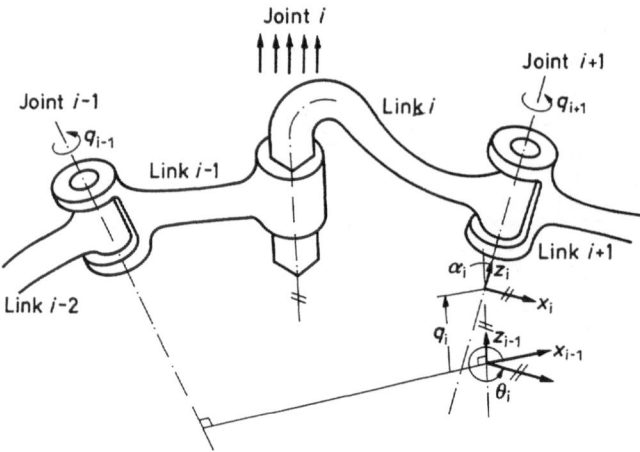

Fig. 2.10. Denavit–Hartenberg parameters for a prismatic joint

Let us now determine transformation matrices between link coordinate systems and distance vectors between their origins. As we have already said, the homogeneous transformation matrices involve both the information in the form of 4×4 matrices (see Eq. (2.5)).

The transformation of system $(i-1)$ into system i can be described by the following set of rotations and translations: rotation about $\vec{z}_{i-1}$ by angle q_i (for a sliding joint by θ_i), translation along $\vec{z}_{i-1}$ for d_i (q_i–for a sliding joint), translation along rotated $\vec{x}_{i-1} = \vec{x}_i$ for a_i and rotation about $\vec{x}_i$ for the twist angle α_i. This sequence of rotations and translations can be presented as a product of the following homogeneous transformation matrices

$$
{}^{i-1}T_i = \begin{bmatrix} \cos q_i & -\sin q_i & 0 & 0 \\ \sin q_i & \cos q_i & 0 & 0 \\ 0 & 0 & 1 & 0 \\ 0 & 0 & 0 & 1 \end{bmatrix} \begin{bmatrix} 1 & 0 & 0 & a_i \\ 0 & 1 & 0 & 0 \\ 0 & 0 & 1 & d_i \\ 0 & 0 & 0 & 1 \end{bmatrix}
$$

$$
\times \begin{bmatrix} 1 & 0 & 0 & 0 \\ 0 & \cos \alpha_i & -\sin \alpha_i & 0 \\ 0 & \sin \alpha_i & \cos \alpha_i & 0 \\ 0 & 0 & 0 & 1 \end{bmatrix} \tag{2.8}
$$

Upon multiplying the matrices on the right-hand side of this equation, we obtain the homogeneous transformation matrix between two successive coordinate systems i and $(i-1)$ for a revolute joint

$$
{}^{i-1}T_i = \begin{bmatrix}
\cos q_i & -\sin q_i \cos \alpha_i & \sin q_i \sin \alpha_i & a_i \cos q_i \\
\sin q_i & \cos q_i \cos \alpha_i & -\cos q_i \sin \alpha_i & a_i \sin q_i \\
0 & \sin \alpha_i & \cos \alpha_i & d_i \\
0 & 0 & 0 & 1
\end{bmatrix}
\tag{2.9}
$$

In the case of a prismatic joint, the link coordinate systems are chosen in such a way that a_i equals zero, d_i becomes joint coordinate q_i, while the angle that was a rotation angle q_i for a revolute joint now becomes a constant angle denoted by θ_i. So the sequence of rotations and translations (Eq. (2.8)) for a prismatic joint yields the transformation matrix in the form

$$
{}^{i-1}T_i = \begin{bmatrix}
\cos \theta_i & -\sin \theta_i \cos \alpha_i & \sin \theta_i \sin \alpha_i & 0 \\
\sin \theta_i & \cos \theta_i \cos \alpha_i & -\cos \theta_i \sin \alpha_i & 0 \\
0 & \sin \alpha_i & \cos \alpha_i & q_i \\
0 & 0 & 0 & 1
\end{bmatrix}
\tag{2.10}
$$

Once the coordinate systems are assigned to the links of a given manipulator and the kinematic parameters α_i, θ_i, d_i, $i = 1, \ldots, n$ are determined, the homogeneous transformation matrices (Eqs. (2.9) and (2.10)) depend only on joint coordinates. As we shall see in the example, angles α_i and θ_i are usually equal to $0°$ or $\pm 90°$ for industrial robots (adjacent joint axes usually parallel or orthogonal). Therefore, the analytical form of elements of ${}^{i-1}T_i$ matrices is usually very simple.

Having obtained the homogeneous transformation matrices between successive coordinate frames, ${}^{i-1}T_i$, $i = 1, \ldots, n$ one can easily obtain the homogeneous transformation between the frame assigned to the last link and the reference coordinate frame, by multiplying these matrices

$$
{}^0T_n = {}^0T_1 \, {}^1T_2 \ldots {}^{n-2}T_{n-1} \, {}^{n-1}T_n
\tag{2.11}
$$

Since the origin of the n-th frame coincides with the manipulator hand tip, the fourth column of ${}^0T_n$ matrix represents the manipulator hand position (Cartesian coordinates) in the reference frame located at manipulator base. The left three columns of ${}^0T_n$ matrix form the rotation matrix between the hand coordinate frame and the reference frame. Once the numerical values of the elements of ${}^0T_n$ matrix are evaluated, given the manipulator and the joint coordinate vector $q = [q_1 \, q_2 \, \ldots \, q_n]^T$, it is possible to determine the Euler angles from the left three columns of ${}^0T_n$. Together with the Cartesian coordinates of the manipulator hand, these angles form the external coordinates vector. Therefore, the evaluation of ${}^0T_n$ matrix, given the joint coordinates, is the main part of the direct kinematic problem solution. The second part – the evaluation of Euler angles, given the ${}^0T_n$ matrix – does not depend on the manipulator structure (this problem will be discussed in detail in Sect. 2.3.3).

Let us now illustrate the direct kinematic problem solution by an example.

Example Illustrating the Direct Kinematic Problem Solution

Solving the direct kinematics will be illustrated by the example of the three degree-of-freedom manipulator shown in Fig. 2.11. The Denavit–Hartenberg kinematic parameters are also listed in this figure.

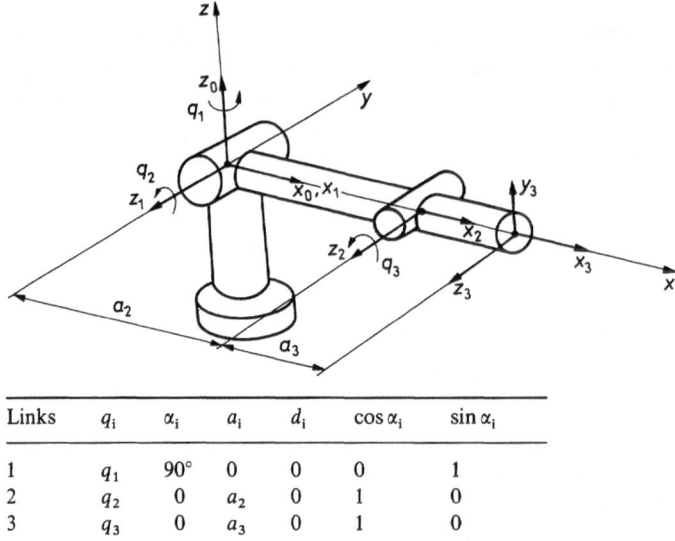

Links	q_i	α_i	a_i	d_i	$\cos \alpha_i$	$\sin \alpha_i$
1	q_1	90°	0	0	0	1
2	q_2	0	a_2	0	1	0
3	q_3	0	a_3	0	1	0

Fig. 2.11. Kinematic scheme of the manipulator with three degrees of freedom

The origin of the reference coordinate frame is located at the point of intersection of the axes of joints 1 and 2. $\vec{z}_i$, $i = 1, \ldots, 3$ axes of link coordinate frames coincide with joint axes. When joint angles have zero values $\vec{x}_i$ axes are parallel. The length of the second and the third link determine the parameters a_2 and a_3, since they correspond to the common normal distances between joint axes. Joint axes 1 and 2 are perpendicular. Therefore, the twist angle about the $\vec{x}_i$ axis is $\alpha_1 = 90°$. The joint axes of the second and the third joint are parallel so that $\alpha_2 = 0°$. The coordinate frame of the last link is chosen to be parallel with the previous frame resulting in $\alpha_3 = 0°$. Since the origins of all the coordinate frames lie in the same plane, i.e. there are no displacements along joint axes, all the d_i parameters are equal to zero.

Bearing in mind the general form (2.9) for the homogeneous transformation matrices for revolute joints and the parameters in the table shown in Fig. 2.11, the homogeneous transformations between adjacent links become

$$
{}^0T_1 =
\begin{bmatrix}
C_1 & 0 & S_1 & 0 \\
S_1 & 0 & -C_1 & 0 \\
0 & 1 & 0 & 0 \\
0 & 0 & 0 & 1
\end{bmatrix}
\tag{2.12}
$$

$$
{}^1T_2 = \begin{bmatrix} C_2 & -S_2 & 0 & a_2C_2 \\ S_2 & C_2 & 0 & a_2S_2 \\ 0 & 0 & 1 & 0 \\ 0 & 0 & 0 & 1 \end{bmatrix}
$$
(2.13)

$$
{}^2T_3 = \begin{bmatrix} C_3 & -S_3 & 0 & a_3C_3 \\ S_3 & C_3 & 0 & a_3S_3 \\ 0 & 0 & 1 & 0 \\ 0 & 0 & 0 & 1 \end{bmatrix}
$$
(2.14)

Here, the following notation has been introduced

$$
\begin{aligned}
C_1 &= \cos q_1, & S_1 &= \sin q_1 \\
C_2 &= \cos q_2, & S_2 &= \sin q_2 \\
C_3 &= \cos q_3, & S_3 &= \sin q_3
\end{aligned}
$$
(2.15)

By multiplying these homogeneous matrices, according to Eq. (2.11), we obtain the homogeneous transformation matrix between the hand coordinate system (n-th system) and the reference frame. Starting from the last link towards the first, we get

$$
{}^0T_3 = {}^0T_1({}^1T_2{}^2T_3)
$$
(2.16)

Thus, we obtain

$$
{}^2T_3 = \begin{bmatrix} C_3 & -S_3 & 0 & a_3C_3 \\ S_3 & C_3 & 0 & a_3S_3 \\ 0 & 0 & 1 & 0 \\ 0 & 0 & 0 & 1 \end{bmatrix}
$$
(2.17)

$$
{}^1T_3 = \begin{bmatrix} C_2C_3-S_2S_3 & -C_2S_3-S_2C_3 & 0 & a_3(C_2C_3-S_2S_3)+a_2C_2 \\ S_2C_3+C_2S_3 & -S_2S_3+C_2C_3 & 0 & a_3(S_2C_3+C_2S_3)+a_2S_2 \\ 0 & 0 & 1 & 0 \\ 0 & 0 & 0 & 1 \end{bmatrix} =
$$

$$
= \begin{bmatrix} C_{23} & -S_{23} & 0 & a_3C_{23}+a_2C_2 \\ S_{23} & C_{23} & 0 & a_3S_{23}+a_2S_2 \\ 0 & 0 & 1 & 0 \\ 0 & 0 & 0 & 1 \end{bmatrix}
$$
(2.18)

$$
^0T_3 = \begin{bmatrix} C_1 C_{23} & -C_1 S_{23} & S_1 & C_1(a_3 C_{23} + a_2 C_2) \\ S_1 C_{23} & -S_1 S_{23} & -C_1 & S_1(a_3 C_{23} + a_2 C_2) \\ S_{23} & C_{23} & 0 & a_3 S_{23} + a_2 S_2 \\ 0 & 0 & 0 & 1 \end{bmatrix} \tag{2.19}
$$

Here, the following abbreviations were introduced

$$
\begin{aligned} S_{23} &= \sin(q_2 + q_3) \\ C_{23} &= \cos(q_2 + q_3) \end{aligned} \tag{2.20}
$$

In this way, by applying the kinematic modelling technique we have obtained the elements of the homogeneous transformation matrix between the hand and the reference frame as explicit analytical functions of joint angles q_1, q_2 and q_3. Now it is possible to evaluate the numerical values of these elements, given the joint angles. As we have already mentioned, the upper left 3×3 submatrix describes the rotation of the third coordinate frame $0_3 x_3 y_3 z_3$ with respect to the reference frame $0xyz$, while the fourth column contains the Cartesian coordinates x, y, z of the manipulator tip in the fixed frame

$$
^0T_3 = \left[\begin{array}{ccc|c} & & & x \\ & ^0A_3 & & y \\ & & & z \\ \hline 0 & 0 & 0 & 1 \end{array} \right] \tag{2.21}
$$

Once the numerical values of the $^0A_3$ matrix are evaluated, the other three external coordinates describing the orientation (Euler angles) may also be determined. Then, the direct kinematic problem is completely solved.

2.3.2 Manipulator Hand Orientation

Manipulator hand orientation with respect to the reference frame is usually specified by three Euler angles (Fig. 2.7). One method for evaluation of Euler angles is the following.

Let us consider two coordinate frames and describe rotation between them. The first system Q_n is assigned to the last manipulator link, and its unit axes are denoted by $\vec{q}_{n1}, \vec{q}_{n2}, \vec{q}_{n3}$ (Fig. 2.12). The transformation matrix ^0A_n which maps vectors from system Q_n to system ϕxyz is

$$
^0A_n = [^0\vec{q}_{n1} \ ^0\vec{q}_{n2} \ ^0\vec{q}_{n3}] = \begin{bmatrix} a_{11} & a_{12} & a_{13} \\ a_{21} & a_{22} & a_{23} \\ a_{31} & a_{32} & a_{33} \end{bmatrix} \tag{2.22}
$$

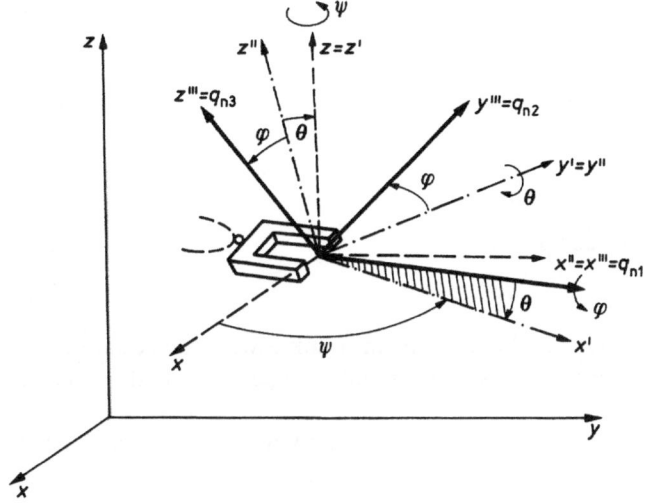

Fig. 2.12. Euler angles between the hand coordinate frame and the fixed reference frame

The rotation of system $0_n x_n y_n z_n$ with respect to $0xyz$ can be described by the following sequence of rotations: a rotation by ψ about z axis (yaw), followed by a rotation by θ about the new y axis (pitch), and rotation by φ about the new x axis (roll) (Fig. 2.12). This sequence of rotations corresponds to the following product of rotation matrices

$$
{}^{0}A_n = \begin{bmatrix} \cos\psi & -\sin\psi & 0 \\ \sin\psi & \cos\psi & 0 \\ 0 & 0 & 1 \end{bmatrix} \begin{bmatrix} \cos\theta & 0 & \sin\theta \\ 0 & 1 & 0 \\ -\sin\theta & 0 & \cos\theta \end{bmatrix} \begin{bmatrix} 1 & 0 & 0 \\ 0 & \cos\varphi & -\sin\varphi \\ 0 & \sin\varphi & \cos\varphi \end{bmatrix} \quad (2.23)
$$

or

$$
{}^{0}A_n = \begin{bmatrix} \cos\psi\cos\theta & \cos\psi\sin\theta\sin\varphi - \sin\psi\cos\varphi & \cos\psi\sin\theta\cos\varphi + \sin\psi\sin\varphi \\ \sin\psi\cos\theta & \sin\psi\sin\theta\sin\varphi + \cos\psi\cos\varphi & \sin\psi\sin\theta\cos\varphi - \cos\psi\sin\varphi \\ -\sin\theta & \cos\theta\sin\varphi & \cos\theta\cos\varphi \end{bmatrix}
$$

$$
(2.24)
$$

By equating the numerical values of a_{jk} elements (Eq. (2.22)), obtained by a kinematic modelling procedure, with the elements from Eq. (2.24), we have

$$
a_{11} = \cos\psi\cos\theta \quad (2.25)
$$

$$
a_{21} = \sin\psi\cos\theta \quad (2.26)
$$

$$a_{31} = -\sin\theta \tag{2.27}$$

$$a_{12} = \cos\psi \sin\theta \sin\varphi - \sin\psi \cos\varphi \tag{2.28}$$

$$a_{22} = \sin\psi \sin\theta \sin\varphi + \cos\psi \cos\varphi \tag{2.29}$$

$$a_{32} = \cos\theta \sin\varphi \tag{2.30}$$

$$a_{13} = \cos\psi \sin\theta \cos\varphi + \sin\psi \sin\varphi \tag{2.31}$$

$$a_{23} = \sin\psi \sin\theta \cos\varphi - \cos\psi \sin\varphi \tag{2.32}$$

$$a_{33} = \cos\theta \cos\varphi \tag{2.33}$$

Evidently, this is a system of 9 equations in 3 unknowns (the equations are not independent since matrix ^0A_n is orthogonal). Angles ψ, θ and φ may be determined in several ways.

By multiplying both sides of Eq. (2.25) by $\sin\psi$ and both sides of Eq. (2.26) by $\cos\psi$, and subtracting these two equations, we obtain

$$a_{11} \sin\psi - a_{21} \cos\psi = 0 \tag{2.34}$$

or

$$\psi = \text{arctg}\,\frac{a_{21}}{a_{11}} + k\pi \tag{2.35}$$

Pitch angle θ can be computed by multiplying Eq. (2.25) by $\cos\psi$ and Eq. (2.26) by $\sin\psi$ and adding these two equations

$$a_{11} \cos\psi + a_{21} \sin\psi = \cos\theta \tag{2.36}$$

The above equation together with Eq. (2.27) yields

$$\theta = \text{arctg}[-a_{31}/(a_{11}\cos\psi + a_{21}\sin\psi)] + 2k\pi \tag{2.37}$$

By multiplying Eq. (2.30) by $\cos\varphi$ and Eq. (2.33) by $\sin\varphi$ and subtracting these two equations, we obtain

$$a_{32} \cos\varphi - a_{33} \sin\varphi = 0 \tag{2.38}$$

or

$$\varphi = \text{arctg}\,\frac{a_{32}}{a_{33}} + 2k\pi \tag{2.39}$$

Problems occurring when denominators in Eqs. (2.35), (2.37), (2.39) are small may be avoided by the use of computer function ATAN2. The only singular situation is when $\theta = k\pi/2$, since in this case $\cos\theta = 0$, resulting in $a_{11} = a_{21} = a_{32} = a_{33} = 0$. This occurs when the manipulator hand is pointing straight up or down and both yaw and roll correspond to the same rotation. Equation (2.35) yields no solution to ψ. In this case the value for ψ can be chosen arbitrarily (e.g. to be equal to the value of ψ from the previous time instant, or to the desired value of ψ). Once the ψ angle is chosen, φ is evaluated from

$$\varphi = \text{arctg}\, \frac{-a_{12}}{a_{22}} - \psi + 2k\pi \;, \quad \text{for } \theta = -k\frac{\pi}{2} \tag{2.40}$$

$$\varphi = \text{arctg}\, \frac{a_{12}}{a_{22}} + \psi + 2k\pi \;, \quad \text{for } \theta = k\frac{\pi}{2} \tag{2.41}$$

The values for integer k are determined following the principle that the change of Euler angles for two successive points in which the direct kinematic problem is solved should be minimal. This criterion follows from the fact that only a continuous evolution of external coordinates can correspond to a continuous manipulator motion (external coordinates must not change for 2π abruptly).

Having obtained matrix ^0T_n and Euler angles, given the joint coordinates, the direct kinematic problem is completely solved.

2.4 Inverse Kinematic Problem

Inverse kinematics involves the computation of the joint coordinates that correspond to a given hand position and orientation specified by the external coordinates vector x_e, i.e. obtaining the inverse transformation

$$q = f^{-1}(x_e) \tag{2.42}$$

The homogeneous transformation matrix ^0T_n between the hand and the reference frame is completely specified, given the external coordinates vector $x_e = [x \; y \; z \; \psi \; \theta \; \varphi \,]^T$ (according to Eqs. (2.21) and (2.24)). Therefore, finding the inverse kinematics solutions is also equivalent to solving the vectorial equation

$$q = f^{-1}(^0T_n) \tag{2.43}$$

In general, the solution to the inverse kinematic problem for non-redundant manipulators is not unique. There exists a finite set of joint coordinate vectors that correspond to the same manipulator hand position and orientation.

Since the function f relating between external and joint coordinates is a non-linear, trigonometric function, finding the inverse kinematics solutions is equivalent to solving the set of non-linear equations. Two different approaches can be recognized. The first is to obtain the inverse vectorial function $f^{-1}(^0T_n)$ in analytical form for each manipulator structure separately [3–5]. The second is to apply some of the methods known from numerical analysis for solving non-linear equations. Both approaches have some advantages and shortcomings. We will discuss them in the following examples.

2.4.1 Analytical Solutions

Obtaining analytical solutions to the inverse kinematic problem for different manipulators is rather a complex problem. It cannot be solved in a general way for an arbitrary manipulator structure. This is due to the fact that the vectorial function f is a complex nonlinear function of n variables.

The analytical inverse kinematics solution (if it is available for a given manipulator) has significant advantages over the numerical solutions. First, the analytical solution is computationally less complex than the numeric one, thus being more convenient for real-time generation of manipulator motion. The second advantage is that the analytical inverse kinematics gives all the solutions (for six degrees-of-freedom manipulators there are eight solutions). The solutions are exact and no numeric errors are accumulated. When deriving the inverse kinematic expressions, the manipulator singularities are easily detected, so they can be treated separately.

Let us illustrate solving inverse kinematics analytically by a simple example. We will consider the three degrees-of-freedom manipulator shown in Fig. 2.11. Its direct kinematics has already been obtained, i.e. the elements of the transformation matrix ${}^{0}T_n = {}^{0}T_3$ are expressed as functions of q_1, q_2 and q_3 (Eqs. (2.12)–(2.19)).

Since we are dealing with a three degrees-of-freedom manipulator in this example, we cannot control its orientation, but only position x, y, z in the reference frame. Equations (2.19) and (2.21) yield

$$x = C_1(a_3 C_{23} + a_2 C_2) \tag{2.44}$$

$$y = S_1(a_3 C_{23} + a_2 C_2) \tag{2.45}$$

$$z = a_3 S_{23} + a_2 S_2 \tag{2.46}$$

By multiplying Eq. (2.44) by S_1 and Eq. (2.45) by C_1 and subtracting these two equations, we obtain

$$S_1 x - C_1 y = 0 \tag{2.47}$$

or

$$q_1 = \operatorname{arctg} \frac{y}{x} + k\pi \tag{2.48}$$

Joint angle q_3 may be evaluated in the following way. By multiplying Eq. (2.44) by C_1 and Eq. (2.45) by S_1 and adding we get

$$C_1 x + S_1 y = a_3 C_{2,3} + a_2 C_2$$
$$z = a_3 S_{2,3} + a_2 S_2 \tag{2.49}$$

Since the angle q_1 has already been determined the left-hand sides of the above equations are known. By squaring both equations and then adding we obtain

$$C_3 = \frac{(C_1 x + S_1 y)^2 + z^2 - a_3^2 - a_2^2}{2a_2 a_3} \tag{2.50}$$

The S_3 variable is obtained from

$$S_3 = \pm \sqrt{1 - C_3^2} \tag{2.51}$$

The signs $+$ and $-$ correspond to the elbows' up and down configuration. The angle q_3 is given by

$$q_3 = \text{arctg} \frac{S_3}{C_3} \tag{2.52}$$

Now, angle q_2 is obtained again from Eq. (2.49):

$$S_2 = \frac{(C_3 a_3 + a_2) z - S_3 a_3 (C_1 x + S_1 y)}{(C_3 a_3 + a_2)^2 + S_3^2 a_3^2} \tag{2.53}$$

$$C_2 = \frac{(C_3 a_3 + a_2)(C_1 x + S_1 y) + S_3 a_3 z}{(C_3 a_3 + a_2)^2 + S_3^2 a_3^2}$$

or

$$q_2 = \text{arctg} \frac{(C_3 a_3 + a_2) z - S_3 a_3 (C_1 x + S_1 y)}{(C_3 a_3 + a_2)(C_1 x + S_1 y) + S_3 a_3 z} \tag{2.54}$$

Having obtained the inverse kinematics analytically the manipulator singularities at which the manipulator becomes degenerate may be easily recognized. It is evident from Eq. (2.48) that angle q_1 is not defined when $x = y = 0$, i.e. when the manipulator tip lies on the z axis. In that case rotation about the first joint axis produces no change in the manipulator tip position, i.e. the first degree of freedom is not active. The second singularity occurs when the manipulator arm is fully stretched so that the manipulator work space boundary is reached and q_3 equals zero. If one would impose a motion towards a point that cannot be reached by the manipulator; the condition

$$x^2 + y^2 + z^2 < (a_2 + a_3)^2 \tag{2.55}$$

or, equivalently,

$$(C_1 x + S_1 y)^2 + z^2 < (a_2 + a_3)^2$$

would not be satisfied. Equation (2.50) would yield $\cos q_3 > 1$ in that case. Therefore, by checking the condition (2.55) we check whether the claimed motion is attainable or not.

In this example (where two joint axes intersect and two were always parallel) it was feasible to solve inverse kinematics analytically and obtain explicit solutions for joint angles q_1, q_2 and q_3. However, even for three degrees-of-freedom manipulators this is not always possible. It was shown in [6] that solving inverse kinematics for three degrees-of-freedom manipulators can be reduced to finding the roots of a fourth degree polynomial in one unknown.

For six degrees-of-freedom manipulators the analytical solution can be obtained if the problem can be separated into two independent sub-problems: evaluating the first three joint coordinates q_1, q_2 and q_3 and evaluating the rest of them q_4, q_5 and q_6. This is possible when the last three joint axes intersect in a single point. Then, knowing the orts of the hand coordinate frame in the reference system, it is feasible to determine the position of the end of the third link in the reference frame. This yields the first three angles q_1, q_2 and q_3. Finally, knowing angles q_1, q_2, q_3 the orientation of the third link is determined and at last angles q_4, q_5 and q_6.

It should be mentioned that the majority of industrial robots produced in the world have such a mechanical structure that the analytical inverse kinematics is obtainable (most of them are of the type shown in Fig. 2.1). However, some non-redundant robots as well as redundant manipulators require the application of purely numerical methods in evaluating joint coordinates, given the manipulator hand position and orientation [7–9].

2.4.2   Numerical Solutions

As we have seen, solving the inverse kinematics is equivalent to seeking for the roots of the set of non-linear equations (2.42), within the work space and the joint coordinates space. Therefore, any of the standard techniques known from numerical analysis may be applied. These procedures are general and do not depend on mechanism configuration. However, as opposed to analytical methods numerical algorithms yield only one solution for joint coordinates, given a vector of external coordinates x_e, which is closest to an initial guess q^0

$$q = f^{-1}(x_e, q^0) \tag{2.56}$$

We will consider only non-redundant manipulators here, i.e. the case when the number of external coordinates is equal to the number of degrees of freedom. Since Newton–Raphson's method is most widely used for inverse kinematics, we will outline it briefly here.

Let us consider the function

$$F(q) = f(q) - x_e \tag{2.57}$$

where $q \in R^n$, $x_e \in R^n$ and $f: R^n \to R^n$ is a continuous, differentiable function which maps joint coordinates into external coordinates.

We should determine the zero of function F which is close to some approximate solution q^k. By expanding function $F(q)$ in Taylor's series and keeping the first two terms only, we obtain

$$F(q) + J(q^k)(q - q^k) = 0 \tag{2.58}$$

where

$$J(q) = \frac{\partial f(q)}{\partial q}, \quad J \in R^{n \times n} \tag{2.59}$$

is the Jacobian matrix of partial derivatives of function f. Equation (2.58) is equivalent to

$$q = q^k - J^{-1}(q^k)(f(q^k) - x_e) = q^k + J^{-1}(q^k)\Delta x_e^k \tag{2.60}$$

where $J^{-1}(q^k)$ is the inverse Jacobian matrix, $\Delta x_e^k = x_e - f(q^k)$ is the increment of external coordinates with respect to the external coordinates which correspond to the approximate solution q^k. If q is replaced by q^{k+1} in Eq. (2.60) an iterative procedure is obtained. The procedure should be ended when the condition $\|\Delta x_e^k\| < \varepsilon$ is satisfied, where ε is a small positive constant. Newton's method has a quadratic convergency. It is clear that it yields a single solution for q, the solution that is closest to q^k.

If the manipulator motion is specified as a continuous trajectory in the external coordinates space, or by a set of points x_e^k, $k = 1, \ldots, N$, so that the increments $\Delta x_e^k = x_e^{k+1} - x_e^k$ are small, the corresponding trajectory in the space of joint coordinates is obtained by a single iteration of Newton's method:

$$q^{k+1} = q^k + J^{-1}(q^k)\Delta x_e^k \tag{2.61}$$

The solution q^{k+1} approximately satisfies $f(q^{k+1}) \cong x_e^{k+1}$ and it is not necessary to check whether the error is less than ε, since it is assumed that the points x_e^k are sufficiently close to each other. However, this method of evaluating the trajectory in the space of joint coordinates, given a trajectory of manipulator hand, inevitably leads to a certain accumulation of the linearization error. In order to avoid this, a compromise between the original Newton–Raphson's algorithm and Eq. (2.61) is applied. Namely, the joint coordinates are still evaluated according to Eq. (2.61) in a single iteration, but instead of evaluating Δx_e^k from $\Delta x_e^k = x_e^{k+1} - x_e^k$, the increment of external coordinates is corrected according to

$$\Delta x_e^k = x_e^{k+1} - f(q^k) \tag{2.62}$$

Instead of dealing with the increments of external and joint coordinates, one can consider the relationship between the velocities

$$\dot{x}_e(t) = J(q(t))\dot{q}(t) \tag{2.63}$$

The joint velocities corresponding to a given velocity vector $\dot{x}_e$ are evaluated from

$$\dot{q}(t) = J^{-1}(q)\dot{x}_e(t) \tag{2.64}$$

This formulation of the inverse kinematics is obviously equivalent to Newton's algorithm (2.60) as applied to an infinitesimal time interval. It is used for rate control of manipulators, where an operator specifies external velocities by means of a joy-stick.

As we can see, solving the inverse kinematics by means of Newton's method requires evaluation of the Jacobian matrix and its inverse. Since this matrix relates to external and joint velocities it is rather important in manipulator motion generation. We will consider its evaluation in more detail in the next section.

The main problems encountered in the application of Newton's method are the following. An initial solution q^0 is required, which is close to the exact solution (otherwise the algorithm diverges). In the vicinity of singularities the trajectory should be modified or specified in joint coordinates, since the inverse Jacobian cannot be obtained. Besides, the computational time for evaluating the Jacobian inverse is long compared to that required for analytical inverse kinematics (if such exists).

2.4.3 The Jacobian Matrix

The Jacobian matrix defined by

$$\dot{x}_e = J(q)\dot{q}$$

correlates between joint rates and the external coordinates velocities. It also correlates between external forces acting on the manipulator hand and the torques transferred to manipulator joints [3]. The Jacobian matrix depends on the type of the external coordinates that have been adopted. However, Jacobian matrices corresponding to spherical, cylindrical coordinates, Euler angles and Euler parameters can easily be obtained if the Jacobian matrix corresponding to Cartesian coordinates and angular velocities of the end-effector is known. Therefore, we will consider the evaluation of the Jacobian defined by

$$\begin{bmatrix} \dot{x} \\ \dot{y} \\ \dot{z} \\ \omega_x \\ \omega_y \\ \omega_z \end{bmatrix} = \begin{bmatrix} J_c(q) \\ J_\omega(q) \end{bmatrix} \begin{bmatrix} \dot{q}_1 \\ \vdots \\ \dot{q}_n \end{bmatrix} = J(q)\dot{q} \ , \quad \begin{matrix} J_c \in R^{3 \times n} \\ J_\omega \in R^{3 \times n} \\ J \in R^{6 \times n} \end{matrix} \tag{2.65}$$

where $\dot{x}$, $\dot{y}$, $\dot{z}$, ω_x, ω_y, ω_z are linear and angular velocities of the manipulator hand in the reference frame.

Let us now consider the evaluation of the sub-matrix J_c. Let us denote by $v^{(i)}$ the part of the total linear velocity $\vec{v} = [\dot{x} \ \dot{y} \ \dot{z}]^T$ which is due to the joint rate $\dot{q}_i$.

Then we have

$$\dot{x}_e = \vec{v} = \sum_{i=1}^{n} \vec{v}^{(i)} \tag{2.66}$$

If the joint i is a revolute one, the component $\vec{v}^{(i)}$ is given by

$$\vec{v}^{(i)} = \vec{z}_{i-1} \times \vec{r}_{i-1,n} \dot{q}_i \tag{2.67}$$

where $\vec{z}_{i-1}$ is the ort of the joint axis, $\vec{r}_{i-1,n}$ is the distance vector between the centre of joint i (origin of coordinate system $(i-1)$) and the manipulator tip (origin of system n), both expressed in the base reference frame (Fig. 2.13).

If joint i is a sliding one, the component $\vec{v}^{(i)}$ of the total linear velocity is given by

$$\vec{v}^{(i)} = \vec{z}_{i-1} \dot{q}_i \tag{2.68}$$

In order to obtain the linear velocity as a function of elements of the homogeneous matrices, we will first express the vectors in Eqs. (2.67) and (2.68) with respect to system $(i-1)$ and then premultiply them by $^0A_{i-1}$ rotation matrix. Since $^{i-1}\vec{z}_{i-1} = [0 \quad 0 \quad 1]^T$ is valid and the distance vectors $^{i-1}\vec{r}_{i-1,n}$ are in fact the fourth columns of the homogeneous transformations $^{i-1}T_n$, $i=1, \ldots, n$,

$$^{i-1}\vec{r}_{i-1,n} = [^{i-1}p_{nx} \quad ^{i-1}p_{ny} \quad ^{i-1}p_{nz}]^T$$

the Jacobian columns are obtained in the form

$$J_c = [j^{(1)} \quad \ldots \quad j^{(n)}] \in R^{3 \times n}$$

$$j^{(i)} = \begin{cases} \begin{bmatrix} -^0a_{(i-1)11}\,^{i-1}p_{ny} + ^0a_{(i-1)12}\,^{i-1}p_{nx} \\ -^0a_{(i-1)21}\,^{i-1}p_{ny} + ^0a_{(i-1)22}\,^{i-1}p_{nx} \\ -^0a_{(i-1)31}\,^{i-1}p_{ny} + ^0a_{(i-1)23}\,^{i-1}p_{nx} \end{bmatrix}, & \text{for revolute joint } i \\[2em] \begin{bmatrix} ^0a_{(i-1)13} \\ ^0a_{(i-1)23} \\ ^0a_{(i-1)33} \end{bmatrix}, & \text{for sliding joint } i \end{cases} \tag{2.69}$$

where $^0a_{(i-1)jk}$ denotes the element (j,k) of matrix $^0A_{i-1}$.

Let us now discuss the evaluation of matrix $J_\omega \in R^{3 \times n}$ relating between angular velocities of the last link, expressed in the base, reference frame, and joint rates. First, we will determine the angular velocity of link i expressed with respect to the reference frame (Fig. 2.14). This velocity can be expressed as the

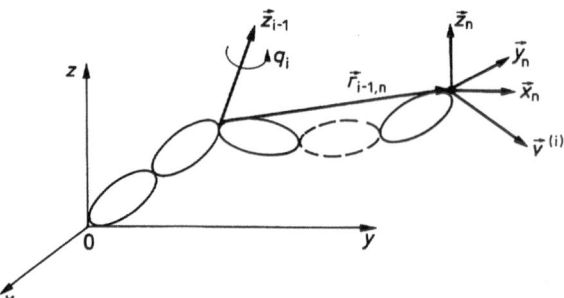

Fig. 2.13. Vectors relevant for evaluating the Jacobian matrix

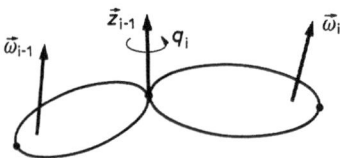

Fig. 2.14. Angular velocity of link i

sum of the angular velocity $\vec{\omega}_i$ of the previous link and the relative angular velocity between link i and $(i-1)$.

For a revolute joint i, we have

$$\vec{\omega}_i = \vec{\omega}_{i-1} + \dot{q}_i \vec{z}_{i-1} , \quad 1 \leqslant i \leqslant n \tag{2.70}$$

where $\vec{z}_{i-1}$ is the ort of the joint axis. All the points belonging to the link have the same angular velocity. In the case of a sliding joint, there is no relative rotation between the links so that the following relation holds

$$\vec{\omega}_i = \vec{\omega}_{i-1} , \quad 1 \leqslant i \leqslant n \tag{2.71}$$

Relations (2.70) and (2.71) may be united into a single equation:

$$\vec{\omega}_i = \vec{\omega}_{i-1} + \dot{q}_i \vec{z}_{i-1}(1 - \xi_i) \tag{2.72}$$

where ξ_i is the indicator whether joint i is a revolute one ($\xi_i = 0$) or a sliding one ($\xi_i = 1$). By successively adding expressions (2.72) for $j = 1, \ldots, n$, a nonrecursive expression for the angular velocity of the last link is obtained

$$\vec{\omega}_n = \sum_{j=1}^{n} \dot{q}_j \vec{z}_{j-1}(1 - \xi_j) \tag{2.73}$$

If the orts of the joints axes in the above relation are expressed with respect to the base reference frame (by premultiplying by $^0A_{j-1}$ matrices) the angular velocity is also expressed in that frame $\vec{\omega}_n = [\omega_x \ \omega_y \ \omega_z]^T$. So we obtain the columns of the Jacobian submatrix J_ω in the form

$$J_\omega = [\vec{j}^{(1)} \ \ldots \ \vec{j}^{(n)}]$$

$$\vec{j}^{\,(i)} = \begin{cases} \begin{bmatrix} {}^0a_{(i-1)13} \\ {}^0a_{(i-1)23} \\ {}^0a_{(i-1)33} \end{bmatrix}, & \text{for revolute joint } i \\[20pt] \begin{bmatrix} 0 \\ 0 \\ 0 \end{bmatrix}, & \text{for prismatic joint } i \end{cases} \tag{2.74}$$

We can see from Eqs. (2.69) and (2.74) that the Jacobian elements depend only on the elements of the homogeneous transformations ${}^0T_i$, $i = 1, \ldots, n$. Having obtained the homogeneous transformations in the analytical form (while solving the direct kinematic problem), we can substitute them into Eqs. (2.69) and (2.74). Thus we obtain the Jacobian matrix with respect to the base reference frame in the analytical form for a specific manipulator structure [1, 3, 10, 11].

The relationship between the joint rates and the rate of changing Euler angles is obtained indirectly using matrix J_ω. Namely, the relation well known from the rigid body kinematics

$$\begin{bmatrix} \omega_x \\ \omega_y \\ \omega_z \end{bmatrix} = \begin{bmatrix} 0 & -\sin\psi & \cos\theta\cos\psi \\ 0 & \cos\psi & \cos\theta\sin\psi \\ 1 & 0 & -\sin\theta \end{bmatrix} \begin{bmatrix} \dot{\psi} \\ \dot{\theta} \\ \dot{\varphi} \end{bmatrix} \tag{2.75}$$

yields the angular velocity in the reference frame, given the rate of changing Euler angles $\dot{\psi}$, $\dot{\theta}$ and $\dot{\varphi}$. Having evaluated ω_x, ω_y, ω_z the corresponding joint velocities are obtained from Eq. (2.65), i.e.

$$\dot{q} = J^{-1}(q) \begin{bmatrix} \dot{x} \\ \dot{y} \\ \dot{z} \\ \omega_x \\ \omega_y \\ \omega_z \end{bmatrix} \tag{2.76}$$

where $J(q)$ is given by

$$J(q) = \left[\begin{array}{c} J_c(q) \\ ------ \\ J_\omega(q) \end{array} \right] \tag{2.77}$$

2.5 Manipulator Path Generation

Having solved the direct and inverse kinematic problem, the primary problem arising in robot control is manipulator motion generation. In this section we shall be concerned with the path generation in the absence of disturbances, i.e. at the level of nominal trajectories [12–18].

The basic property which the motion generation algorithms should possess is provision for the functionality of motion, i.e. to provide for continuity and smoothness of motion between the user-specified end-effector positions and orientations. Beside this basic requirement, additional properties are also needed to increase the adaptivity and flexibility of robots. For example, motion generation algorithms should provide for obstacle avoidance, the treatment of various sensors such as cameras, tactile sensors, force sensors, proximity sensors, working on-moving conveyors, and so on.

Manipulator motion generation usually requires:

1. specification of the set of points (knots) either in the space of external or joint coordinates, which the manipulator end-effector should pass through, and
2. defining the trajectories between these points.

The selection of points the manipulator is to pass through is mostly determined by technological requirements, the distribution of handled objects, tools, obstacles, etc. The points are usually specified with respect to the fixed coordinate frame at the manipulator base. However, it is possible to specify them with respect to several frames attached to various objects and machines in the manipulator environment [3]. Then, the external coordinates have first to be transformed into the external coordinates with respect to the base reference frame, before the joint coordinates can be solved. In this case the points do not have to be learned again if the objects or machines change their location. Only the new position of the frame attached to the object is to be changed.

In modern robot control systems manipulation tasks are specified by means of robot languages and teaching boxes (robot languages will be discussed in detail in Chap. 6). In the process of teaching robots very often the rate control mode is used, where the operator specifies the manipulator path by specifying the rate of changing joint or external coordinates. The kinematic module transforms the external coordinates into joint trajectories and feeds up the servo system. The specification of robot tasks by means of teaching boxes will be described in Chapter 5.

Motion synthesis between user-specified points can be done in many ways. The simplest way is to generate a straight-line motion stopping at each point. The straight line may be interpolated in joint coordinates between the initial and the final points q^0, $q^F \in Q \subset R^n$, given the time period T

$$q(t) = q^0 + \lambda(t)(q^F - q^0) , \qquad 0 \leqslant t \leqslant T \tag{2.78}$$

where $\lambda(t) \in [0, 1]$ is the scalar parameter defining the velocity and acceleration profile along the trajectory. It can be chosen in different ways. Figure 2.15 shows the rectangular, parabolic, trapezoidal and square-sine-function-like acceleration profiles. Profiles which have a smaller jerk (third derivative) are desirable. However, they yield higher velocities and accelerations along the trajectory.

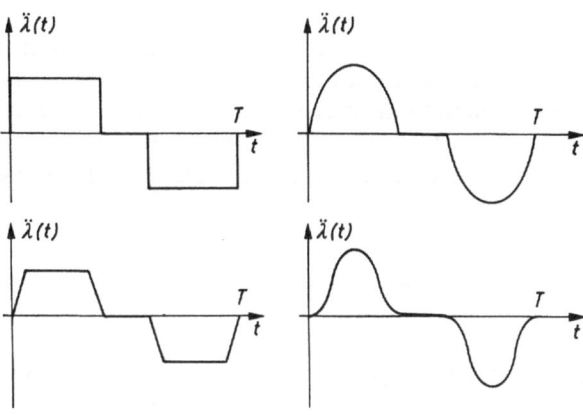

Fig. 2.15. Acceleration profiles

Linear motion in joint coordinates (Eq. (2.78)) obviously does not produce straight-line motion in Cartesian space. However, joint interpolated motion is used if high speed motions are required and straight-line motion is not very important. It is also used to move the manipulator from one configuration into another, when a kinematic degeneracy is located between them.

Cartesian interpolated motion between two points x_e^0 and x_e^F

$$x_e(t) = x_e^0 + \lambda(t)(x_e^F - x_e^0) , \qquad 0 \leqslant t \leqslant T \tag{2.79}$$

may be performed using the same acceleration profiles as shown in Fig. 2.15. In this case, however, the manipulator control system has to provide for joint angles at servo rate by solving the inverse kinematics. If external sensors are to be included in order to modify the trajectory, the motion generation has to be done in real time. Then, the time required for obtaining the inverse kinematics solution is critical, especially for high-speed robots (e.g. 5 m/s). If this time is several times greater than the time required by the servosystem, the cubic spline interpolation is used to generate continuous joint positions and velocities [13].

Very often it is not necessary to stop at each user-supplied point. Such motion may be realized in various ways. One of them is shown in Fig. 2.16 [3].

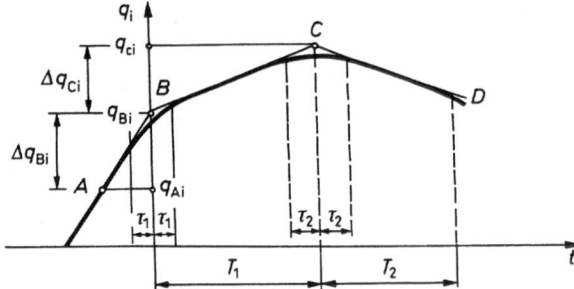

Fig. 2.16. Time history of joint coordinate q_i

Here, the quadratic change of acceleration in the vicinity of the transition points provides for continuity of velocity and acceleration. It is applied to the tasks where it is sufficient to pass close enough to the user-supplied points. According to the notation introduced in Fig. 2.16, the time history of joint position $q_i(t)$, velocity $\dot{q}_i(t)$ and acceleration $\ddot{q}_i(t)$ in time interval $-\tau_1 \leqslant t \leqslant \tau_1$, can be expressed as

$$q_i(t) = \left[\left(\Delta q_{ci} \frac{\tau_1}{T_1} - \Delta q_{Bi} \right)(2 - h)h^2 + 2\Delta q_{Bi} \right] h + q_{Ai}$$

$$\dot{q}_i(t) = \left[\left(\Delta q_{ci} \frac{\tau_1}{T_1} - \Delta q_{Bi} \right)(1.5 - h) 2h^2 + \Delta q_{Bi} \right] \frac{1}{\tau_1} \qquad (2.80)$$

$$\ddot{q}_i(t) = \left(\Delta q_{ci} \frac{\tau_1}{T_1} - \Delta q_{Bi} \right)(1 - h) \frac{3h}{\tau_1^2}$$

$$h = \frac{t + \tau_1}{2\tau_1} , \qquad -\tau_1 \leqslant T \leqslant \tau_1$$

while for the intermediate part of the trajectory the velocity is constant

$$q_i(t) = \Delta q_{ci} \frac{t}{T_1} + q_{Bi} , \qquad \tau_1 \leqslant t \leqslant T_1 - \tau_2 \qquad (2.81)$$

The time function $q(t)$, $-\tau_1 \leqslant t \leqslant T_1 - \tau_2$ depends solely on the variables related only to the three points A, B and C on the trajectory and does not depend on the other trajectory segments. This feature is convenient in real time motion synthesis.

One of the classical methods of manipulator path generation which provides for continuity of position, velocity and acceleration on the entire trajectory is the use of cubic spline functions [16,18]. Given the points q_i^k, $k=0, \ldots, N$ for the i-th joint coordinate, $i=1, \ldots, n$ and the corresponding time instants t_k, the time history of the i-th coordinate, $t_k \leqslant t \leqslant t_{k+1}$, is given by

$$q_i(t) = \omega_i^k \frac{(t_{k+1}-t)^3}{6h_k} + \omega_i^{k+1} \frac{(t-t_k)^3}{6h_k} + \left(q_i^k - \frac{\omega_i^k h_k^2}{6} \right) \frac{t_{k+1}-t}{h_k} +$$

$$+ \left(q_i^{k+1} - \frac{\omega_i^{k+1} h_k^3}{6} \right) \frac{t-t_k}{h_k}, \qquad t_k \leqslant t \leqslant t_{k+1} \qquad (2.82)$$

where $h_k = t_{k+1} - t_k$. This function satisfies boundary conditions

$$q_i(t_k) = q_i^k, \qquad k=0, \ldots, N \quad i=1, \ldots, n$$

$$\dot{q}_i(t_0) = \dot{q}_i(T) = 0, \qquad\qquad i=1, \ldots, n$$

Variables $\omega_i^k = \ddot{q}_i(t_k)$ are evaluated as the solution to the following linear system of $N+1$ equations

$$2\omega_i^0 + \omega_i^1 = 6 \frac{q_i^1 - q_i^0}{h_0^2}$$

$$h_{k-1} \omega_i^{k-1} + 2(h_k + h_{k-1}) \omega_i^k + h_k \omega_i^{k+1} = 6 \left[\frac{q_i^{k+1} - q_i^k}{h_k} - \frac{q_i^k - q_i^{k-1}}{h_{k-1}} \right],$$

$$k=1, \ldots, N-1 \qquad (2.83)$$

$$\omega_i^{N-1} + 2\omega_i^N = -6 \frac{q_i^N - q_i^{N-1}}{h_{N-1}^2}$$

The main drawback of this algorithm is that it requires all the joint coordinates q^k, $k=0, \ldots, n$, corresponding to the user-specified knots x_e^k, to be evaluated before the trajectory is executed. This prevents the use of such algorithms in on-line motion generation if the number of points is large.

If the selected Cartesian knots are sequentially generated in real time (for example, from a path planner or a vision system), it is not acceptable to allow the arm to rest until all the knots are determined. One piece of a joint trajectory can be determined once a few knots are known. In this situation, X-spline function [17], one kind of cubic spline which requires only local knot information for curve fitting, can be used [12]. However, this method does not provide for continuity of acceleration at time instants t_k. If the continuity of acceleration is to be preserved, then a quartic spline function of the type

$$q(t) = q^k + a_1^k (t-t_k) + a_2^k (t-t_k)^2 + a_3^k (t-t_k)^3 +$$

$$+ a_4^k (t-t_k)^4, \qquad t_k \leqslant t \leqslant t_{k+1}, \quad k=0, \ldots, n-1 \qquad (2.84)$$

can be used [12]. The evaluation of coefficients requires only the knowledge of knots q^k and q^{k+1}.

Industrial tasks sometimes require motion along geometrically defined curves, such as circles, ellipses, etc. In this case it is convenient to introduce coordinate frames in which the curve can easily be described and then describe the curve in parametric form of this system (procedurally defined motion).

The above mentioned techniques are some of the motion generation algorithms which are purely kinematic. On the other hand motion can be optimized with respect to some dynamic criteria, such as actuator energy consumption, execution time, etc. in order to improve robot efficiency. Very important aspects of manipulator path generation today are the planning of manipulator paths at a higher control level to avoid obstacles and adapting the execution of the task from information sensors.

2.6   Conclusion

In this chapter the main principles of manipulator kinematic modelling have been presented. It includes Denavit–Hartenberg kinematic notation, solving for manipulator hand position and orientation in the reference frame, numerical and analytical methods for solving the inverse kinematics, the Jacobian matrix evaluation and the basic algorithms for manipulator path generation.

References

[1] Vukobratović, M., Kircanski, M., "Kinematics and Trajectory Synthesis of Manipulation Robots, Series": Scientific Fundamentals of Robotics 3", Springer-Verlag, 1985.

[2] Denavit, J., Hartenberg, R.S., "Kinematic Notation for Lower-Pair Mechanisms Based on Matrices", ASME J. of Applied Mechanics, June 1955, pp. 215–221.

[3] Paul, R., Robot Manipulators: Mathematics, Programming, and Control, The MIT Press, 1981.

[4] Lee, C.S.G., Ziegler, M., "A Geometric Approach in Solving the Inverse Kinematics of PUMA Robots", Proc. 13th ISIR, Chicago, 1983.

[5] Featherstone, R., "Position and Velocity Transformations Between Robot End-Effector Coordinates and Joint Angles", The Int. J. of Robotics Research, Vol. 2, No. 2, Summer 1983.

[6] Pieper, D.L., Roth, B., "The Kinematics of Manipulators Under Computer Control", Proc. II Int. Congress on Theory of Machines and Mechanisms, Vol. 2., Zakopane, Poland, 1969.

[7] Klien, Ch.A., Ching-Hsian Huang, "Review of Pseudo-Inverse Control for Use with Kinematically Redundant Manipulators", IEEE Trans. on Syst., Man, and Cyber., Vol. SMC-13, No. 3, March/April 1983.

[8] Wampler, C.W., "Manipulator Inverse Kinematic Solutions Based on Vector Formulations and Damped Least-Square Methods", IEEE Trans. on Syst., Man, and Cyber., Vol. SMC-16, No. 1, January/February 1986.

[9] Goldenberg, A.A., Benhabib, B., Fenton, R., "A Complete Generalized Solution to the Inverse Kinematics of Robots", IEEE J. of Robotics and Automation, Vol. RA-1, No. 1, March 1985.

[10] Kircanski, M., Vukobratović, M., "Computer-Aided Generation of Manipulator Kinematic Models in Symbolic Form", Proc. of 15th ISIR, Tokyo, 1985.

[11] Kircanski, M., Vukobratović, M., "A New Program Package for Generating Symbolic Kinematic Models of Arbitrary Serial-Link Manipulators", Proc. of 16th ISIR, Brussels, Belgium, 1986.

[12] Lin, Ch.Sh., Chang, P.R., "Joint Trajectories of Mechanical Manipulators for Cartesian Path Approximation", IEEE Trans. on Syst., Man, and Cyber., Vol. SMC-13, No. 6, Nov./Dec. 1983, pp. 1094–1102.

[13] Chand, S., Doty, K., "On-Line Polynomial Trajectories for Robot Manipulators", Int. J. of Robotics Research, Vol. 4, No. 2, Summer 1985, pp. 38–48.

[14] Paul, R., Zhang Hong, "Robot Motion Trajectory Specification and Generation", II Int. Symp. on Robotics Research, Kyoto, 1984, pp. 373–380.

[15] Castain, R., Paul, R., "An On-Line Dynamic Trajectory Generator", Int. J. of Robotics Research, Vol. 3, No. 1, Spring 1984, pp. 68–72.

[16] Vereschagin, A.F., Generozov, V.L., "Planning of Manipulator Trajectories" (in Russian), Teknicheskaya Kibernetika AN USSR, No. 2, 1978.

[17] Behforooz, H., Papamichael, N, Worsey, A.J., "A Class of Piecewise-Cubic Interpolatory Polynomials", J. Inst. Math. Appl., Vol. 25, 1980, pp. 53–65.

[18] Lin, Ch.Sh., Chang, P.R., "Formulation and Optimization of Cubic Polynomial Joint Trajectories for Industrial Robots", IEEE Trans. on Automatic Control, Vol. AC-28, No. 12, December 1983, pp. 1066–1074.

Chapter 3

Dynamics and Dynamic Analysis of Manipulation Robots

3.1 Introduction

The modern development of robotic mechanisms is rapidly increasing. A linear composite speed of manipulation robots now achieves 5 m/s and angular speed of their particular links (segments) surpasses 8 rad/s. This is not only characteristic of robots with small reachability, but also of robots with larger manipulator possibilities. Such an increase leads to a significant dynamic effect in robot mechanisms, necessitating a careful study of their dynamics. Adequate dynamic models of manipulation robots can be used for robot mechanism design, optimal choice of its actuators, and also for modern robot controller design. A mathematical model derivation of the dynamics of artropoidal robot configuration, frequently used today in industrial practice, is presented below.

3.2 Mathematical Model of Manipulation Robot Dynamics

Forming a mathematical model of large-space mechanisms "by hand" incurs unavoidable errors. Therefore, automatic forming of differential equations of robot mechanism motion by a computer plays a central role in the study of effective dynamics and algorithm synthesis for any robot systems control. Whilst endeavouring to introduce the reader to robotics, a modern procedure for forming mathematical robot models is presented through the accessibility of the computer.

By using digital computers a general algorithm can be found in the forming of mathematical models of active mechanisms. Such an algorithm has to satisfy the following demands.

With information on the kinematic scheme, parameters and the type of problem of the mechanism which has to be solved the algorithm has to:

1. determine the "home" position of the kinematic scheme;
2. calculate the positions, velocities and accelerations of mechanism's segments;

3. assemble differential equations of motion;
4. integrate formed equations with imposed specific conditions on the basic iterative procedure, or to calculate driving forces for a prescribed motion.

The first phase of such an automatic procedure is to determine the "home" mechanism's position or assembly mechanism. In this case, by changing data in input, each mechanism structure change can be simply performed. The rest of the procedure is automatically carried out and the algorithm forms a mechanism movement, or calculates its driving forces. Due to the automatic mechanism assembly, the very delicate selection problem of the robot's mechanism kinematic scheme is easily solved. This would be an inconceivable task, if a mathematical model of the mechanism in analytical form for each possible configuration had to be performed "by hand". Thus the automatic forming of mathematical models of the robot's mechanism motion represents a decisive step in the systematic study of robot mechanics and the synthesis of algorithm control in the various tasks of applied robotics. Systematization of methods for automatic generation of mathematical models can be achieved by adopting various criteria. It is customary that the adopted criterion is founded on the laws of mechanics according to which the method was formulated. According to this criterion, three basic groups of methods are distinguished:

1. Methods based on the second order Lagrange's equations.
2. Methods based on Newton–Euler's dynamic equations (method of general theorems of mechanics)
3. Methods based on Appel's equations.

In this book, it was decided to use Newton–Euler's equations as the most suitable way of demonstrating the physical sense of the problem. These as-

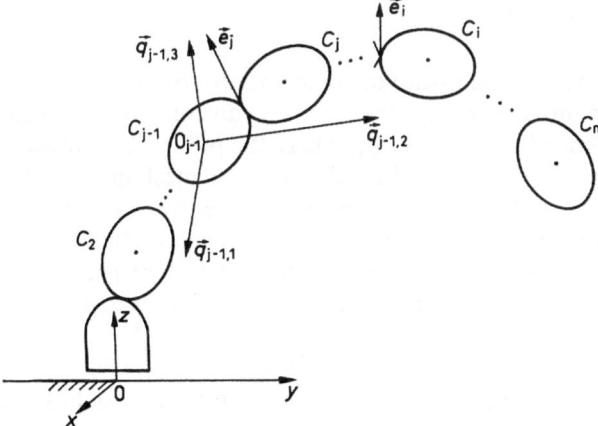

Fig. 3.1. Model of robot mechanism with n links and n joints

sumptions were made:

(a)   links which are modelled by rigid bodies, ignoring the effects of elasticity;
(b)   the kinematic chain of robotic manipulator is neither branched nor closed;
(c)   there are no kinematically coupled degrees of freedom.

Let us consider the model of the robot mechanism shown in Fig. 3.1. The model consists of n rigid bodies which represent the links of the mechanism chain. These links are interconnected by prismatic (sliding) or rotational joints. For a precise derivation of differential equations of the robot dynamics, a number of basic definitions are introduced.

Configuration

Configuration of the mechanism is represented by an arranged n-tuple $(I_1, \ldots, I_n)$, such that for each $i \in N = \{1, \ldots, n\}$, $I_i \in \{R, T\}$, where R denotes a rotational and T a prismatic joint.

To illustrate this, the configuration $RTT-RRR$ stands for a mechanism with six joints, where the second and third joints are prismatic, and the remaining ones are rotational.

Link (segment) of Mechanism

A link is defined by an arranged set of parameters $C_i(K_i, D_i)$, where K_i represents a set of kinematic, D_i a set of dynamic parameters, and $i \in N$ index of mechanism link.

The sets K_i and D_i may be defined in various ways. In this method, these sets are assumed to have the form

$$K_i = (Q_i, \tilde{R}_i, \tilde{E}_i) , \qquad D_i = (m_i, \underline{J}_i)$$

where: $Q_i = [\vec{q}_{i1}, \vec{q}_{i2}, \vec{q}_{i3}]$ the internal (local) orthonormal coordinate system is attached to the mechanism link i, $\tilde{R}_i = \{\vec{r}_{ik}\}$ the set of the distance vectors from points Z_{ik} to the origin of coordinate system Q_i, where the points Z_{ik} represent the centres of joint connections between the i-th and k-th mechanism links, $k \in \{1, \ldots, k_i\}$, $\tilde{E}_i = \{\vec{e}_{ik}\}$ the set of the unit vectors of joint axes by which the i-th mechanism link C_i is connected to the remaining links C_k in points Z_{ik}, m_i-mass of the i-th link, and $\underline{J}_i$ inertia tensor of the i-th link defined with respect to the local system Q_i. Frequently we accept that the axes of coordinate system Q_i are aligned along the principal axes of inertia, so the inertia tensor $\underline{J}_i$ reduces to three moments of inertia $\underline{J}_i = (J_{i1}, J_{i2}, J_{i3})$. In addition, the origin of the coordinate system Q_i is accepted as coinciding with the centre of mass of link C_i. An example of a robot link is shown in Fig. 3.2.

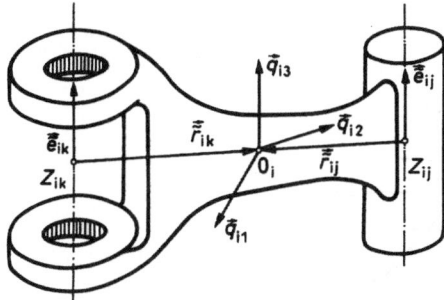

Fig. 3.2. Mechanism link C_i with two joints (Z_{ik} and Z_{ij}) with the centre of mass in point O_i

Kinematic Pair

A kinematic pair P_{ik} represents a set of two adjacent mechanism links $\{C_i, C_k\}$ interconnected by a joint in point Z_{ik}.

The notion of class and subclass of a kinematic pair is introduced depending on the type of joint connection, which, in our consideration, is of a rotational type a common joint type of contemporary robots.

Kinematic Chain

A kinematic chain Λ_n is a set of n interconnected kinematic pairs, $\Lambda_n = \{P_{ik}\}$, $i \in N$, $k \in N$.

According to the structure of connections, chains are classified into simple, complex, open and closed. A chain in which no link C_i, $\forall i \in N$, enters into more than two kinematic pairs is said to be a simple kinematic chain. On the other hand, a complex kinematic chain contains at least one link C_i, $i \in N$ which enters into more than two kinematic pairs. An open kinematic chain possesses at least one link C_i, $i \in N$, which belongs to one kinematic pair only. If each link C_i, $i \in N$ enters into at least two kinematic pairs, the chain is said to be closed.

Joint Coordinates

Scalar quantities which determine, in a unique manner, the relative position of the links of kinematic pair $P_{ik} = \{C_i, C_k\}$ are referred to as manipulator joint coordinates q_{ik}^l. The superscript $l \in \{1, \ldots, s\}$, where $s = 6 - j$, is the number of degrees of freedom, and j the class of pair P_{ik}. In this paragraph we only considered kinematic pairs of the V class, i.e. to pairs with only one rotational degree of freedom of the relative motion. Figure 3.3 shows a rotational kinematic pair whose joint angle (coordinate) is denoted by q_i. For a rotational kinematic pair the corresponding joint (internal) coordinate q_{ik}^l is defined as the rotation angle in a joint around axis $\vec{e}_i$ and can be seen as the angle between the projections of the vectors $-\vec{r}_{i-1,i}$ and $\vec{r}_{ii}$ to the plane vertical to $\vec{e}_i$.

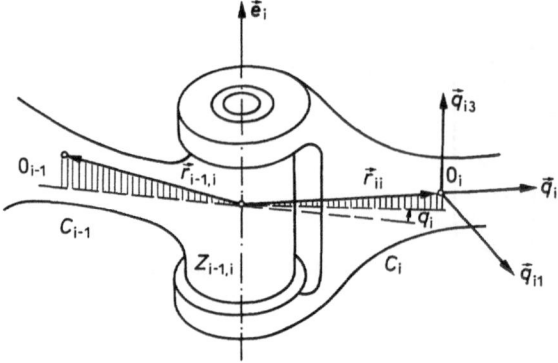

Fig. 3.3. Joint coordinate of a revolute kinematic pair $\{C_{i-1}, C_i\}$

External Coordinates

Scalar quantities x_{ek}, $k \in \{1, \ldots, m\}$, which determine the position and (partially or completely) the orientation of the n-th link of chain Λ_n with respect to a reference coordinate system are said to be external coordinates of any mechanism. The specification of external coordinates, and of their number $m \leqslant 6$, depends on the type of manipulation task, which is mentioned in Chapter 2 of this book.

Active Mechanism

An active mechanism represents a system comprising: (a) a mechanical part which may be modelled by an appropriate kinematic chain, and (b) a set of actuators through which driving torques (forces) are realized.

To form the mathematical model of an active mechanism, it is necessary to identify the parameters of the kinematic chain and actuators, form the dynamic model of mechanism, and form the models of actuators.

The dynamic model of an open active mechanism represents a set of n non-linear differential equations which describe the system motion in the space of joint coordinates through the action of driving forces.

The method which should be presented is a recursive numerical one for forming dynamic models of open active mechanisms by computers. This method consists of the following stages:

1. determination of the "home" position;
2. determination of the actual position of any mechanism links in the reference frame;
3. a kinematic stage;
4. a dynamic stage.

Stage 1: Determination of "home" Position – Mechanism Assembly

It is assumed that the active mechanism under consideration may be described by a simple, open kinematic chain Λ_n consisting of a set of kinematic pairs $\{C_{i-1}, C_i\}$, $i \in N$. In fact the kinematic pair $\{C_0, C_1\}$ represents the first link of the mechanism which is connected to support C_0 by joint Z_{10}. It is necessary to determine the position of all links in the reference frame, under the supposition that all joint coordinates equal zero, $q_i = 0$, $i \in N$.

Let us assume all kinematic parameters $K_j^0 = (Q_j^0, R_j^0, E_j^0)$ for $j \in (1, \ldots, i-1)$ to be known, where the superscript 0 means that $q_j = 0$, $\forall j \in N$.

The task of determining the "home" position is now reduced to determining $K_i^0 = (Q_i^0, R_i^0, E_i^0)$. The sets R_i and E_i in the case of a simple kinematic chain have the following form:

$$R_i = \{\vec{r}_{ii}, \vec{r}_{i,i+1}\} \quad \text{and} \quad E_i = \{\vec{e}_i, \vec{e}_{i+1}\}$$

where $\vec{e}_i$ and $\vec{e}_{i+1}$ stand for vectors $\vec{e}_{ii}$ and $\vec{e}_{i,i+1}$.

On the basis of the assumption that K_{i-1}^0 is known, it follows that $\vec{e}_i^0$ and $\vec{r}_{i-1,i}^0$ are also known. Further, let us note the set of three orthogonal unit vectors in points $Z_{i-1,i}$: $\vec{e}_i^0$, $\vec{a}_i^0$ and $\vec{e}_i^0 \times \vec{a}_i^0$, where $\vec{a}_i^0 = -\text{ort}(\vec{e}_i^0 \times (\vec{r}_{i-1,i}^0 \times \vec{e}_i^0))$, whose components are known in the reference coordinate system (Fig. 3.4). It should be noticed that ort($\cdot$) denotes the unit vector of ($\cdot$).

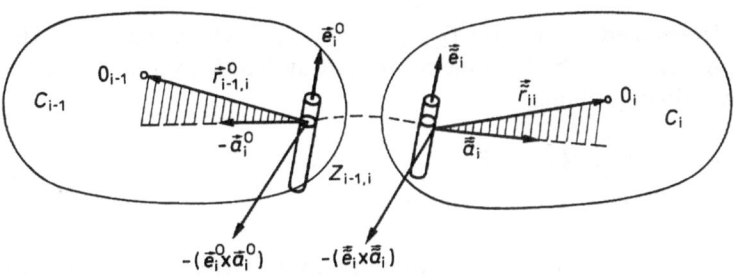

Fig. 3.4. Determination of the "home" position

On the other hand, let us note the set of vectors, $\vec{\tilde{e}}_i$, $\vec{\tilde{a}}_i$, and $\vec{\tilde{e}}_i \times \vec{\tilde{a}}_i$, where $\vec{\tilde{a}}_i = -\text{ort}(\vec{\tilde{e}}_i \times (\vec{\tilde{r}}_{ii} \times \vec{\tilde{e}}_i))$, whose components are known with respect to the local coordinate system Q_i. Under the condition $q_i = 0$, these two sets of vectors coincide. Since $Q_i^0 = [\vec{q}_{i1}^0 \; \vec{q}_{i2}^0 \; \vec{q}_{i3}^0]$ represents the transformation matrix of the i-th link C_i joint coordinate system into the reference system, it follows that

$$\vec{e}_i^0 = Q_i^0 \vec{\tilde{e}}_i \,, \quad \vec{a}_i^0 = Q_i^0 \vec{\tilde{a}}_i \,, \quad \vec{e}_i^0 \times \vec{a}_i^0 = Q_i^0 (\vec{\tilde{e}}_i \times \vec{\tilde{a}}_i) \tag{3.1}$$

and this completely determines the matrix Q_i^0:

$$Q_i^0 = [\vec{e}_i^0 \; \vec{a}_i^0 \; \vec{e}_i^0 \times \vec{a}_i^0] [\vec{\tilde{e}}_i \; \vec{\tilde{a}}_i \; \vec{\tilde{e}}_i \times \vec{\tilde{a}}_i]^\mathsf{T} \tag{3.2}$$

Let us note that the matrix transformation has been used instead of its inverse, due to the orthogonality of vectors $\vec{e}_i$, $\vec{a}_i$ and $\vec{e}_i \times \vec{a}_i$.

It is now necessary to determine the remaining elements of set K_i^0, i.e. sets R_i^0 and E_i^0. Since the transformation matrix Q_i^0 has been determined, the following relations evidently hold

$$\vec{r}_{ii}^0 = Q_i^0 \vec{r}_{ii} \; , \qquad \vec{r}_{i,i+1}^0 = Q_i^0 \vec{r}_{i,i+1} \; , \qquad \vec{e}_{i+1}^0 = Q_i^0 \vec{e}_{i+1} \tag{3.3}$$

Recursiveness required for calculating Q_i^0 for $\forall i \in N$, and of all vectors contained in R_i^0 and E_i^0, has thus been established.

The derivation of the transformation matrix is presented in Appendix A.3.1.

Stage 2: Mechanism Position

This stage consists of determining the elements of the set of kinematic variables $K_i = (Q_i, R_i, E_i)$, where the mechanism position is determined by joint co-ordinates q_i, $i \in N$.

Applying the theorem of finite rotations[†] (Rodrigues' formula) for the case of rotational joints, one obtains

$$\vec{q}_{ij} = \begin{bmatrix} \vec{e}_i \times (\vec{q}_{ij}^0 \times \vec{e}_i) \\ (\vec{e}_i \times \vec{q}_{ij}^0) \\ (\vec{e}_i \cdot \vec{q}_{ij}^0)\vec{e}_i \end{bmatrix}^{\mathrm{T}} \begin{bmatrix} \cos q_i \\ \sin q_i \\ 1 \end{bmatrix} \; , \qquad j = 1, 2, 3 \tag{3.4}$$

We have thus determined the matrix $Q_i = [\vec{q}_{i1}, \vec{q}_{i2}, \vec{q}_{i3}]$ which represents the transformation matrix from the i-th link coordinate system S_i into the reference system. Now, we directly obtain

$$\begin{aligned} R_i &= \{\vec{r}_{ii}, \vec{r}_{i,i+1}\} = \{Q_i \vec{r}_{ii}, Q_i \vec{r}_{i,i+1}\} \\ E_i &= \{\vec{e}_i, \vec{e}_{i+1}\} = \{\vec{e}_i, Q_i \vec{e}_{i+1}\} \end{aligned} \tag{3.5}$$

where $\vec{r}_{ij} = \vec{r}_{i-1,j} - \vec{r}_{i-1,i} + \vec{r}_{ii}$, $j \in I$, $i \in N$, evidently holds. Determination of $\vec{e}_{i+1}$ provides for the recursiveness of calculating the transformation matrix Q_i according to Eq. (3.4), and thus the elements of sets, Eq. (3.5).

Stage 3: Mechanism Kinematics

This stage consists of forming the set of kinematic quantities $\alpha_i = \{\Omega_i, W_i\}$, where

$$\Omega_i = \{\vec{\omega}_i\} \; , \qquad W_i = \{\vec{\varepsilon}_i, \vec{w}_i\}$$

[†] Instead of Denavit–Hartenberg's coordinates which have been used in Chapter 2, here we use Rodrigues' formula of finite rotations which is somewhat more favourable for forming dynamic equations of the robot. Namely, by applying Rodrigues' formula joint (local) coordinate systems can be coincided with the principal axes of inertia, so that the inertia tensor reduces to three principal moments of inertia.

with $\vec{\omega}_i$ as the angular velocity of the i-th link, $\vec{\varepsilon}_i$ the angular acceleration of the i-th link, and $\vec{w}_i$ the linear acceleration of the i-th link.

These kinematic quantities are functions of the velocities and accelerations of joint coordinates $\dot{q}_i$ and $\ddot{q}_i$ and variables formed in the preceding stage: $K_i = (Q_i, R_i, E_i)$. All relations considered in this stage are recursive. Applying the basic theorems of rigid body kinematics, one obtains

$$
\begin{aligned}
\vec{\omega}_i &= \vec{\omega}_{i-1} + \dot{q}_i \vec{e}_i \\
\vec{\varepsilon}_i &= \vec{\varepsilon}_{i-1} + \dot{q}_i \vec{\omega}_{i-1} \times \vec{e}_i + \ddot{q}_i \vec{e}_i \\
\vec{w}_i &= \vec{w}_{i-1} - \vec{\varepsilon}_{i-1} \times \vec{r}_{i-1,i} - \vec{\omega}_{i-1} \times (\vec{\omega}_{i-1} \times \vec{r}_{i-1,i}) + \\
&\quad + \vec{\varepsilon}_i \times \vec{r}_{ii} + \vec{\omega}_i \times (\vec{\omega}_i \times \vec{r}_{ii})
\end{aligned}
\tag{3.6}
$$

The basic relations of rigid body kinematics are given in Appendix A.3.2 of this chapter.

It follows that angular and linear accelerations are functions of the second derivatives of joint coordinates $\ddot{q}_j$, $j \in I$. However, to form the matrices of the dynamic model of mechanism, expressions $\vec{\varepsilon}_i$ and $\vec{w}_i$ should be rearranged so that accelerations $\ddot{q}_j$, $j \in N$, figure explicitly

$$
\begin{aligned}
\vec{\varepsilon}_i &= [\vec{\alpha}_{i1} \ldots \vec{\alpha}_{ii} \, 0 \ldots 0] \ddot{q} + \vec{\theta}_i + \vec{\varepsilon}_0 \\
\vec{w}_i &= [\vec{\beta}_{i1} \ldots \vec{\beta}_{ii} \, 0 \ldots 0] \ddot{q} + \vec{\eta}_i + \vec{w}_0
\end{aligned}
\tag{3.7}
$$

where $\ddot{q} = [\ddot{q}_1 \ldots \ddot{q}_n]^T$ and $\vec{\varepsilon}_0$, $\vec{w}_0$ are the angular and linear acceleration of the basic mechanism link. The set of kinematic quantities W_i thus reduces to

$$
W_i = \{\vec{\alpha}_{ij}, \vec{\beta}_{ij}, \vec{\theta}_i, \vec{\eta}_i, j \in I\}
$$

On the basis of Eqs. (3.6) and (3.7) we can write the following relations:

$$
\begin{aligned}
\sum_{j=1}^{i} \vec{\alpha}_{ij} \ddot{q}_j + \vec{\theta}_i + \vec{\varepsilon}_0 &= \sum_{j=1}^{i-1} \vec{\alpha}_{i-1} \ddot{q}_j + (\ddot{q}_i \vec{e}_i + \dot{q}_i \omega_{i-1} \times \vec{e}_i) + \\
&\quad + \vec{\theta}_{i-1} + \vec{\varepsilon}_0 \\[2ex]
\sum_{j=1}^{i} \vec{\beta}_{ij} \ddot{q}_j + \vec{\eta}_i + \vec{w}_0 &= \sum_{j=1}^{i-1} \vec{\beta}_{i-1,j} \ddot{q}_j + \vec{\eta}_{i-1} + \vec{w}_0 - \\
&\quad - \left(\sum_{j=1}^{i-1} \vec{\alpha}_{i-1,j} \ddot{q}_j + \vec{\theta}_{i-1} + \vec{\varepsilon}_0 \right) \times \vec{r}_{i-1,i} - \\
&\quad - \vec{\omega}_{i-1} \times (\vec{\omega}_{i-1} \times \vec{r}_{i-1,i}) + \left(\sum_{j=1}^{i-1} \vec{\alpha}_{i-1} \ddot{q}_j + \vec{\theta}_{i-1} + \vec{\varepsilon}_0 \right) \times \vec{r}_{ii} + \\
&\quad + (\ddot{q}_i \vec{e}_i + \dot{q}_i \vec{\omega}_{i-1} \times \vec{e}_i) \times \vec{r}_{ii} + \vec{\omega}_i \times (\vec{\omega}_i \times \vec{r}_{ii})
\end{aligned}
$$

Equalizing the coefficients with equal j, one obtains

$$\vec{\alpha}_{ij} = \vec{\alpha}_{i-1,j} , \quad j \neq i$$

$$\vec{\alpha}_{ij} = \vec{e}_j , \quad j \in I , \quad i \in N$$

$$\vec{\theta}_i = \vec{\theta}_{i-1} + \dot{q}_i(\vec{\omega}_{i-1} \times \vec{e}_i) , \quad \vec{\theta}_0 = 0 , \quad i \in N \tag{3.8}$$

$$\vec{\beta}_{ij} = \vec{\beta}_{i-1,j} + \vec{\alpha}_{i-1,j} \times (\vec{r}_{ii} - \vec{r}_{i-1,i}) , \quad j \in \{1, 2, \ldots, i-1\} , \quad i \in N$$

$$\vec{\beta}_{ii} = (\vec{e}_i \times \vec{r}_{ii})$$

$$\vec{\eta}_i = \vec{\eta}_{i-1} + \vec{\theta}_{i-1} \times (\vec{r}_{ii} - \vec{r}_{i-1,i}) + \dot{q}_i(\vec{\omega}_{i-1} \times \vec{e}_i) \times \vec{r}_{ii} +$$

$$+ \vec{\gamma}_{ii} - \vec{\gamma}_{i-1,i} - \vec{\varepsilon}_0 \times (\vec{r}_{i-1,i} - \vec{r}_{ii})$$

$$\vec{\eta}_0 = 0 , \quad \vec{\gamma}_{ij} = \vec{\omega}_i \times (\vec{\omega}_i \times \vec{r}_{ij}) , \quad j \in I , \quad i \in N$$

In this way recursive relations for determining all necessary vector coefficients $\vec{\alpha}_{ij}$, $\vec{\beta}_{ij}$, $\vec{\theta}_i$ and $\vec{\eta}_i$ are obtained. For instance, the vector coefficient $\vec{\eta}_i$ for the i-th kinematic pair is expressed via the vector coefficient of the $(i\text{-}1)$-th pair. If we now express $\vec{\eta}_{i-1}$ via $\vec{\eta}_{i-2}$, and so on to the first pair, one obtains

$$\vec{\eta}_i = \vec{\eta}_i(\vec{\varepsilon}_0 = 0) + \vec{\varepsilon}_0 \times \sum_{k=1}^{i} (\vec{r}_{kk} - \vec{r}_{k-1,k}) = \vec{\eta}_i(\vec{\varepsilon}_0 = 0) + \vec{\varepsilon}_0 \times \vec{r}_{i,1}$$

where $\vec{\eta}_i(\vec{\varepsilon}_0 = 0)$ is the vector coefficient $\vec{\eta}_i$ for case $\vec{\varepsilon}_0 = 0$, and $\vec{r}_{i,1}$ is the radius vector from the first joint to the centre of mass of the i-th segment. Using these coefficients, requested accelerations (linear and angular) can be found in the function of relative (generalized) accelerations.

Stage 4: Mechanism Dynamics

This stage includes evaluation of dynamic quantities $\Delta_i = \{F_i, M_i\}$, $i \in N$, where

$$F_i = \{\vec{F}_i\} , \quad M_i = \{\vec{M}_i\}$$

with

$\vec{F}_i$ the inertial force in the centre of mass of the i-th link, and

$\vec{M}_i$ the moment of the inertial force of the i-th link.

The inertial force may be calculated using the second Newton's law:

$$\vec{F}_i = -m_i\vec{w}_i = [\vec{a}_{i1} \ldots \vec{a}_{ii} \, 0 \ldots 0]\ddot{q} + \vec{a}_i^0 - m_i\vec{w}_0 \tag{3.9}$$

where m_i is the mass of the i-th mechanism link. Comparing this with Eq. (3.7) we obtain

$$\vec{a}_{ij} = -m_i\vec{\beta}_{ij} , \quad \vec{a}_i^0 = -m_i\vec{\eta}_i \tag{3.10}$$

The moments of inertial forces are determined from Euler's dynamic equations (see Appendix A.3.3):

$$\tilde{M}_i^1 = -J_{i1}\tilde{\varepsilon}_i^1 + (J_{i2}-J_{i3})\tilde{\omega}_i^2\tilde{\omega}_i^3$$

$$\tilde{M}_i^2 = -J_{i2}\tilde{\varepsilon}_i^2 + (J_{i3}-J_{i1})\tilde{\omega}_i^3\tilde{\omega}_i^1 \qquad (3.11)$$

$$\tilde{M}_i^3 = -J_{i3}\tilde{\varepsilon}_i^3 + (J_{i1}-J_{i2})\tilde{\omega}_i^1\tilde{\omega}_i^2$$

where $\sim$ denotes projections of the vectors $\vec{M}$, $\vec{\tilde{\varepsilon}}$ and $\vec{\tilde{\omega}}$ to the local (mobile) coordinate system. The superscript indicates the projection number, i.e. $\tilde{\varepsilon}_i^j = \vec{\varepsilon}_i \cdot \vec{q}_{ij}$, or $\tilde{M}_i^j = \vec{M}_i \cdot q_{ij}$, $i \in I, j = 1, 2, 3$.

Moving to the immobile system, a matrix transformation Q_i of the i-th segment is used:

$$\vec{M}_i = Q_i \vec{\tilde{M}}_i \qquad (3.12)$$

Multiplying by Q_i the left and right side of the expression (3.11) and substituting acceleration projections $\tilde{\varepsilon}_i^j$ with their expressions (3.7), the moment projections of inertial forces to the axes of absolute system are obtained:

$$M_i^j = -[(Q_i^{j1}J_{i1}q_{i1}^1 + Q_i^{j2}J_{i2}q_{i2}^1 + Q_i^{j3}J_{i3}q_{i3}^1)\varepsilon_i^1 +$$
$$+ (Q_i^{j1}J_{i1}q_{i1}^2 + Q_i^{j2}J_{i2}q_{i2}^2 + Q_i^{j3}J_{i3}q_{i3}^2)\varepsilon_i^2 + \qquad (3.13)$$
$$+ (Q_i^{j1}J_{i1}q_{i1}^3 + Q_i^{j2}J_{i2}q_{i2}^3 + Q_i^{j3}J_{i3}q_{i3}^3)\varepsilon_i^3] + \lambda_i^j , \quad j = 1, 2, 3$$

where

$$\vec{\lambda}_i = \begin{bmatrix} \lambda_i^1 \\ \lambda_i^2 \\ \lambda_i^3 \end{bmatrix} = Q_i \begin{bmatrix} (\vec{\omega}_i \cdot \vec{q}_{i2})(\vec{\omega}_i \cdot \vec{q}_{i3}) \cdot (J_{i2}-J_{i3}) \\ (\vec{\omega}_i \cdot \vec{q}_{i3})(\vec{\omega}_i \cdot \vec{q}_{i1}) \cdot (J_{i3}-J_{i1}) \\ (\vec{\omega}_i \cdot \vec{q}_{i1})(\vec{\omega}_i \cdot \vec{q}_{i2}) \cdot (J_{i1}-J_{i2}) \end{bmatrix} \qquad (3.14)$$

Since $Q_i^{jk} = q_{ik}^j$, an expression for the moment of inertial forces of the i-th segment may be written in the form

$$\vec{M}_i = -T_i\vec{\varepsilon}_i + \vec{\lambda}_i \qquad (3.15)$$

where T_i is the matrix 3×3 whose elements are

$$T_i^{jk} = \sum_{l=1}^{3} Q_i^{jl}J_{il}q_{il}^k = \sum_{l=1}^{3} q_{il}^j q_{il}^k J_{il} \qquad (3.16)$$

Substituting Eq. (3.7) into Eq. (3.15), one obtains

$$\vec{M}_i = \sum_{j=1}^{i} \vec{b}_{ij}\ddot{q}_j + \vec{b}_i^0 - T_i\vec{\varepsilon}_0 \qquad (3.17)$$

where $\vec{b}_{ij} = -T_i\vec{\alpha}_{ij}$, $\vec{b}_i^0 = -T_i\vec{\theta}_i + \vec{\lambda}_i$.

Apart from the inertial forces and moments mechanism, links are also acted upon by external forces and moments $\vec{G}_i$, $\vec{M}_i^G$, $i \in N$. Thus the total forces and moments may be expressed in the form

$$\vec{F}_i^u = \vec{F}_i + \vec{G}_i \, , \qquad \vec{M}_i^u = \vec{M}_i + \vec{M}_i^G \tag{3.18}$$

Introducing Eq. (3.9), (3.17) into Eq. (3.18), one obtains

$$\vec{F}_j^u = \sum_{k=1}^{j} \vec{a}_{jk} \ddot{q}_k + \vec{a}_j^0 - m_j \vec{w}_0 + \vec{G}_j$$

$$\vec{M}_j^u = \sum_{k=1}^{j} \vec{b}_{jk} \ddot{q}_k + \vec{b}_j^0 - T_j \vec{\varepsilon}_0 + \vec{M}_j^G \tag{3.19}$$

Let a kinematic chain be fictitiously disconnected in the i-th joint. Consider then the equilibrium of the mechanism free end (Fig. 3.5). The action of the "rejected" mechanism part is substituted by a force $\vec{R}_i$ and moment $\vec{M}_i^*$. This force and moment will be termed as the total reaction force at the i-th joint. In determining the overall reaction the external forces $\vec{F}_j^u$ and moments $\vec{M}_j^u$ ($j = 1, i+1, \ldots, n$) are reduced to the center of the i-th joint. Reactions of drives of all subsequent joints need not be taken into account since each drive acts upon two adjacent links with forces and moments of equal magnitude but opposite sense which are thus annulled in the summation process.

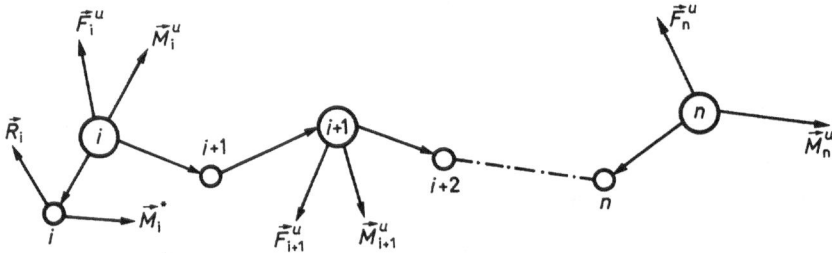

Fig. 3.5. Free chain equilibrium

Then, the total reactions may be presented in the following form

$$\vec{R}_i = -\sum_{j=i}^{n} \vec{F}_j^u = -\sum_{j=i}^{n} \left(\sum_{k=1}^{j} \vec{a}_{jk} \ddot{q}_k + \vec{a}_j^0 - m_j \vec{w}_0 + \vec{G}_j \right)$$

$$\vec{M}_i^* = -\sum_{j=i}^{n} \left(\vec{M}_j^u + \vec{r}_{ji} \times \vec{F}_j^u \right) = -\sum_{j=i}^{n} \left[\sum_{k=1}^{j} (\vec{b}_{jk} + \vec{r}_{ji} \times \vec{a}_{jk}) \ddot{q}_k \right.$$

$$\left. + \vec{r}_{ji} \times \vec{a}_j^0 - m_j \vec{r}_{ji} \times \vec{w}_0 + \vec{b}_j^0 - T_j \vec{\varepsilon}_0 + \vec{M}_j^G + \vec{r}_{ji} \times \vec{G}_j \right] \tag{3.20}$$

where n is number of the chain links.

To determine the reactions it is appropriate to establish the recursive relations between the reactions in two neighbouring joints. From Fig. 3.6, where one mechanism link is presented, it is obviously

$$\vec{R}_i = \vec{R}_{i+1} - \vec{F}_i^u \tag{3.21}$$
$$\vec{M}_i^* = \vec{M}_{i+1}^* + (\vec{r}_{ii} - \vec{r}_{i+1,i}) \times \vec{R}_{i+1} + \vec{r}_{ii} \times \vec{F}_i^u - \vec{M}_i^u$$

Going along the chain from the last to the first link according to Eq. (3.21) all the reactions can be determined.

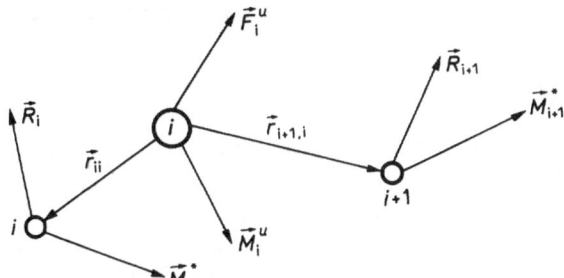

Fig. 3.6. Reaction of the i-th link

Disassemble total reaction in the i-th joint in two components: one parallel with vector $\vec{e}_i$, and the second, perpendicular to the joint axis. The perpendicular components cannot cause mechanism movement, they are only acting as load on the joint, and are necessary when considering the friction forces. One of the parallel components (moment for linear kinematic pairs and forces for rotational pairs) also contributes to the friction forces. In order to maintain the mechanism immobile, the other parallel component of the reaction must be balanced by a reaction at the i-th joint. If the joint is rotational driving torque is

$$P_i = \vec{M}_i^* \cdot \vec{e}_i \tag{3.22}$$

Using relations (3.22) and (3.20), it is possible to determine the necessary forces and moments in each joint of the robot mechanism. Thus, we obtain

$$P_i^M = -\vec{e}_i \cdot \sum_{k=1}^{n} \cdot \sum_{j=\max(i,k)}^{n} (\vec{b}_{jk} + \vec{r}_{ji} \times \vec{a}_{jk}) \ddot{q}_k - \vec{e}_i \cdot \left[\sum_{j=i}^{n} \vec{r}_{ji} \times (\vec{a}_j^0 + \right.$$
$$\left. + \vec{G}_j - m_j \vec{w}_0) + (\vec{b}_j^0 - T_j \vec{\varepsilon}_0) \right]$$

This expression may be written in matrix form

$$P = H(q, \theta)\ddot{q} + h(q, \dot{q}, \theta) \tag{3.23}$$

where $P = [P_1 \ldots P_n]^T$ is the vector of driving torques, $q = [q_1 \ldots q_n]^T$ is the vector of joint (internal) coordinates, $\theta = [\theta_1 \ldots \theta_n]^T$ is the vector of geometrical and dynamic parameters. According to the above expressions, matrix elements (3.23) have the form

$$[H_{ik}] = \left[-\vec{e}_i \cdot \sum_{j=\max(i,k)}^{n} (\vec{b}_{jk} + \vec{r}_{ji} \times \vec{a}_{jk}) \right]$$ (3.24)

$$[h_i] = \left[-\vec{e}_i \cdot \sum_{j=i}^{n} \vec{r}_{ji} \times (\vec{a}_j^0 + \vec{G}_j - m_j \vec{w}_0) + (\vec{b}_j^0 - T_j \vec{\varepsilon}_0) \right]$$

In order to understand more the presented computer algorithms in Appendix A.3.4 a mathematical model of the dynamics of one robot is formed manually.

3.2.1 Mathematical Models of Actuators

It is well known that manipulation systems and robots generally consist of the mechanical part S^M and actuators S^i which are powered mechanism degrees-of-freedom. As previously stated, the mathematical model of the dynamics of the mechanical system part S^M can be expressed in the form of Eq. (3.23). Generally, three actuator types are used for robotic mechanisms. These are the electro-mechanical, electro-hydraulic and electro-pneumatic actuators. The electro-pneumatic actuators are applicable for manipulator mechanisms, the other types being applicable for robot mechanisms. Recently the payload of industrial robots with electromechanical actuators has grown to 100 kg and more.

Generally, actuator models can be presented in the following form

$$S^i: \quad \dot{x}^i = A^i x^i + b^i N(u_i) + f^i P_i$$ (3.25)

where

A^i = subsystem matrix,
x^i = state vector of the i-th actuator,
b^i = input distribution vector,
$N(u_i)$ = amplitude saturation nonlinearity,

$$N(u_i) = \begin{cases} -u_i^m & \text{for} \quad u_i^m < -u_i^m \\ u_i & \text{for} \quad -u_i^m \leqslant u_i \leqslant u_i^m \\ u_i^m & \text{for} \quad u_i > u_i^m \end{cases}$$

f^i = load distribution vector, and
P_i = vector of driving torque (force) of the i-th degree of freedom

The total system dynamic model S is obtained by uniting the model of the system mechanical part S^m and the actuators model S^i. In order to avoid matrix calculations and to reduce necessary computing time, we proceed as follows since the model of the total system is sufficiently complex. We start from the differential equations which describe the behaviour of electro-mechanical actuator (D.C. motor electrical scheme is presented in Fig. 3.7) in the following way:

$$L_r \dot{i}_r + R_r i_r + C_E \dot{q} = u \tag{3.26}$$
$$-C_M i_r + J_r \ddot{q} + B_C \dot{q} = -P$$

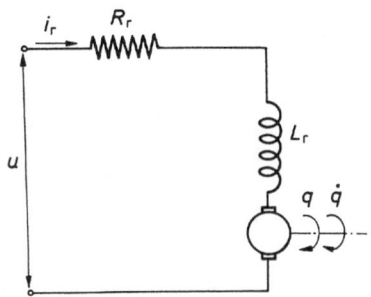

Fig. 3.7. Electrical scheme of D.C. motor

where

$L_r[H]$	= rotor inductivity,
$i_r[mA]$	= rotor current,
$R_r[\Omega]$	= rotor resistance,
$C_E[V/\text{rad/s}]$	= proportionality constant of the electromotor force,
$q[\text{rad}]$	= rotational angle of the motor output shaft
$u[V]$	= voltage on the rotor of motor,
$C_M[\text{Nm/mA}]$	= moment proportionality constant,
$J_r[\text{kg m}^2]$	= rotor moment of inertia J_M reduced to the output shaft
$B_C[\text{Nm/rad/s}]$	= viscous friction coefficient, and
$P[\text{Nm}]$	= moment of the external motor load

Here, it has to be noticed that

$$C_E = C_e N_V , \qquad C_M = C_m N_M$$
$$B_C = B_c N_V N_M , \qquad J_r = J_M N_V N_M$$

where N_v and N_M are the gear speed ratio and gear torque ratio, respectively; C_e, C_m, B_c are particular catalogue values.

Dividing the first equation from Eq. (3.26) with L_r, and the second equation with J_r, we obtain the following relationship

$$\dot{i}_r = -\frac{C_E}{L_r}\dot{q} - \frac{R_r}{L_r}i_r + \frac{u}{L_r} \tag{3.27}$$

$$\ddot{q} = -\frac{B_C}{J_r}\dot{q} + \frac{C_M}{J_r}i_r - \frac{P^\dagger}{J_r} \tag{3.28}$$

Adopting the state vector of the robot subsystem in form $x^i = (q_i, \dot{q}_i, i_r^i)^T$ we obtain all the matrix A^i elements and components of the vectors b^i and f^i from Eq. (3.25).

$$A^i = \begin{bmatrix} 0 & 1 & 0 \\ 0 & -\dfrac{B_C}{J_r} & \dfrac{C_M}{J_r} \\ 0 & -\dfrac{C_E}{L_r} & -\dfrac{R_r}{L_r} \end{bmatrix}, \quad b^i = \begin{bmatrix} 0 \\ 0 \\ \dfrac{1}{L_r} \end{bmatrix}, \quad f^i = \begin{bmatrix} 0 \\ -\dfrac{1}{J_r} \\ 0 \end{bmatrix} \tag{3.29}$$

The mathematical models of the actuators, as shown, are in the form of the system of the third order linear differential equation with constant coefficients. However, by neglecting the motor rotor inductivity ($L_r \approx 0$) and viscous friction ($B_C \approx 0$) the models of the second order are used more frequently, because they are simpler for calculation. It can be simply proven that in this case the subsystem matrix of the actuator A^i and the vectors b^i and f^i are in the form of

$$A^i = \begin{bmatrix} 0 & 1 \\ 0 & -\dfrac{C_M C_E}{J_r R_r} \end{bmatrix}, \quad b^i = \begin{bmatrix} 0 \\ \dfrac{C_M}{J_r R_r} \end{bmatrix}, \quad f^i = \begin{bmatrix} 0 \\ -\dfrac{1}{J_r} \end{bmatrix} \tag{3.30}$$

Neglecting the rotor inductivity only and solving the first equation of (3.26) for i_r we obtain

$$i_r^i = (u_i - C_E^i \dot{q}_i)/R_r^i \tag{3.31}$$

where index i denotes the joint number of manipulator.

Substituting Eq. (3.31) into Eq. (3.28) we obtain

$$\ddot{q}_i = -\frac{B_C^i}{J_r^i}\dot{q}_i + \frac{C_M^i}{J_r^i}(u_i - C_E^i \dot{q}_i)\frac{1}{R_r^i} - \frac{P_i}{J_r^i} =$$

$$= -\left(\frac{B_C^i}{J_r^i} + \frac{C_M^i C_E^i}{J_r^i R_r^i}\right)\dot{q}_i + \frac{C_M^i}{J_r^i R_r^i}u_i - \frac{P_i}{J_r^i} \tag{3.32}$$

$\dagger$ When the angular motion of robot's mechanism realized by D.C. electro-motor or vane hydraulic actuator, then a load M_i^* which acts on the actuator equals to the driving torque P_i of the robot mechanism.

From Eq. (3.32) u_i can be easily calculated:

$$u_i = \frac{J_r^i R_r^i}{C_M^i} \ddot{q}_i + \left(\frac{R_r^i B_C^i}{C_M^i} + C_E^i \right) \dot{q}_i + \frac{R_r^i}{C_M^i} P_i$$

i.e. for all manipulation systems u can be expressed in the form

$$u = \frac{J_r R_r}{C_M} \ddot{q} + \left(\frac{R_r B_C}{C_M} + C_E \right) \dot{q} + \frac{R_r}{C_M} P \ . \tag{3.33}$$

where: $u = [u_1 \ldots u_n]^T$, $\ddot{q} = [\ddot{q}_1 \ldots \ddot{q}_n]^T$, $\dot{q} = [\dot{q}_1 \ldots \dot{q}_n]^T$, $P = [P_1 \ldots P_n]^T$.

If, instead of driving torques P, we substitute Eq. (3.23) into Eq. (3.33), we obtain

$$u = \frac{J_r R_r}{C_M} \ddot{q} + \left(\frac{R_r B_C}{C_M} + C_E \right) \dot{q} + \frac{R_r}{C_M} (H\ddot{q} + h)$$

This relationship may be written in a more suitable form as

$$\frac{J_r R_r}{C_M} \ddot{q} + \frac{R_r}{C_M} H\ddot{q} = u - \left(\frac{R_r B_C}{C_M} + C_E \right) \dot{q} - \frac{R_r}{C_M} h$$

If we multiply the previous relationship by C_M and divide by J_r, we can obtain the following form

$$J_r \ddot{q} + H\ddot{q} = \frac{C_M}{R_r} u - \left(B_C + \frac{C_E C_M}{R_r} \right) \dot{q} - h$$

or

$$(H + J_r I_n)\ddot{q} = \frac{C_M}{R_r} u - \left(B_C + \frac{C_E C_M}{R_r} \right) \dot{q} - h$$

where I_n is a unit matrix of the n-th order. Solving this equation, we obtain

$$\ddot{q} = (H + J_r I_n)^{-1} \left[\frac{C_M}{R_r} u - \left(B_C + \frac{C_E C_M}{R_r} \right) \dot{q} - h \right] \tag{3.34}$$

After two integrations of the expression (3.34) we obtain a law of the change of internal coordinates. In Eqs. (3.32)–(3.34) equal actuators were assumed.

3.2.2 Trajectories Synthesis and Dynamic Analysis of Manipulation Robots

Mathematical models (3.23) give us the possibility to solve two basic problems of dynamics: direct and inverse. The direct problem comprises a necessary driving forces and moments calculation in mechanism joints, for realizing the prescribed motion. The inverse problem comprises calculating the motion for given driving forces or moments.

Practical robotic tasks are solved through a direct problem approach, i.e. for prescribed robot motion (known joints trajectories), corresponding driving forces need to be calculated. However, in practice a manipulation task, as a rule, is not prescribed in the generalized (internal) mechanism coordinates, but via the so-called "external quantities" X; for example: tip manipulator motion law and gripper orienting in external space. In order to prescribe a manipulation task correctly, it is necessary to have unique dependence between the external quantities (vector X) and the generalized (internal) coordinates (vector q). In the case where unique dependence does not exist, for example if there is an excess in mechanism's degrees of freedom (system redundancy), this represents a special problem which will not be considered here.

Let us designate by η the function which transforms the generalized coordinates q into external ones X,

$$X = \eta(q) \tag{3.35}$$

where q and X are n-dimensional vectors. The function η is one-place and can always be determined (not explicitly but as a computational algorithm). The problem lies in the difficulty of calculating q from such a system of equations (3.35) resulting from the impossibility to express q either explicitly, or even approximately, numerically because of the complexity of the system which has to be solved.

To form the mathematical models, it is necessary to know q and $\dot{q}$ in each time instant, but if we want to determine the drives P, we have to know $\ddot{q}$, too.

Let us explain the procedure for forming this model on the basis of a prescribed profile (trajectories) of external quantities X and known states q, $\dot{q}$.

By double differentiation,

$$\dot{X} = \frac{\partial \eta}{\partial q} \dot{q} \tag{3.36}$$

$$\ddot{X} = \frac{\partial \eta}{\partial q} \ddot{q} + \frac{\partial^2 \eta}{\partial q^2} \dot{q}^2 \tag{3.37}$$

Let us introduce the notation $X^a = \ddot{X}$, $J = \dfrac{\partial \eta}{\partial q}$, $A = \dfrac{\partial^2 \eta}{\partial q^2} \dot{q}^2$. Equation (3.37) then becomes

$$X^a = J\ddot{q} + A \tag{3.38}$$

where J is an $n \times n$ Jacobian matrix, n the number of degrees of freedom of manipulation mechanism, A is an $n \times 1$ adjoint matrix. The matrices A and J are functions of state q, $\dot{q}$. Hence, it is necessary to prescribe X^a in a series of time instants. X^a denotes an acceleration profile of the tip of manipulation mechanism. Note that the acceleration profile X^a may be simply obtained from the velocity profile, which is very often prescribed on the basis of technological demands of the manipulation task. Note also that the acceleration in external coordinates of any manipulation mechanism is prescribed by the linear acceleration vector of the centre of mass of the gripper or its tip ($\vec{w}$) and by the angular acceleration vector of the gripper (last link) $\vec{\varepsilon}$, $X^a = \begin{bmatrix} w \\ \varepsilon \end{bmatrix}$. With the assumption that the Jacobian matrix J is nonsingular, the required generalized accelerations $\ddot{q}$ may be calculated in the following way:

$$\ddot{q} = J^{-1}(X^a - A) \tag{3.39}$$

Then, using a mathematical model of the dynamics (Eq. (3.23)) the driving forces (or moments) can be calculated. The presented procedure, suitable for computer realization, is shown by block-scheme in Fig. 3.8. Since in practical tasks of industrial robotics prescribing the trajectories or velocity profile of the tip of manipulation mechanism is more frequently the case, we will shortly consider a problem of robot trajectories synthesis by transferring the manipulator tip along

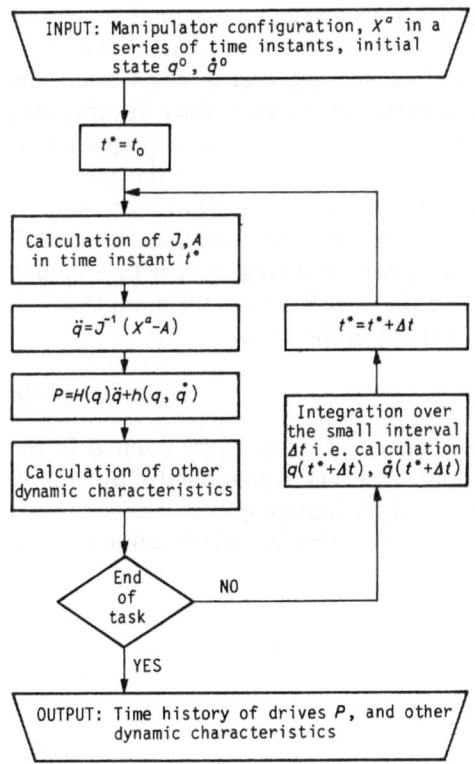

Fig. 3.8. Block-scheme of the algorithm

a prescribed trajectory and transferring the working object, while maintaining some given orientation along the prescribed trajectory.

3.2.2.1 Transfer of Manipulator Tip along Prescribed Trajectory

Consider the most simple manipulation grasp: transferring the manipulator tip with gripper and working object along a given trajectory in space. In principle, this task may be defined as a demand to transfer the manipulator tip from an initial position to a given point without prescribing the manner of its motion. In this case, the trajectory synthesis task can be solved by minimizing some criteria.

However, as different obstacles exist in the working and technological environment of the robot, its motion is limited, making optimization unsuitable. It is therefore more suitable to perform any synthesis of robot trajectory using functional requirements, bearing in mind the constraints limitations of working space, in any given manipulation task. Here, as well as in other parts of the book, we restrict our consideration to non-redundant manipulation mechanisms, i.e. to mechanisms with rigid links and having up to six degrees-of-freedom.

Since any point in a working space (which is limited by kinematical manipulator possibilities) may be obtained with only three degrees-of-freedom, it is sufficient to observe just the minimal robot configuration to attain this. Hence, we restrict our consideration here to observing just three degrees-of-freedom, the other three degrees-of-freedom (gripper) we consider as immobile (q_4, q_5, q_6 = const.). We consider the problem of synthesizing the nominal trajectory $x^0(t)$ and programmed control. It should be noticed, that in cases where the actuators do not change their subsystem order, the vector of the i-th subsystem consists of two coordinates, $x^i = (q_i \dot{q}_i)^T$. In this case, there is a unique correspondence between the angles of the three degrees-of-freedom of the anthropomorphic manipulator and the tip coordinates of minimal configuration (point D in Fig. 3.9) in Cartesian coordinates. In this way, it is obvious that the point D can be moved in advance along a desired prescribed trajectory with a unique change of angles. The relationship between the coordinates of the point $D(x, y, z)$ and the angular coordinates can be presented in the form

$$\vec{S} = (x, y, z)^T = f(q) = f(q_1, q_2, q_3) \tag{3.40}$$

In order to calculate the manipulator angle trajectories, when point D in the space is given, let us consider the small increments of the manipulator tip motion along a prescribed trajectory $\Delta \vec{S} = (\Delta x, \Delta y, \Delta z)^T$. Assuming that the increments are sufficiently small, it is possible to write the following relationship between the angles and the Cartesian coordinates

$$\left(\frac{\partial f_i}{\partial q_1}\right)\Delta q_1 + \left(\frac{\partial f_i}{\partial q_2}\right)\Delta q_2 + \left(\frac{\partial f_i}{\partial q_3}\right)\Delta q_3 = \Delta S_i , \quad i = 1, 2, 3 \tag{3.41}$$

i.e.

$$A\Delta q = \Delta \vec{S} , \quad \Delta q = (\Delta q_1, \Delta q_2, \Delta q_3)^T \tag{3.42}$$

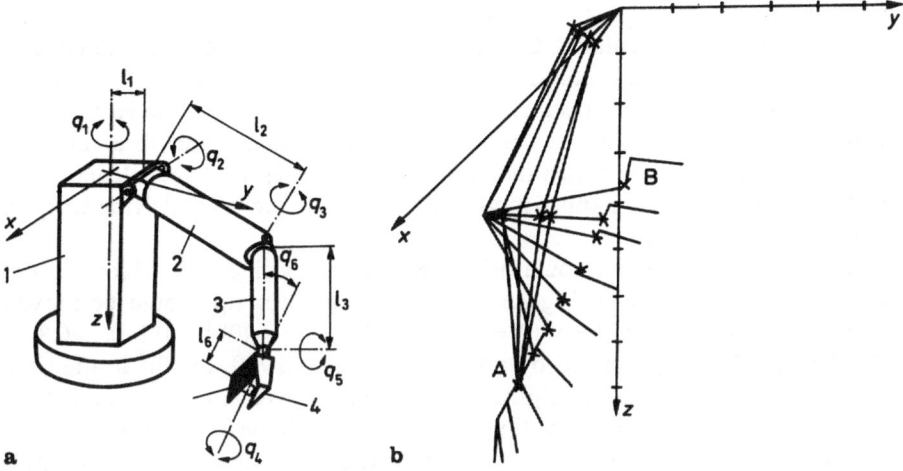

Fig. 3.9. a Mechanical scheme of anthropomorphic manipulator with six degrees of freedom, **b** manipulator task

where elements of the matrix A

$$a_{ij} = \frac{\partial f_i}{\partial q_j} , \quad i, j = 1, 2, 3$$

are calculated for the previous point on the trajectory. Since

$$\Delta q = q(t_1) - q(t_{1-1}) , \quad \Delta \vec{S} = \vec{S}(t_1) - \vec{S}(t_{1-1})$$

the elements of the matrix A are calculated for $q(t_{1-1})$ and, on the basis of Eq. (3.42), the angles on the next point of the tip D trajectory are calculated.

The algorithm for synthesizing the prescribed (functional) kinematics of the minimal (basic) configuration of the manipulation mechanism is as follows. The trajectory $\vec{S}^0(t)$ along which the point D should be moved is prescribed with a given velocity distribution $\dot{\vec{S}}^0(t)$, starting from the initial position of the manipulator (with angles $q^0(0)$ and coordinates $\vec{S}^0(0)$). For the point $q^0(0)$ the matrix A is calculated from Eq. (3.42) and the desired $\dot{\vec{S}}^0(t)$. Thus,

$$\dot{q}^0(0) = A^{-1}(0) \, \dot{\vec{S}}^0(t) \quad \text{if} \quad \Delta t \to 0 \tag{3.43}$$

assuming that the matrix A is non-singular.

Let us consider sufficiently short time intervals $\Delta t = t_1 - t_{1-1}$, during which the values of the matrix A elements are not changing significantly. For sufficiently small Δt we may assume that

$$q^0(t_1) = q^0(t_{1-1}) + \dot{q}^0(t_{1-1}) \Delta t \tag{3.44}$$

In this manner, the angles for the next point on the desired trajectory $\vec{S}^0(t)$ of the manipulator tip are calculated. Using new $q^0(t_1)$, the matrix A is calculated, and then, from Eq. (3.43), the necessary angular velocity $\dot{q}^0(t_1)$ is calculated, etc. Thus, one obtains the nominal trajectories for all coordinates of the state vector $x^0(t)$, corresponding to the minimal configuration (three angles and three angular velocities are calculated). When the nominal trajectories $x^0(t)$ are calculated on the basis of the robot dynamics model, Eq. (3.23), the nominal driving torques $P^0(t)$ can be calculated.

The motion of the manipulator UMS-1 for a specific task which we are considering is presented in Fig. 3.9(b). The manipulator tip should be moved from the point A, defined by $x^0(0)=(q_1(0), \dot{q}_1(0), \ldots, \dot{q}_3(0))^T=(0.1, 0, -0.8, 0, 0.1, 0)$, to the point B, defined by $x^0(\tau)=(-0.4, 0, -0.9, 0, 1.9, 0)^T$ in the time interval $\tau_s=\tau=1.8$ s. The manipulator tip should move along a straight line between the points A and B. The tip acceleration should be constant and change its sign once during the movement. Namely, the manipulator tip acceleration should be

$$a_{tip}=\begin{cases} a_{max} & \text{if } \vec{S}^0(t)\leqslant 0.5(\vec{S}^0(\tau)+\vec{S}^0(0)) \\ -a_{max} & \text{if } \vec{S}^0(t)>0.5\ (\vec{S}^0(\tau)+\vec{S}^0(0)) \end{cases} \qquad (3.45)$$

where $a_{max}=\dfrac{4\ \text{dist}}{\tau^2}$, $\text{dist}=\|\vec{S}^0(\tau)-\vec{S}^0(0)\|$.

The results of the nominal dynamics synthesis for this particular motion with the manipulator UMS-1 are presented in Figs. 3.10 and 3.11. In Fig. 3.10 the nominal trajectories are presented for all three angles of the manipulator, and in Fig. 3.11 the corresponding driving torques are presented.

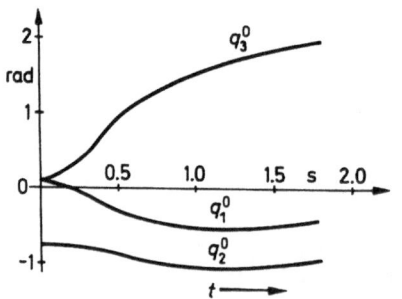

Fig. 3.10. Nominal trajectories of minimal configuration

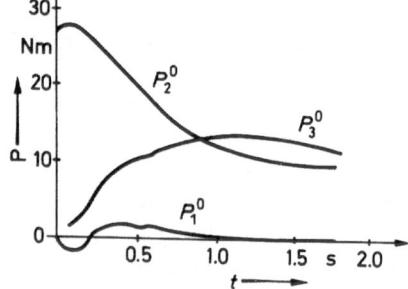

Fig. 3.11. Nominal driving torques of minimal configuration

3.2.2.2 Transfer of Working Object with Desired Orientation along Prescribed Trajectory

This task represents an extension of the task previously discussed. Here, in addition to transferring the manipulator tip (gripper), with or without a working object, along a prescribed trajectory in space, it is necessary to achieve a particular orientation of the working object during (or at the end of) transfer, i.e. to ensure that some axis of the working object has a defined orientation with respect to the axes of the absolute coordinate system. In order to realize this task three degrees of freedom of the minimal configuration of the manipulator are not sufficient. Hence, for solving this task, we consider the manipulator with six degrees of freedom (Fig. 3.9).

Let us consider now a problem of the nominal trajectory synthesis $x^0(t)$, i.e. a desired motion of the object with a prescribed orientation to the coordinate system axes. Determining the trajectories of the six angles and six angular velocities of the manipulator, when the trajectory and the orientation of the working object is given in space, is not simple. However, if the system is decoupled into two subsystems, so that one subsystem consists of the first three degrees of freedom (manipulator minimal configuration), and the other consists of three degrees of freedom of the gripper, then the problem of nominal trajectory synthesis $q^0(t)$ and $\dot{q}^0(t)$ becomes simpler. In this case, the state vector x is divided into two state vectors of the subsystems, $x_0^0 = (x_1^T, x_2^T, x_3^T)^T$, $X_x^0 = (x_4^T, x_5^T, x_6^T)^T$.

Since the manipulator tip trajectory is defined, all three angles and all three angular velocities for the first subsystem (subsystem for positioning) are uniquely defined. The calculation can be performed in the manner previously described, since the task for the first subsystem is reduced in the same way as the manipulator tip motion along a prescribed trajectory in space.

When the nominal trajectories of the subsystem for positioning x_0^0 are calculated, we have to synthesize the nominal trajectories for the gripper subsystem. This subsystem should provide the desired orientation of the object during its transfer along the prescribed trajectory. Prescribing the object orientation during its transfer can be performed by the trajectory prescribing of the object tip with respect to the manipulator tip (gripper joint).

The object tip coordinates in the Cartesian coordinate system are functions of all six manipulator angles:

$$(S_{p_i}) = f_{p_i}(q_1, q_2, \ldots, q_6) , \quad i = 1, 2, 3 \tag{3.46}$$

where $(S_{p_i}) = (x_p, y_p, z_p)^T$ are the coordinates of the object tip. Since with Eqs. (3.41–3.44) the trajectories of the first three degrees of freedom are defined by q_i^0, $i = 1, 2, 3$, we obtain

$$(S_{p_i}) = f_{p_i}(q_1^0, q_2^0, q_3^0, q_4, q_5, q_6) , \quad i = 1, 2, 3 \tag{3.47}$$

i.e. the coordinates of the object tip are functions of the three angles of the gripper. Thus, there exists a unique relation between the coordinates of the object tip (or gripper) and the three angles of the gripper.

Based on Eq. (3.46), one can write

$$[A_{p1} \mid A_{p2}] \left\{ \frac{\Delta q_0^0}{\Delta q_x^0} \right\} = \Delta \vec{S}_p$$

where the elements of the matrices A_{p1} and A_{p2} are given by

$$a_{ij}^{p1} = \frac{\partial f_{pi}}{\partial q_j}, \qquad a_{ij}^{p2} = \frac{\partial f_{pi}}{\partial q_{j+3}}, \qquad i, j = 1, 2, 3$$

Since the orientation of the object is defined in space, the object tip trajectory $\vec{S}_p^0(t)$ is defined as well as the trajectory of the object tip velocity $\dot{\vec{S}}_p^0(t)$. As with the calculation of the minimal configuration trajectories, we shall observe small increments of the object tip movement along the trajectory $\vec{S}_p^0(t)$, during which the matrices A_{p1} and A_{p2} do not change significantly. Let us consider the same time intervals $\Delta t_1 = t_1 - t_{1-1}$ as in the synthesis of the minimal configuration trajectories. Since the position of the manipulator and the gripper at instant t_1 are known, the matrices $A_{p1}(t_1)$, $A_{p2}(t_1)$ can be calculated and since $\dot{q}_0^0(t_1)$ is already known, we get

$$\dot{q}_x^0(t_1) = A_{p2}(t_1)^{-1} (\dot{\vec{S}}_p^0(t_1) - A_{p1}(t_1) \dot{q}_0^0(t_1)) \qquad (3.48)$$

assuming that the matrix A_{p2} is non-singular. In this manner, the angular velocities of the gripper and the angle trajectories of the gripper are calculated, so that the desired orientation of the object is provided.

If the matrix A_{p2} is singular, the singular points should be carefully studied.

In this way, the nominal angles $q^0(t)$ and nominal velocities $\dot{q}^0(t)$ can be calculated for both functional subsystems (positioning subsystem and orienting subsystem).

From the prescribed velocity profile of the manipulator tip (Eq. (3.45)), the nominal, generalized accelerations $\ddot{q}^0(t)$ for all six degrees of freedom can be calculated. Furthermore, on the basis of the model of the robot mechanism dynamics (Eq. (3.23)) the nominal driving torques $P^0(t)$ in robot joints can be calculated.

Figures 3.12 and 3.13 present the results of the nominal trajectories synthesis and driving torques for one concrete task for the manipulation system UMS-1 (Fig. 3.9). The working object should move from position A to position B (Fig. 3.12e). The object position at point A is defined by the manipulator coordinates (or gripper joint) with $\vec{S}_A(0) = (0.425, 0.167, 0.571)^T$, while the gripper centre of gravity coordinates are $\vec{S}_{pA}^0(0) = (0.455, 0.203, 0.530)^T$. The position of the object at the point B is given by the manipulator tip coordinates $\vec{S}_B^0(\tau) = (0.367, 0.365, 0.355)^T$ $[m]$, while the coordinates of the gripper centre of gravity are $\vec{S}_{pB}^0(\tau) = (0.391, 0.410, 0.410)^T$ $[m]$. The manipulator angles corresponding to initial position are $q_0^0(0) = (0, 1.1, 0.5, 0, 0, 0)^T$ $[rad]$, while the manipulator angles for point B are $q_x^0(\tau) = (0.08, 0.94, 1.25, 0, 0.78, -1.50)^T$ $[rad]$.

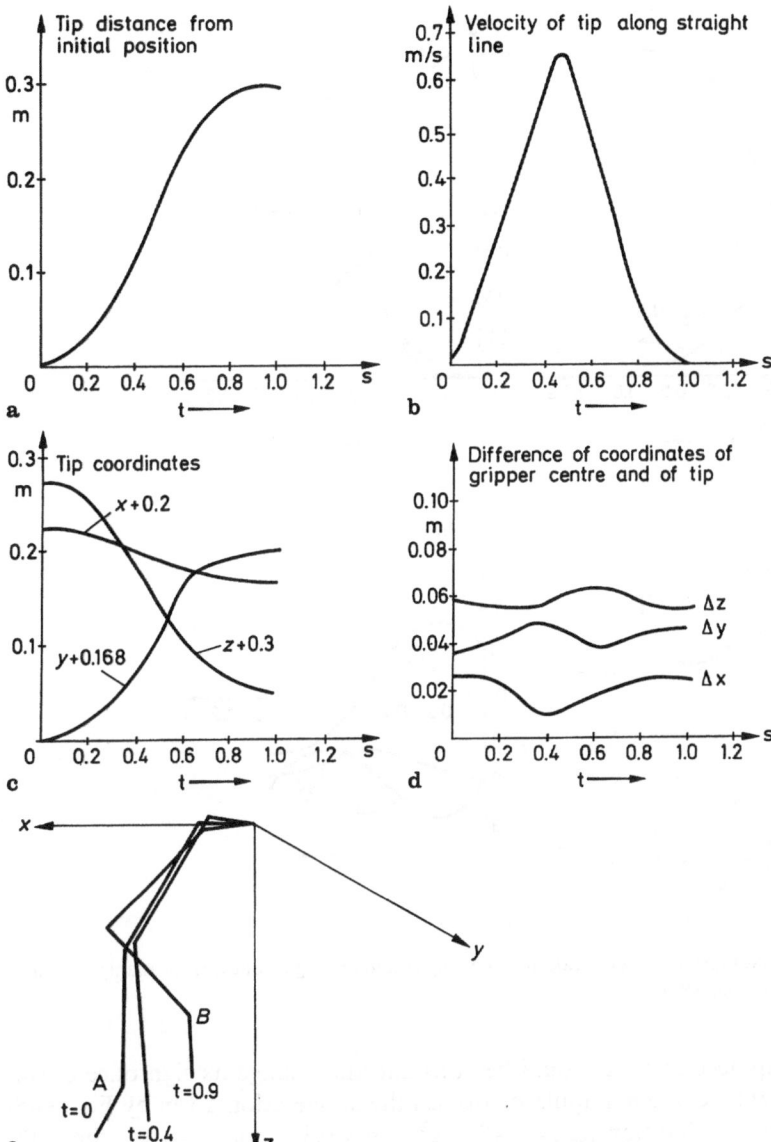

Fig. 3.12 a–e. Nominal dynamics: coordinates of tip and gripper centre of gravity (UMS-1)

The movement of the object should be performed in $\tau = \tau_s = 0.9\,\text{s}$. The working object should be moved in such a way that the tip of the manipulator should move along a straight line between points $A(\vec{S}^0_A(0))$ and $B(\vec{S}^0_B(\tau))$ and the gripper centre of gravity should also move along a straight line between its terminal positions $A(\vec{S}_{pA}(0))$ and $B(\vec{S}_{pB}(\tau))$. Also, it is required that the ma-

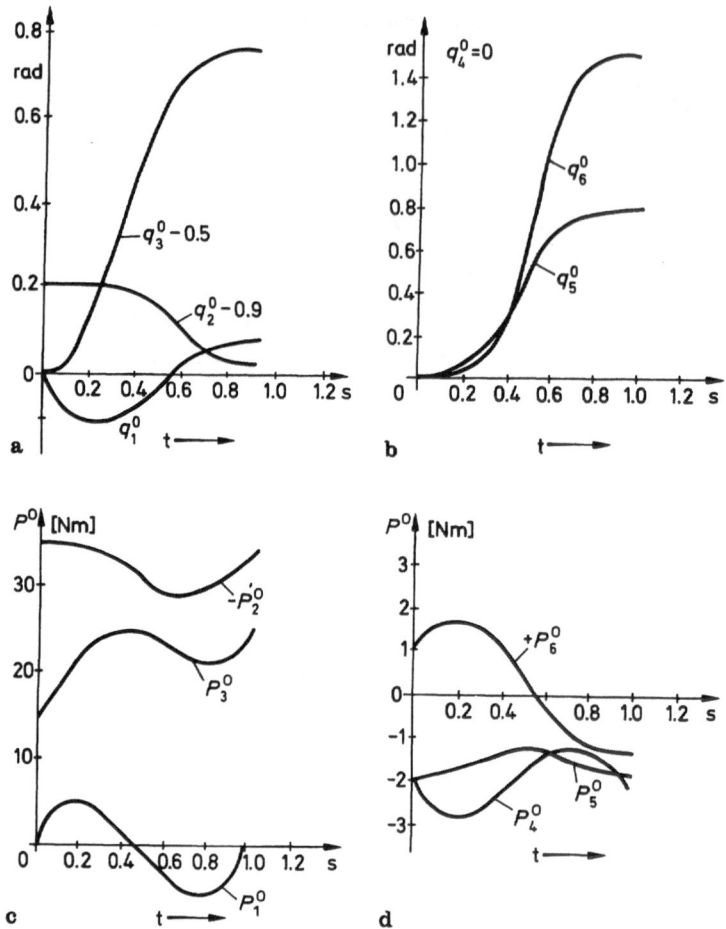

Fig. 3.13 a–d. Nominal dynamics: trajectories of angles and driving torques for the manipulation system UMS-1 (Fig. 3.9a)

nipulator tip acceleration should be constant and change its sign once during the movement, i.e. the manipulator tip has the acceleration given by Eq. (3.45), adopting the acceleration $a_{max} = 1.48$ m/s^2 (dist $= 0.3$ m and $\tau = 0.9$ s). Since the conditions of the object orientation during its transfer are defined so that only two gripper degrees of freedom are sufficient to define it, the matrix A_{p2} in Eq. (3.48) is singular. For that reason, one degree of freedom of the gripper has to be fixed (in this case $q_4(t) = 0$), and the conditions of the gripper orientation are achieved by the other two degrees of freedom of the subsystem S_x. The angular velocities $\dot{q}_5^{(t)}$ and $\dot{q}_6^{(t)}$ of the fifth and sixth degrees of freedom are calculated according to Eq. (3.48), but, in this case, the matrix $A_{p2}(t)$ has the dimension 2×2 and $A_{p1}(t)$ matrix has the dimension 2×3.

Figure 3.12(a) shows the distance of the manipulator tip from the initial position A in the time domain. Figure 3.12(b) shows the desired velocity of the tip object during the motion as in Eq. (3.45). The tip coordinates in the Cartesian absolute coordinate system are presented in Fig. 3.12(c). It is required that the angular velocities in the initial and terminal position have to be zero, $\dot{q}^0(0) = 0$, $\dot{q}^0(\tau) = 0$. Differences between the gripper centre of gravity coordinates and the coordinates of the minimal (basic) configuration tip are presented in Fig. 3.12(d), representing the gripper (object) orientation change during motion. Figure 3.13(a) shows the nominal trajectories of the angles of the positioning subsystem for the given task. Figure 3.13(b) shows the nominal trajectories of the gripper angles (orienting subsystem). With the nominal trajectories for all state coordinates $x^0(t)$, the nominal driving torques are calculated using the mathematical model (Eq. (3.23)) which are presented in Figs. 3.13(c) and (3.13)(d).

3.2.3   Calculation of Other Dynamic Characteristics

In Sects. 3.2, 3.2.1 and 3.2.2 it was shown how driving torques (forces) $P(t)$ which produce the desired motion of a manipulator, i.e. its trajectories of the state vector $(q(t), \dot{q}(t))$, are calculated. However, we are interested in other dynamic characteristics which we will now consider.

3.2.3.1   Diagrams of Torque vs. r.p.m.

One useful characteristic of the algorithm presented in the previous text is the diagram of torque vs. rotation speed of motor. Since the rotation speed of a motor is usually expressed in terms of revolutions per minute (r.p.m.), we will now consider torque–r.p.m. diagram. This diagram can be computed for each joint, i.e. for each motor (actuator), since we consider the robot mechanisms powered by one actuator in each joint. Here, as has been emphasized in (3.2.1), we are only considering D.C. electromotors. The algorithm for dynamic analysis, presented in 3.2.2, calculates the driving torques P_i and internal (generalized) velocities $\dot{q}_i$ for each joint of the considered mechanism. The r.p.m for the i-th joint is $n_i = \dfrac{60}{2\pi}\dot{q}_i$ ($\dot{q}_i$ is expressed in [rad/s]), so each time instant gives one point of the $P_i - n_i$ diagram. This diagram is valid for the shaft of the joint considered. For the moment we are only interested in the diagram of the motor itself. The reducer in the joint must also be taken into account. Let us consider a reducer with the speed reduction ratio equal to N. The motor r.p.m. is

$$n_i^M = N_i n_i = N_i \frac{60}{2\pi}\dot{q}_i \tag{3.49}$$

and the necessary driving torque is

$$P_i^m = \frac{P_i}{N_i \eta_i} \tag{3.50}$$

where η_i is the mechanical efficiency of the reducer. The motor diagram $P_i^m - n_i^m$ is obtained by calculating Eqs. (3.49) and (3.50) in time instants $t_0, t_1, t_2, \ldots$. One example of a torque–r.p.m. diagram is shown in Fig. 3.14.

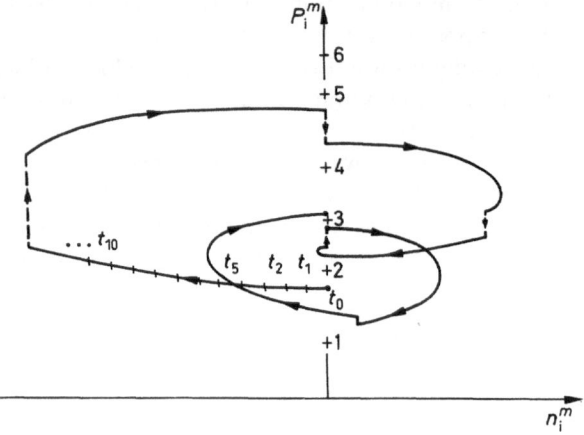

Fig. 3.14. A torque r.p.m. diagram

Such diagrams are very useful during the synthesis and choice of D.C. servosystems. The producer gives the $P_{max}^m - n^m$ motor characteristic in the catalogue, where P_{max}^n is the maximal motor torque at motor r.p.m. $= n^m$. By comparing the necessary characteristic obtained by means of the algorithm described, with the one from the catalogue we can decide whether the chosen motor suits its application. The use of these characteristics will be discussed further in Sect. 3.2.4.

3.2.3.2   Calculation of the Power Needed and the Energy Consumed

When choosing actuators the power needs in each mechanism joint should be considered. For the i-th joint the power needed is obtained by $Q_i = P_i \dot{q}_i$. However, the power produced by the motor has to be larger because of the power loss in the reducer. So the necessary motor power is $Q_i^m = P_i \dot{q}_i / \eta_i$, where η_i is the mechanical efficiency of the reducer. It should be pointed out that this is the output mechanical power of the motor. The energy consumption may also be easily computed. Let us consider the i-th joint and the corresponding actuator. Let $E_i^{(k)}$ be the energy consumed in the first k time steps, and let ΔE_i^k be the energy consumed in the k-th time step (time interval Δt_k). The total energy consumed by the i-th joint actuator is calculated in such a way that during the time-iterative procedure of dynamic analysis, summation of the energy at each step Δt_k is performed

$$E_i^{(k)} = E_i^{(k-1)} + \Delta E_i^k \ . \tag{3.51}$$

To calculate ΔE_i^k, we adopt the medium drive value on the interval

$$P_{i\,med}^k = \tfrac{1}{2}(P_i^{k-1} + P_i^k) \qquad (3.52)$$

where the upper index indicates the k-th time instant. Using Eq. (3.52), one obtains

$$\Delta E_i^k = P_{i\,med}^k \cdot \Delta q_i^k \qquad (3.53)$$

where $\Delta q_i^k = q_i^k - q_i^{k-1}$.

It should be stressed that this discussion on energy has only dealt with the mechanical power and energy. If we want to calculate the energy which has to be taken from an energy source (for instance from an electric battery) then we should take the energy lost in actuators into consideration. In this paragraph we only give some ideas of such energy consumption calculation. If a manipulator is driven by D.C. electromotors then the energy loss follows from the resistance and friction effects. The power required from a source can be computed as $Q = ui_r$ where u is the control voltage and i_r is the rotor current. In a time step Δt_k the energy increment is

$$\Delta E_i^k = u_i^k i_{r_i}^k \Delta t_k = Q_i \Delta t_k$$

where the lower index represents the number of joints and the upper index indicates the k-th time instant. The energy consumed by the whole manipulator is the sum of energies consumed by the actuators.

3.2.4   Testing of Dynamic Characteristics

Knowledge of dynamic characteristics is very useful in design processing and in the application of robot systems. Hence, we will present here the algorithm for testing any relevant dynamic characteristics. It is an automatic procedure based on choosing manipulator parameters, which calculates the relevant dynamic characteristics and tests whether they satisfy any conditions imposed. We thus immediately have the answer as to whether the chosen parameters are correct. If they are not correction is made.

In principle, any dynamic characteristic calculated can be tested. Here, we mention only the most relevant tests.

3.2.4.1   Tests of a D.C. Electromotor

Suppose that we have chosen a D.C. electromotor as the actuator for the i-th manipulator joint. The mathematical model of such an actuator is presented in Sect. 3.2.1. Here, we consider the tests of the D.C. motors because they are frequently used with today's robot mechanisms.

All the calculations refer to some defined manipulation task with a prescribed execution time.

Test 1

We can find the $P_{max}^m - n^m$ characteristic of the chosen motor in the catalogue. It is the diagram of maximal torque depending on motor r.p.m. If the diagram is not given directly, it can be constructed from the data given in the catalogue. With robot systems, we often use permanent magnet D.C. motors. For such motors the $P_{max}^m - n^m$ characteristic has a polygonal form (straight lines in Fig. 3.15). In the catalogue we sometimes find the value of maximal motor torque (corresponding to the point A) and the maximal rotation speed (point B) (Fig. 3.15). These two values define the maximal characteristic (assuming a straight line). Let the value of maximal torque for $n^m \to 0$ (point A) be marked by P_M^m and let the value of rotation speed for $P^m \to 0$ (point B) be marked by n_M^m. The torque P_M^m is often called stall torque, and the rotation speed n_M^m is called no-load speed. When this speed is expressed in terms of r.p.m. then it is marked by n_M^m and if it is expressed in terms of [rad/s] then we mark it by ω_M^m. It should be said that there is sometimes a difference between the real value of maximal torque (P_{Mr}^m) and its theoretical value (P_M^m); the real value of P_{Mr}^m is less than P_M^m.

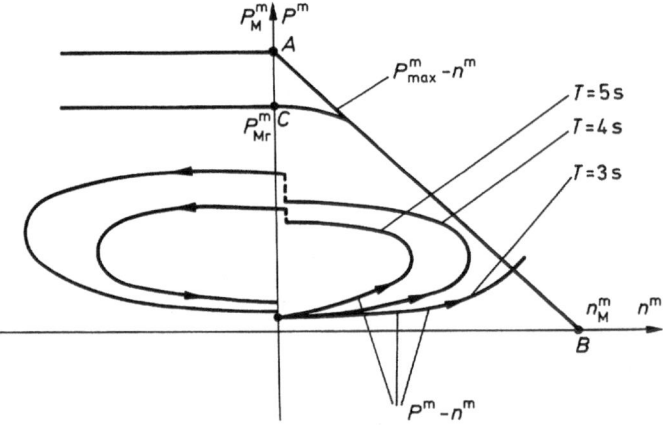

Fig. 3.15. P^m–n^m diagrams

In such a case the maximal characteristic $P_{max}^m - n^m$ has an upper bound P_{Mr}^m (point C in Fig. 3.15). This characteristic defines the domain. The real $P^m - n^m$ characteristics must be wholly within this domain. In each iteration a new point of the diagram is obtained. The algorithm checks whether it is within the permissible domain. If it is, a new iteration starts, if not, the algorithm signals that there is a violation of the constraint. This constraint then follows the mathematical model of the D.C. actuator, which was discussed in Sect. 3.2.1, and given here only as a simplified derivation of the torque–speed constraint.

According to Ohm's law it is

$$u = R_r i_r + C_E \dot{q}_m \tag{3.54}$$

where u is input voltage, R_r is the rotor resistance, i_r is the rotor current, C_E is the constant of electromotor force, and $\dot{q}_m$ is the rotation speed [rad/s]. If we ignore the rotor acceleration effects and friction term, the motor output torque is

$$P^m = C_M i_r \tag{3.55}$$

where C_M is the torque constant. Combining Eq. (3.54) and (3.55) we obtain

$$P^m = \frac{C_M}{R_r} u - \frac{C_M C_E}{R_r} \dot{q}_m \tag{3.56}$$

If the rotation speed is expressed in terms of r.p.m. ($n^m = \frac{60}{2\pi} \dot{q}_m$) then

$$P^m = \frac{C_M}{R_r} u - \frac{C_M C_E 2\pi}{R_r 60} n^m \tag{3.57}$$

Let us introduce the constraint of the maximal input voltage u_{max}. Then, from Eqs. (3.56) and (3.57) it follows

$$\frac{P^m_{max}}{P^m_M} + \frac{n^m}{n^m_M} = 1 \tag{3.58}$$

where $P^m_M = \frac{u_{max} C_M}{R_r}$ is the stall torque and $n^m_M = \frac{60}{2\pi} \cdot \frac{u_{max}}{C_E}$ is the no-load speed. This constraint of maximal input voltage can be represented by a straight line in the $P^m - n^m$ plane ((1) in Fig. 3.16). We use this constraint in the quadrants I and III of the $P^m - n^m$ plane. For the quadrants II and IV we introduce the constraint

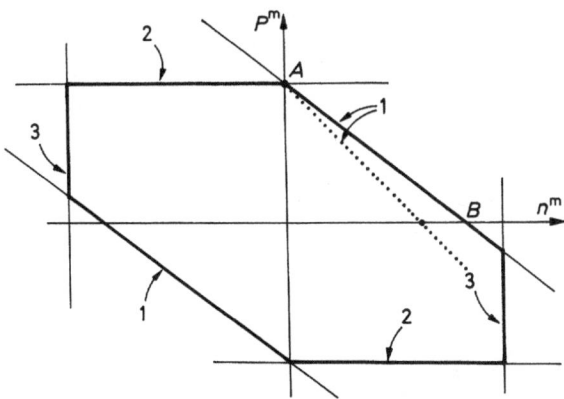

Fig. 3.16. Constraints of D.C. motor

of maximal rotor current in order to keep this current smaller than the stall current value: $|i| \leqslant i_M = u_{max}/R_r$. This constraint is represented by a horizontal line (2) in the $P^m - n^m$ plane. Finally, we introduce the constraint of maxima in the speed allowed n_{max} ((3) in Fig. 3.16); for instance, $n_{max} = n_M^m$.

If the viscous friction is not ignored then the no-load speed becomes

$$\omega_M^m = u_{max} \bigg/ \left(C_E + \frac{R_r B_C}{C_M} \right),$$ where B_C is the viscous friction coefficient. This modi-

fied constraint is represented by dotted line in Fig. 3.16.

One example of the constraint is shown in Fig. 3.15. Straight lines represent the constraint $P_{max}^m - n^m$. It can therefore be concluded that the $P^m - n^m$ diagrams spread when the working speed increases, i.e. the execution time T decreases. For $T = 5s$ and $T = 4s$ the diagrams are wholly within the permissible domain which means that the actuators chosen can produce the manipulator work at this speed. For $T = 3s$ the diagram extends beyond the permissible domain i.e. the constraint is violated, so therefore the motor cannot produce the manipulator work at that speed.

Test 2

Another test can be made of the necessary motor power. The algorithm computes the power needed in each joint, i.e. $Q_i^m = P_i \dot{q}_i/\eta_i$. By comparing this function (for the joint considered) with the maximal power which can be produced by the chosen motor we conclude whether the motor is chosen correctly.

Here we will give only a short presentation of the power test procedure. Besides the power $Q_i = P_i \dot{q}_i$ which is a product of both the torque and rotation speed, we introduce a notation of dynamic power (or acceleration power) DQ as a product of the torque and acceleration, i.e. $DQ_i = P_i \ddot{q}_i$. A very useful characteristic is a diagram connecting the power and dynamic power. Each time instant gives one point having the coordinates Q^m and DQ^m. In this way, at the end of the manipulation task we obtain the $Q^m - DQ^m$ characteristic (Fig. 3.17). This

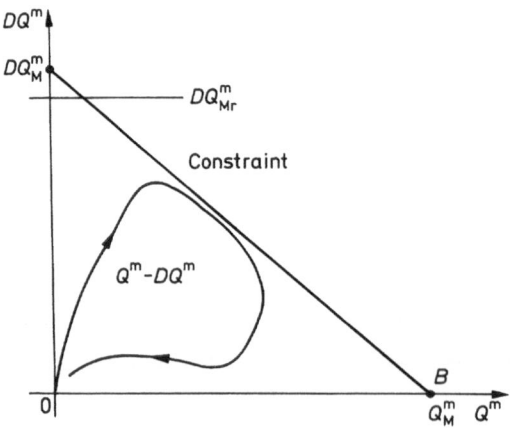

Fig. 3.17. $Q^m - DQ^m$ diagram

characteristic has to be within the domain defined by a straight line connecting the maximal values Q_M^m and DQ_M^m (points B and A respectively, in Fig. 3.17). If the diagram violates this constraint then the test is negative. Q_M^m is the maximal motor power and can be expressed in the form $Q_M^m = P_M^m \omega_M^m/4 = (P_M^m)^2 R_r/4C_M C_E$ where P_M^m is the stall torque of motor, ω_M^m is the no-load speed, R_r is the rotor resistance, and C_M and C_E the constants of torque and electromotor force. Maximal value DQ_M^m can be obtained by $DQ_M^m = Q_M^m/T_{em}$. T_{em} is called the electromechanical constant and has the form $T_{em} = J_r R_r/C_M C_E$, where J_r is the rotor moment of inertia. Note that the real value of maximal dynamic power DQ_{Mr}^m can be less than the theoretical value DQ_M^m. This difference appears because the real and theoretical driving torques are different (Fig. 3.15).

Test 3

Finally, we introduce the test of motor heating. For this, the method of equivalent torque is used. We compute this equivalent motor torque by

$$
P_{eq}^m = \sqrt{\frac{\sum_k (P^{mk})^2 \Delta t_k}{\sum_k \Delta t_k}}
\tag{3.59}
$$

where P^{mk} is motor torque at the k-th time instant and Δt_k is time increment. The summation is performed over the whole manipulation task or over a finite time interval corresponding to a part of the manipulation task. The equivalent torque is now tested against the nominal motor torque P_{nom}^m. If $P_{eq}^m \leqslant P_{nom}^m$ the test is positive. If $P_{eq}^m > P_{nom}^m$ then there is overheating of the motor and the test is negative.

These are the main tests when choosing a D.C. actuator. Some other tests which include rotor current, control voltage and other control characteristics can also be applied.

3.2.4.2 Choice of Optimal Design Parameters

Mathematical models of the robot mechanism dynamics can be used for determining adequate dimensions and optimal choices for robot parameters.

It is first necessary to choose optimal criteria for efficient, practical use of industrial robots; both operation speed and energy consumption have to be taken into consideration. For optimization and evaluation of these manipulation mechanisms the criteria needed can be defined as:

(a) work speed criterion (or time criterion);
(b) energy consumption criterion;
(c) combination of (a) and (b).

We will briefly explain the essence of these criteria. An example only using the speed criterion follows.

1. *Velocity (time) criterion.* Let T denote the time of the set manipulation task. Optimization regarding the velocity criterion is reduced to determining that manipulator configuration which permits the greatest work speed, i.e. the least time T.
2. *Energy criterion.* Let E denote the total energy the manipulator uses in performing the test task. Optimization means the minimization of E, i.e. finding the configuration which ensures minimal consumption of energy.
3. *Combined criterion.* It presents the combination of two previous criteria, taking into account both the work velocity and energy consumption.

These criteria give the possibility for appropriate choosing optimal parameters in the manipulator design process. However, these criteria can be also used for evaluating and comparing different manipulation robots on the market. The criteria (a) and (b) can be used in optimization procedures, and the combined criterion (c) can be used for evaluating and comparing robot systems.

Limitations

The choice of optimal parameters has to be made within the limits of these adopted criteria. Limitation of reachability is of a formal nature and it should be understood more as a certain kinematic equalization of the manipulators.

1. *Limitation of drives.* Driving motors in the robot joints should produce the driving forces (torques) necessary for any given manipulation task.
2. *Limitation of stress.* Stresses in the manipulator segments should not exceed their permitted values.
3. *Limitation of rigidity.* Positioning and orienting deviations due to elastic deformations of the manipulator segments should not exceed their permitted values.

Example

It is given that the manipulation robot UMS-1 (Fig. 3.9) has to transfer a payload of 5 kilograms along trajectory $ABCA$ (Fig. 3.18), maintaining the manipulator gripper all the time its initial orientation in space. Let the manipulator run the trajectory parts of AB, BC and CA in equal time with triangular velocity profile (Fig. 3.19). Let T denote the execution time of the whole task.

Robot segments are assumed in the form of circular tubes with a constant ratio of inner and outer radius $\psi = r/R = 0.75$. It is also possible to adopt other forms of cross-section of the robot segment. As material for segment production the light alloy $AlMg_3$ was adopted. Data about specific density (ρ), permitted bending stress (σd) and torsion stress (τd), Young's modulus (E_j) are:

$$\rho = 2.700\,\text{kg/m}^3 \ , \qquad \sigma d = \frac{130}{k}\,daN/\text{mm}^2 \ , \qquad \tau d = \frac{125}{k}\,daN/\text{mm}^2 \ ,$$

$$E_j = 7.848 \cdot 10^{10}\,\text{N/m}^2 \ , \qquad k = 5 \text{ the safety coefficient.}$$

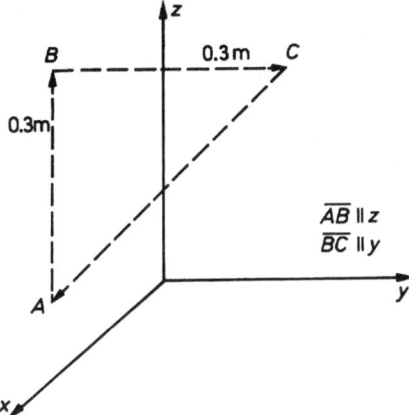

Fig. 3.18. Trajectory of the object transfer

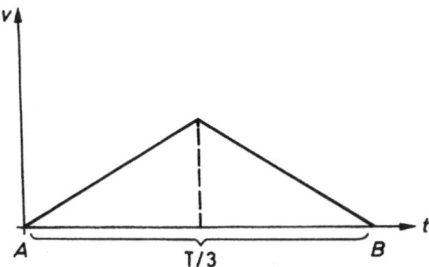

Fig. 3.19. Triangular velocity profile

D.C. permanent magnet INDOX motors, Frame 23, type 2315-p20-0 produced by Indiana General were chosen as driving motors for the manipulator joints. The reduction ratio is 100 in rotational joints.

Optimization procedure using the speed criterion was adopted.

The procedure is the following: for a selected value of R a series of simulations is performed successively reducing the task execution time, T. This continues until the drive limitation (II) is violated, i.e. until the driving motors cannot produce the manipulator working speed. Then R is reduced and the procedure repeated. Thus, the curve $T_{min}(R)$, i.e. minimal execution time depending on R is obtained. Let us first consider the limitations (I)–(II) only. The procedure of reducing R is repeated until limitation (I) is violated, i.e. until the stresses in segments exceed the permitted values. If further reduction of R is required, T has to be increased. Consequently, minimum time (maximum velocity) appears in both limitations (I) and (II). Figure 3.20 illustrates these results (for the anthropomorphic manipulator in Fig. 3.9), i.e. the curve $T_{min}(R)$ and limitations (I) and (II). The minimum appears at the point M_1 designated by a circle. Let us now introduce the limitation of elasticity (III). We impose the condition that the manipulator tip linear deviation due to segment elasticity be less than 0.001 m. In this case we consider only the quasi-static deflexion due to nominal dynamics. By introducing limitation (III) the permitted domain is narrowed and the minimum point moves into position M_2, designated by a square in Fig. 3.20. The coordinates $(R_{opt}^T, T_{min}^{abs})$ correspond to the point M_2. The dotted line in Figure 3.20 represents the corresponding energy consumptions. The abrupt decline of energy consumption to the left of M_1 is due to the abrupt increase of working time T, i.e. to a velocity decline.

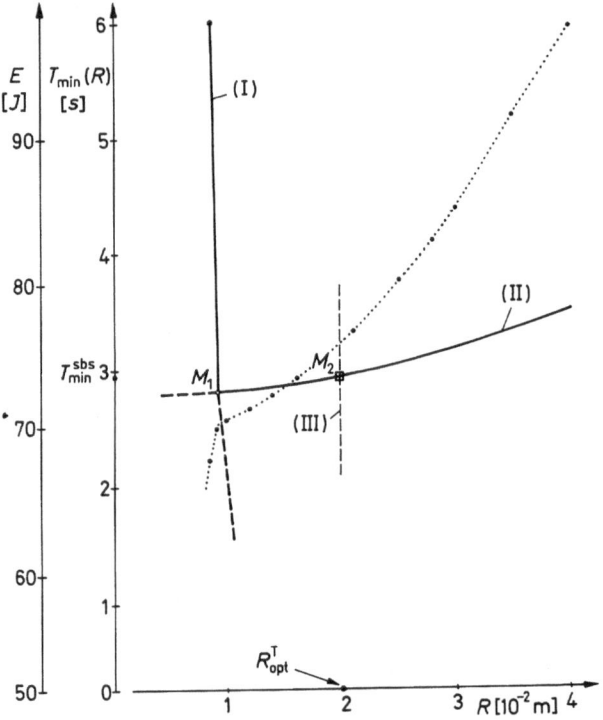

Fig. 3.20. Dependence $T_{min}(R)$ for the anthropomorphic manipulator in Fig. 3.9a

3.3 Dynamics of Manipulation Robots in Conditions of Mechanical Vibrations Impact

As it is known from industrial practice in working conditions of robots that the vibrations are often present, transmitted through the basis to the carrying structure. In general, manipulation robots are rigidly connected to their base with no dampening vibrations. The question of any vibration impact on positioning and orienting precision is therefore ever present, particularly in conditions where high precision is required.

3.3.1 Basis Vibrations

Generally, mechanical vibrations are periodic and aperiodic. Periodic vibrations are generated by machines and devices which rotate uniformly. Those are, for example, electric engines, pumps, compressors, etc. Aperiodic vibrations arise from machines with linear motion which often finishes with an impact. Such a motion is produced by the blacksmith's hammer, or shock vibration machines, etc. In machines which rotate uniformly, rotational masses (rotors) are balanced theoretically, that is, the exact coincidence of a rotating path gravity axes with its geometric axes is almost never possible. Hence, the centrifugal forces are unavoidable and are transmitted through the carrying structure to the basis. With large frequencies these forces can be considerable, although the eccentricity is still negligibly small. Thus generated centrifugal forces present the simplest form of the excitation periodical forces whose change can be presented by a simple sine function. A somewhat more complex change of centrifugal force appears with machines which have a rotational and return linear motion. In working conditions forces of an impulse type are also often present. However, the most frequent case of time-dependent excitation forces is in the form of random function. One of such real displacement plots which is recorded on tape is shown in Fig. 3.21. Figure 3.22 presents an acceleration plot in vertical direction because, as we shall later show, a calculation of the manipulator tip deviation from a nominal trajectory in conditions of mechanical vibrations is based on the acceleration known of the initial (basis) segment of the robot's mechanism.

It is clear that a random acceleration signal could be approximatively presented by "white noise" in the time interval when impulse (shock) accelerations are not apparent. This supposition denotes a density spectrum of the random signal is constant and frequency independent. It is very useful in system state estimation and in the other dynamic analysis.

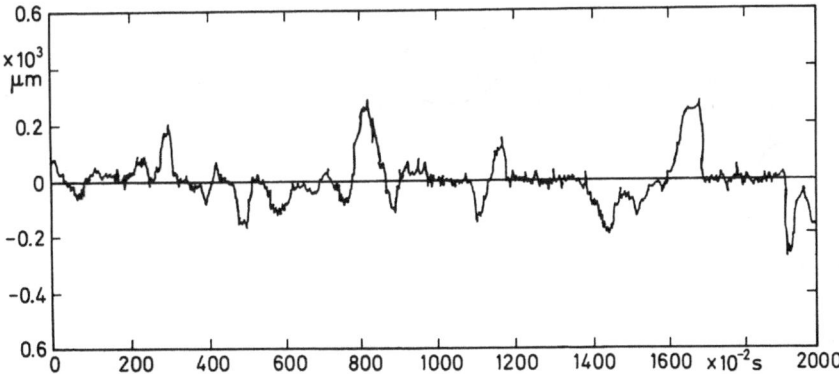

Fig. 3.21. Displacement in vertical direction

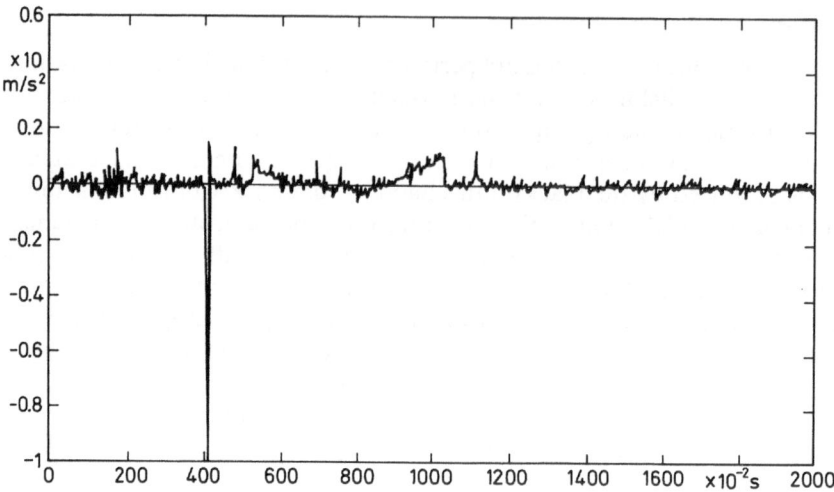

Fig. 3.22. Acceleration in vertical direction

3.3.2   Forming Differential Equations of Motion

To obtain differential equations of the manipulator robot's motion under mechanical vibration impact we suppose that the mechanism of the robot is rigidly connected to the basis which is vibrating. The vibrations are presented by linar $\vec{w}_0$ and angular $\vec{\varepsilon}_0$ acceleration vectors (Fig. 3.23). Also we assume that the robot mechanism segments are rigid bodies connected to one other by joint links which permit its relative motion. The neighbouring mechanism segments are connected by rotational joints with one degree of freedom.

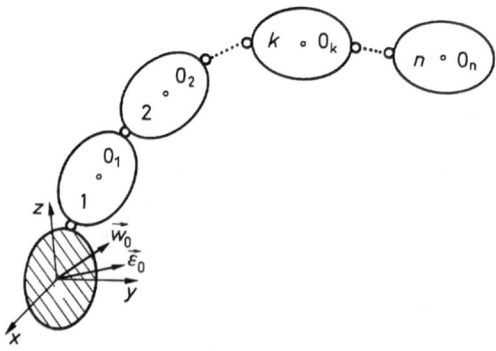

Fig. 3.23. An open kinematic chain

We shall consider the motion of the manipulation mechanism with respect to the external fixed Cartesian coordinate system $Oxyz$ which is connected to the gravity centre of its basis segment. The geometric parameters and parameters which define kinematic and dynamic characteristics of the robot are considered in Sect. 3.2.

By including both accelerations, linear w_0 and angular ε_0, into Eq. (3.23) the mathematical model of the mechanism robot dynamics under vibration impact could be obtained. This model can be presented in following matrix form:

$$H(q)\ddot{q} + h_1(q)w_0 + h_2(q)\varepsilon_0 + \zeta(q, \dot{q}) = P \qquad (3.60)$$

where:

H is the inertial matrix,

ζ is the vector equal to the vector h in Eq. (3.23),

h_1 and h_2 are the matrices associated to the linear acceleration w_0 and angular acceleration ε_0, respectively.

By including the dynamics of actuators which power particular mechanism joints we can establish the complete dynamic robot model which enables us to calculate its exact dynamics in a perturbed working regime. This perturbed regime comes from mechanical vibrations which can be present as noticed in Sect. 3.3.1, by "white noise". Using the complete dynamic model for a real manipulation robot, the manipulator deviation from nominal trajectory can be calculated, the results of which are presented in the next example.

Example

We observe the motion of the manipulator UMS-2 which is shown in Fig. 3.24. The geometric manipulator parameters are presented in Table 3.1, the dynamic parameters in Table 3.2, and the actuator parameters in Table 3.3.

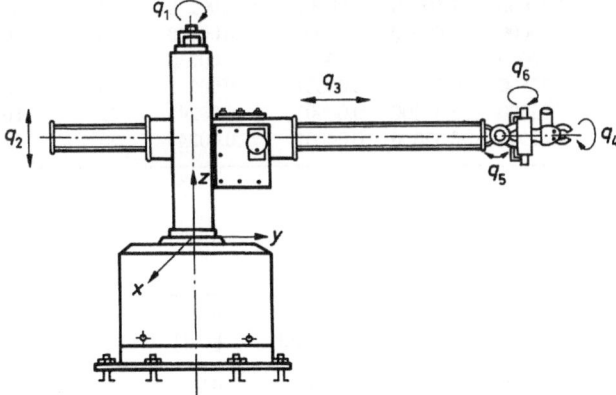

Fig. 3.24. Manipulator UMS-2

Table 3.1.   Manipulator geometric parameters

Joint	Type of joint[†]	Unit vectors of joint axes $\vec{e}_i$	Vectors $\vec{r}_{ii}[m]$	Vectors $\vec{r}_{i,i+1}[m]$
1	R	0, 0, 1	0, 0, 0.188	0, 0, −0.192
2	T	0, 0, 1	0, 0, 0.01	0, 0, −0.01
3	T	0, 1, 0	0, 0.01, 0	0, −0.44, 0
4	R	0, 1, 0	0, 0.025, 0	0, −0.025, 0
5	R	1, 0, 0	0, 0.025, 0	0, −0.025, 0
6	R	0, 0, 1	0, 0.025, 0	0, −0.025, 0

[†] R – rotational, T – prismatic

Table 3.2.   Manipulator dynamic parameters

Segment	Mass $m_i[kg]$	Moment of inertia [kg m²]		
		J_{xi}	J_{yi}	J_{zi}
1	10.	0	0	0.0294
2	7.	0.055	0	0.055
3	4.15′		$J_S = 0.318$,	$J_N = 0.318$
4	0.5	0.00015	0.00010	0.00015
5	0.5	0.00015	0.00010	0.00015
6	0.5	0.00015	0.00010	0.00015

Table 3.3.   Actuator parameters

Actuator	$B_C\left[\dfrac{Nm}{rad/s}\right]$	$C_M\left[\dfrac{Nm}{A}\right]$	$C_E\left[\dfrac{V}{rad/s}\right]$	$R_r[\Omega]$	$J_M[kg\,m^2]$	$L_r[H]$	$N_V[-]$	$N_M[-]$
1	0.00580	1.50000	1.43000	1.60000	0.00003	0.00230	31.17	31.17
2	40.30000	125.40000	120.30000	1.60000	0.00003	0.00230	2616	2616
3	14.50000	75.50000	72.20000	1.60000	0.00003	0.00230	1570	1570
4	0.00750	1.36000	0.32500	3.32000	0.00001	0.00245	25	10
5	0.00750	1.36000	0.32500	3.32000	0.00001	0.00245	25	10
6	0.00750	1.36000	0.32500	3.32000	0.00001	0.00245	25	10

Table 3.4.   Initial and final positions

Degree of freedom	q_1 [rad]	q_2 [m]	q_3 [m]	q_4 [rad]	q_5 [rad]	q_6 [rad]
Initial position	0.2	0.	0.03	0.3	0.5	0.2
Final position	0.6	0.1	0.12	0.8	1.0	0.7

The manipulator motion is observed in internal coordinates whose initial and final positions are presented in Table 3.4.

The motion between the initial and final positions is performed over a straight-line segment with an adopted parabolic velocity profile. The internal coordinates and velocities can be presented in the form of

$$
\begin{aligned}
q(t) &= q^0 + \lambda(t)(q^F - q^0) \\
\dot{q}(t) &= \dot{\lambda}(t)(q^F - q^0)
\end{aligned}
\quad , \quad 0 \leqslant t \leqslant T
\tag{3.61}
$$

where

q^0 is the initial position,

q^F is the final position,

$\lambda(t) \in [0, 1]$ is the scalar parameter which defines the velocity change of the internal coordinates,

$$
\lambda(t) = \frac{6}{T^2}\left(\frac{1}{2} - \frac{1}{3}\frac{t}{T}\right)t^2
$$

T is the movement duration time.

The motion is simulated under conditions of the vibration impact which are presented by linear $\vec{w}_0$ and angular $\vec{\varepsilon}_0$ acceleration vectors. It is assumed that on the manipulator, only the linear acceleration w_{0z} in vertical direction and the angular acceleration ε_{0z} around that axis are operating. The accelerations are shown by a simple sine function in the following form:

$$
w_{0z} = -e_1\Omega^2 \sin\Omega t \ , \qquad \varepsilon_{0z} = -e_r\Omega^2 \sin\Omega t
\tag{3.62}
$$

where e_1 – maximal linear amplitude of basis,

$\quad e_r$ – maximal angular amplitude of basis,

$\quad \Omega$ – angular frequency.

This form, as we have seen in Sect. 3.3.1, corresponds to vibrations which appear due to centrifugal forces in machines with an unbalanced rotor.

For the following adopted values

$$
e_1 = 0.001 \ [\text{m}] \ , \qquad e_r = 0.001 \ [\text{rad}] \ , \qquad \Omega = 100 \ [\text{s}^{-1}]
\tag{3.63}
$$

deviation from the nominal trajectory in internal coordinates space for the excitation case, Eq. (3.62), is presented in Fig. 3.25. Only the deviation from nominal values of the linear coordinate q_2 is presented, coinciding with the direction of the linear acceleration w_{0z}. Solid line corresponds to the nominal position, and the dotted one to the real position in instant t. Using the known relation between the internal and external coordinates (given in Sect. 3.2.2), the deviation of the manipulator tip from the nominal trajectory can be calculated in the external coordinates space. By introducing the subsystem feedbacks, as well as force-feedbacks, it is possible to achieve satisfactory tracking of the nominal trajectories. For understandable reasons, the problem of maintaining

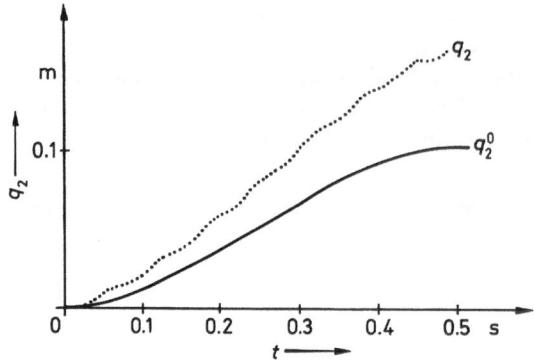

Fig. 3.25. Position deviation q_2 from nominal trajectory

robot accuracy in mechanical vibrations will not be considered in this book. This presents a special control problem; the reader is referred to Reference [1].

3.4 Dynamics of Robots Under Action of External Reaction Forces

It is well known from industrial practice in many manipulation tasks that manipulation robots change their kinematic structure, i.e. from open to closed kinematic chains. The dynamics of these open chains, under conditions of mechanical vibration impact, is presented in Sect. 3.3. Here we will consider closed chain dynamics of robots with constrained gripper motion. These problems often appear in practical manipulation tasks, such as grinding, polishing, writing, drawing, assembling and bilateral manipulation.

The dynamic model of closed chains enables to calculate the relative gripper motion of imposed constraints and the reactions of these constraints. However, by using the robot dynamic model we can consider non-stationary constraints as a more complicated case together with the constraints associated with friction. Furthermore, we can solve the collision problem between the manipulator and constraint. In our consideration we will only present the mathematical model of the robot dynamics which can calculate its nominal dynamics, i.e. we only calculate the necessary driving torques which realize the desired robot's motion. Using the robots dynamics model one practical case of the robot's constrained motion is solved.

3.4.1 Practical Cases of Constrained Gripper Motion

Here we want to present some typical examples of constrained gripper motion which often appear in practical use. Let us now consider a manipulation writing or drawing task on a given surface (Fig. 3.26). It is clear that here there appears a surface type constraint which we will consider in detail in Sect. 3.4.3. While the manipulator gripper motion with the device for "writing" is above the surface the kinematic chain is open. The chain becomes closed when contact appears between the surface and the manipulator gripper.

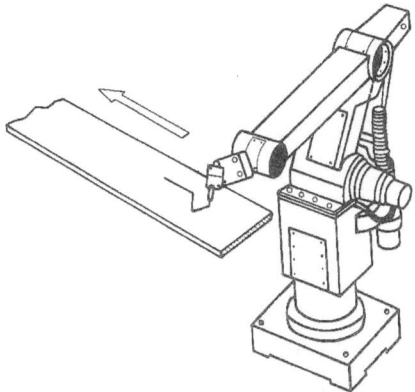

Fig. 3.26. Writing task

The second example of constrained gripper motion is a grinding task (Fig. 3.27). The necessary theoretical considerations for this type of constraint are demonstrated in Sect. 3.4.3 where the numerical example of the grinding process with a six degrees-of-freedom manipulator is also solved.

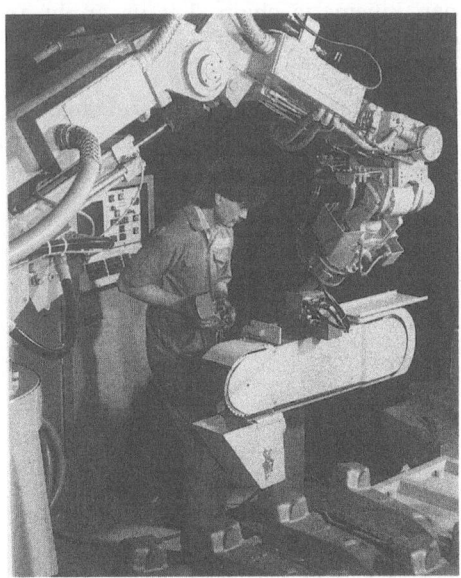

Fig. 3.27. Manipulation grinding task

The next example is very typical in industrial practice. This is an assembly task. Figure 3.28 shows the manipulation peg-in-hole assembly task. The assembly problem is often the subject of the work of many researchers. For instance, in [3] the dynamic analysis and control synthesis using force feedback in the peg-in-hole assembly task are presented. In [4] the complete dynamic analysis for non-stationary constraint is derived and one example of the assembly simulation is given.

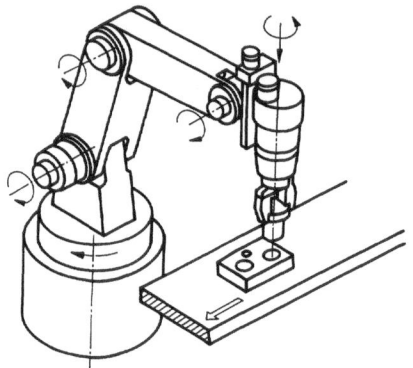

Fig. 3.28. Manipulation assembly task

3.4.2   Mathematical Model of Manipulator with Constraints of Gripper Motion

In order to present the mathematical model of the robot closed configuration we consider first a manipulator as an open kinematic chain. Let us apply an external force $\vec{F}_A$ acting on the gripper at point A and external moment $\vec{M}_A$ (Fig. 3.29).

We assume that the mechanism is a set of n segments with n degrees of freedom. Also, we assume that the mechanism position is determined by a vector of generalized coordinates, q. In a completely analogous way as

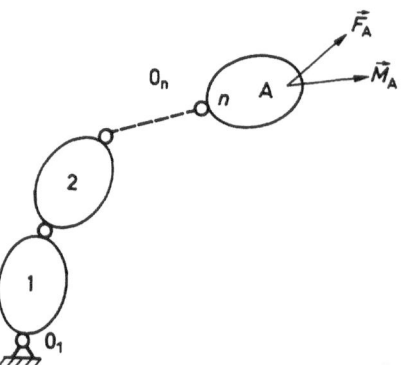

Fig. 3.29. Open kinematic chain under action of external force $\vec{F}_A$ and moment $\vec{M}_A$

presented in Sect. (3.2) the dynamic model of such a mechanism can be presented in the following matrix form

$$H\ddot{q}+h=P+D_1 F_A+D_2 M_A \tag{3.64}$$

where $H=H(q):(n \times n)$ is the inertial matrix, $h=h(q,\dot{q}):(n \times 1)$ the vector consisting of gravity, centrifugal and Coriolis forces, $P=P(t):(n \times 1)$ the vector of driving torques (forces) in joints, $\ddot{q}:(n \times 1)$ the acceleration vector in internal coordinates, D_1 and $D_2:(n \times 3)$ the matrices associated to the external force F_A and external moment M_A, respectively.

It is easy to show that the matrices D_1 and D_2 are

$$D_{1\,(n \times 3)}=\begin{bmatrix} d_{11}^T \\ \cdot \\ \cdot \\ \cdot \\ d_{1n}^T \end{bmatrix}, \quad \vec{d}_{1i}=\vec{e}_i \times \vec{r}_{iA} \tag{3.65}$$

$$D_{2\,(n \times 3)}=\begin{bmatrix} d_{21}^T \\ \cdot \\ \cdot \\ \cdot \\ d_{2n}^T \end{bmatrix}, \quad \vec{d}_{2i}=\vec{e}_i \tag{3.66}$$

In Eq. (3.65) $\vec{r}_{iA}$ denotes a vector from the axes of the i-th mechanism joint to the force acting point. Using the notation from Section 3.2, $\vec{r}_{iA}$ can be presented in the form

$$\vec{r}_{iA}=\overrightarrow{O_i A}=\sum_{k=i}^{n-1}(\vec{r}_{kk}-\vec{r}_{k,\,k+1})+\vec{r}_{nn}+\vec{p} \tag{3.67}$$

where $\vec{p}$ is the vector from the centre of mass of the last segment to the force acting point.

The dynamic model (Eq. (3.64)) can be written in a more suitable form as

$$H\ddot{q}+h=P+DR_A \tag{3.68}$$

where:

$$D_{(n \times 6)}=[D_{1\,(n \times 3)} \,\vdots\, D_{2\,(n \times 3)}]$$

$$R_{A\,(6 \times 1)}=\begin{bmatrix} F_{A\,(3 \times 1)} \\ \hline M_{A\,(3 \times 1)} \end{bmatrix} \tag{3.69}$$

In order to define a functional robot motion we introduce the generalized position vector x_e. For a robot with six degrees of freedom we adopt the position vector in the form

$$x_e = [x_A \quad y_A \quad z_A \quad \theta \quad \varphi \quad \psi]^T \tag{3.70}$$

where x_A, y_A, z_A are Cartesian coordinates of the gripper point which determine its position and θ, φ, ψ are Euler's angles which determine its orientation (Fig. 3.30).

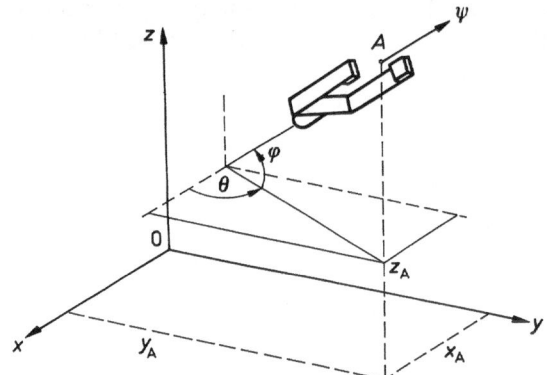

Fig. 3.30. External coordinates

Let us consider a manipulation task in which the gripper cannot move freely, its motion being imposed by this constraint. Here we have a closed chain. Imposed gripper motion reduces the number of degrees of freedom (d.o.f.). Let n_r be this reduced number of d.o.f. If n is the number of d.o.f., it holds that $n_r \leqslant n$. The equality holds when there is no constraint.

We now introduce n_r free and independent parameters $u_1, \ldots, u_n$ which define the constrained position of the gripper. The reduced position vector X_r is introduced

$$X_r = [u_1 \ldots u_n]^T \tag{3.71}$$

We express the constrained motion by the second-order Jacobian form connecting the position vector (Eq. (3.70)) and the reduced position vector (Eq. (3.71)):

$$\ddot{x}_e = J_r \ddot{X}_r + A_r \tag{3.72}$$

where J_r is the reduced Jacobian form and A_r is the associated reduced vector, the dimensions of which are $J_r(n \times n_r)$, $A_r(n \times 1)$. For motion without constraints, see Sect. 3.2.

$$\ddot{x}_e = J\ddot{q} + A \tag{3.73}$$

Combining Eqs. (3.72) and (3.73) we obtain

$$\ddot{q} = J^{-1} J_r \ddot{X}_r + J^{-1}(A_r - A) \tag{3.74}$$

Before substituting Eq. (3.74) into Eq. (3.68) it is noted that the constraints produce reactions, forces and moments which are already introduced into the dynamic model (Eq. (3.68)).

Depending on the constraint imposed and the manipulator configuration, there are some conditions which should be satisfied by the six-component reaction R_A. Namely, there are $6-(n-n_r)$ scalar conditions which can be expressed in matrix form:

$$ER_A = 0 \tag{3.75}$$

where E is a matrix of dimensions $(6-n+n_r) \times 6$. Now, Eqs. (3.68), (3.74) and (3.75) define the complete mathematical model of a closed chain configuration.

When we want to calculate the nominal dynamics (driving torques in the mechanism joints), we assume that the forces which we want to realize during motion are given. Thus F_A, M_A and accordingly R_A, are known. These values must be prescribed so that they satisfy Eq. (3.75). Now, the necessary driving torques (forces) can be calculated using matrix Eq. (3.68). However, if we want to calculate the unknown motion and reactions, then substituting Eq. (3.74) into Eq. (3.68), we obtain

$$HJ^{-1}J_r \ddot{X}_r + h - DR_A = P - HJ^{-1}(A_r - A) \tag{3.76}$$

Combining Eqs. (3.76) and (3.75), the following matrix equation can be reached:

$$\begin{bmatrix} HJ^{-1}J_r & -D \\ \hline 0 & E \end{bmatrix} \begin{bmatrix} \ddot{X}_r \\ \hline R_A \end{bmatrix} + \begin{bmatrix} h \\ \hline 0 \end{bmatrix} = \begin{bmatrix} P \\ \hline 0 \end{bmatrix} + \begin{bmatrix} -HJ^{-1}(A_r - A) \\ \hline 0 \end{bmatrix} \tag{3.77}$$

which can be solved for $\ddot{X}_r$ and R_A. Using Eq. (3.74), we can calculate $\ddot{q}$, i.e. the unknown motion.

The presented theoretical considerations will be applied to the case of the surface type constraint.

3.4.3 Gripper Moving Along a Surface

Let us consider a manipulator with a gripper which cannot move freely but its point A is forced to move along a given surface (Fig. 3.31). The relative position of the gripper point A with respect to a surface we define by means of two parameters, u_1 and u_2. This leads us to the parametric form of the moving surface (non-stationary constraint):

$$x = f_x(u_1, u_2, t) , \quad y = f_y(u_1, u_2, t) , \quad z = f_z(u_1, u_2, t) \tag{3.78}$$

Now, $n_r = n - 1$ and therefore the reduced position vector X_r, for a manipulator with six degrees of freedom, is

$$X_r = [u_1 u_2 \theta \rho \psi]^T \tag{3.79}$$

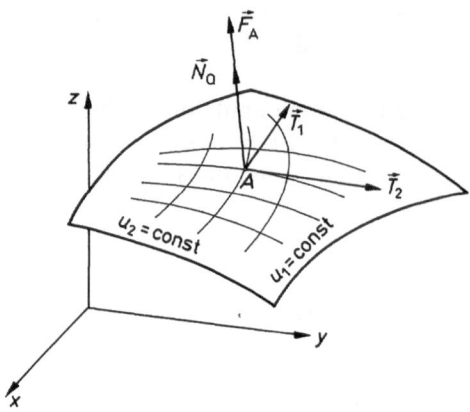

Fig. 3.31. Point A on a moving surface

i.e. $u_3 = \theta$, $u_4 = \rho$, $u_5 = \psi$.

The first and the second derivative of Eq. (3.78) gives

$$\dot{x} = \frac{\partial f_x}{\partial u_1}\dot{u}_1 + \frac{\partial f_x}{\partial u_2}\dot{u}_2 + \frac{\partial f_x}{\partial t}$$

$$\dot{y} = \frac{\partial f_y}{\partial u_1}\dot{u}_1 + \frac{\partial f_y}{\partial u_2}\dot{u}_2 + \frac{\partial f_y}{\partial t} \qquad (3.80)$$

$$\dot{z} = \frac{\partial f_z}{\partial u_1}\dot{u}_1 + \frac{\partial f_z}{\partial u_2}\dot{u}_2 + \frac{\partial f_z}{\partial t}$$

and

$$\ddot{x} = \frac{\partial^2 f_x}{\partial u_1^2}\dot{u}_1^2 + 2\frac{\partial^2 f_x}{\partial u_1 \partial u_2}\dot{u}_1\dot{u}_2 + \frac{\partial^2 f_x}{\partial u_2^2}\dot{u}_2^2 + 2\frac{\partial^2 f_x}{\partial u_1 \partial t}\dot{u}_1 +$$

$$+ 2\frac{\partial^2 f_x}{\partial u_2 \partial t}\dot{u}_2 + \frac{\partial^2 f_x}{\partial t^2} + \frac{\partial f_x}{\partial u_1}\ddot{u}_1 + \frac{\partial f_x}{\partial u_2}\ddot{u}_2 = \alpha_x + \frac{\partial f_x}{\partial u_1}\ddot{u}_1 + \frac{\partial f_x}{\partial u_2}\ddot{u}_2$$

$$\ddot{y} = \frac{\partial^2 f_y}{\partial u_1^2}\dot{u}_1^2 + 2\frac{\partial^2 f_y}{\partial u_1 \partial u_2}\dot{u}_1\dot{u}_2 + \frac{\partial^2 f_y}{\partial u_2^2}\dot{u}_2^2 + 2\frac{\partial^2 f_y}{\partial u_1 \partial t}\dot{u}_1 + \qquad (3.81)$$

$$+ 2\frac{\partial^2 f_y}{\partial u_2 \partial t}\dot{u}_2 + \frac{\partial^2 f_y}{\partial t^2} + \frac{\partial f_y}{\partial u_1}\ddot{u}_1 + \frac{\partial f_y}{\partial u_2}\ddot{u}_2 = \alpha_y + \frac{\partial f_y}{\partial u_1}\ddot{u}_1 + \frac{\partial f_y}{\partial u_2}\ddot{u}_2$$

$$\ddot{z} = \frac{\partial^2 f_z}{\partial u_1^2}\dot{u}_1^2 + 2\frac{\partial^2 f_z}{\partial u_1 \partial u_2}\dot{u}_1\dot{u}_2 + \frac{\partial^2 f_z}{\partial u_2^2}\dot{u}_2^2 + 2\frac{\partial^2 f_z}{\partial u_1 \partial t}\dot{u}_1 +$$

$$+ 2\frac{\partial^2 f_z}{\partial u_2 \partial t}\dot{u}_2 + \frac{\partial^2 f_z}{\partial t^2} + \frac{\partial f_z}{\partial u_1}\ddot{u}_1 + \frac{\partial f_z}{\partial u_2}\ddot{u}_2 = \alpha_z + \frac{\partial f_z}{\partial u_1}\ddot{u}_1 + \frac{\partial f_z}{\partial u_2}\ddot{u}_2$$

The three equations from (3.81) can be written together in the matrix form

$$
\begin{bmatrix} \ddot{x}_A \\ \ddot{y}_A \\ \ddot{z}_A \end{bmatrix} = \begin{bmatrix} \dfrac{\partial f_x}{\partial u_1} & \dfrac{\partial f_x}{\partial u_2} \\ \dfrac{\partial f_y}{\partial u_1} & \dfrac{\partial f_y}{\partial u_2} \\ \dfrac{\partial f_z}{\partial u_1} & \dfrac{\partial f_z}{\partial u_2} \end{bmatrix} \begin{bmatrix} \ddot{u}_1 \\ \ddot{u}_2 \end{bmatrix} + \begin{bmatrix} \alpha_x \\ \alpha_y \\ \alpha_z \end{bmatrix}
\tag{3.82}
$$

Now, the Jacobian form (3.72) can be obtained. The reduced Jacobian and the reduced associated vector are

$$
J_r = \begin{bmatrix} \dfrac{\partial f_x}{\partial u_1} & \dfrac{\partial f_x}{\partial u_2} & \vline & \\ \dfrac{\partial f_y}{\partial u_1} & \dfrac{\partial f_y}{\partial u_2} & \vline & 0_{(3 \times (n-3))} \\ \dfrac{\partial f_z}{\partial u_1} & \dfrac{\partial f_z}{\partial u_2} & \vline & \\ \hline 0_{((n-3) \times 2)} & & \vline & I_{((n-3) \times (n-3))} \end{bmatrix} , \quad A_r = \begin{bmatrix} \alpha_x \\ \alpha_y \\ \alpha_z \\ \hline 0_{((n-3) \times 1)} \end{bmatrix}
\tag{3.83}
$$

where α_x, α_y, α_z are defined as in Eq. (3.81) and I is a unit matrix of the corresponding dimension.

Let us consider the reactions R_A. It is clear that there exists a reaction force $\vec{F}_A$ perpendicular to the surface and the reaction moment $\vec{M}_A$ equals zero (Fig. 3.31). During a gripper motion along a surface there appear friction forces which due to limited space are not considered in this book.

If we define two tangents

$$
\vec{T}_1 = \left\{ \frac{\partial f_x}{\partial u_1}, \frac{\partial f_y}{\partial u_1}, \frac{\partial f_z}{\partial u_1} \right\}
\tag{3.84}
$$

$$
\vec{T}_2 = \left\{ \frac{\partial f_x}{\partial u_2}, \frac{\partial f_y}{\partial u_2}, \frac{\partial f_z}{\partial u_2} \right\}
$$

having the unit vectors

$$
\vec{T}_{01} = \vec{T}_1 / |\vec{T}_1| , \quad \vec{T}_{02} = \vec{T}_2 / |\vec{T}_2|
\tag{3.85}
$$

then it holds that $\vec{F}_A \perp \vec{T}_{01}$ and $\vec{F}_A \perp \vec{T}_{02}$ or

$$
\vec{T}_{01} \cdot \vec{F}_A = 0 , \quad \vec{T}_{02} \cdot \vec{F}_A = 0
\tag{3.86}
$$

The conditions (3.86) together with the condition $\vec{M}_A = 0$ can be written in the form of eq. (3.75) where

$$
E = \begin{bmatrix}
T_{01}^T & | & 0_{(1 \times 3)} \\
\text{-----} & | & \text{-----} \\
T_{02}^T & | & 0_{(1 \times 3)} \\
\text{-----} & | & \text{-----} \\
0_{(3 \times 3)} & | & I_{(3 \times 3)}
\end{bmatrix}_{(5 \times 6)}
\tag{3.87}
$$

Now, all the elements of the dynamic model (Eq. (3.77)) are determined and the model can be solved. For a manipulator with six degrees of freedom the nominal dynamics are calculated and the results are presented in the next Section.

Numerical Example

Let us consider the manipulation robot during grinding of the working object which is moved with constant velocity of 0.1 m/s (Fig. 3.32). During the period T_1 the manipulator moves freely, i.e. the manipulator with a cutting tool moves towards the working object. Let us assume that the contact between the gripper

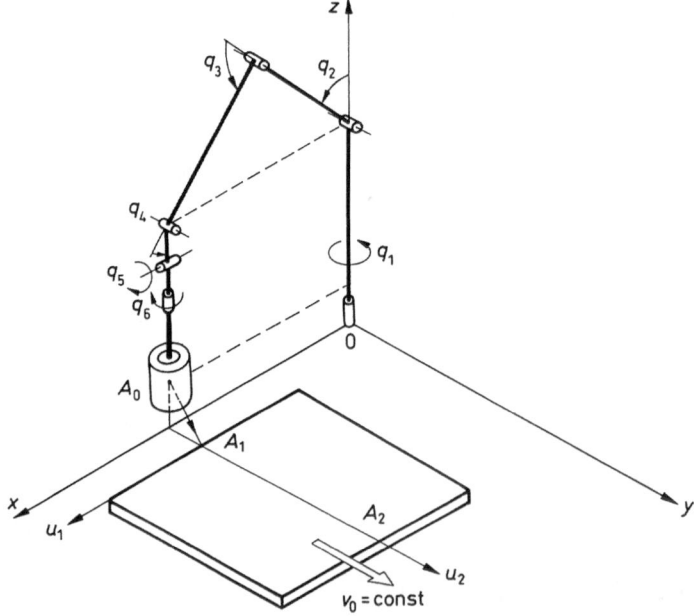

Fig. 3.32. Scheme of manipulation grinding task

(cutting tool) and the working object is impactless and that the end of period T_1 is, at the same time, the start of the interval T_2 (point A_1 in Fig. 3.32). The motion from A_1 to A_2 is constrained, i.e. the manipulator is considered as a closed chain.

The relative manipulator motion with respect to the surface, as mentioned before, is defined by means of two parameters, u_1 and u_2. The parameter u_1 is constant during the grinding process and the parameter u_2 can be defined in the form:

$$u_2(t) = \frac{L}{T}t = \frac{0.4}{2}t = 0.2t \tag{3.88}$$

where $L[m]$ is the cutting length and $T[s]$ is the total grinding time.

Following the theory explained above, the reduced Jacobian form and the reduced associated vector are

$$
J_r = \begin{bmatrix}
\dfrac{\partial f_x}{\partial u_1} & \dfrac{\partial f_x}{\partial u_2} & \\[2mm]
\dfrac{\partial f_y}{\partial u_1} & \dfrac{\partial f_y}{\partial u_2} & 0_{(3\times(n-3))} \\[2mm]
\dfrac{\partial f_z}{\partial u_1} & \dfrac{\partial f_z}{\partial u_2} & \\[2mm]
\hline
0_{((n-3)\times 2)} & I_{((n-3)\times(n-3))}
\end{bmatrix}
=
\begin{bmatrix}
1 & 0 & | & 0 & 0 & 0 \\
0 & 1 & | & 0 & 0 & 0 \\
0 & 0 & | & 0 & 0 & 0 \\
\hline
0 & 0 & | & 1 & 0 & 0 \\
0 & 0 & | & 0 & 1 & 0 \\
0 & 0 & | & 0 & 0 & 1
\end{bmatrix}
\tag{3.89}
$$

$$
A_r = \begin{bmatrix}
\alpha_x \\
\alpha_y \\
\alpha_z \\
\hline
0_{((n-3)\times 1)}
\end{bmatrix}
=
\begin{bmatrix}
0 \\
0 \\
0 \\
\hline
0 \\
0 \\
0
\end{bmatrix}
\tag{3.90}
$$

The tangents $\vec{T}_1$ and $\vec{T}_2$, defined with Eq. (3.84), are

$$\vec{T}_1 = \left\{ \frac{\partial f_x}{\partial u_1}, \frac{\partial f_y}{\partial u_1}, \frac{\partial f_z}{\partial u_1} \right\} = \{1, 0, 0\}$$

$$\vec{T}_2 = \left\{ \frac{\partial f_x}{\partial u_2}, \frac{\partial f_y}{\partial u_2}, \frac{\partial f_z}{\partial u_2} \right\} = \{0, 1, 0\}$$

$$\tag{3.91}$$

and the unit vectors (Eq. (3.85)) are

$$\vec{T}_{01} = \vec{T}_1 / |\vec{T}_1| = \{1, 0, 0\}$$
$$\vec{T}_{02} = \vec{T}_2 / |\vec{T}_2| = \{0, 1, 0\} \ . \tag{3.92}$$

On the basis of the cutting theory, it is known that during grinding, two perpendicular components appear, radial F_r and tangential F_t. It is also known that their ratio is $1.5 \div 3$. In this example we adopt $F_r/F_t = 2$. The reaction force $\vec{F}_A$ (cutting force F_r) satisfies the condition (3.86), and the matrix E (3.87) is

$$
E = \begin{bmatrix} T^{\mathrm{T}}_{01} & 0_{(1 \times 3)} \\ \hline T^{\mathrm{T}}_{02} & 0_{(1 \times 3)} \\ \hline 0_{(3 \times 3)} & I_{(3 \times 3)} \end{bmatrix} = \left[\begin{array}{ccc|ccc} 1 & 0 & 0 & 0 & 0 & 0 \\ 0 & 1 & 0 & 0 & 0 & 0 \\ \hline 0 & 0 & 0 & 1 & 0 & 0 \\ 0 & 0 & 0 & 0 & 1 & 0 \\ 0 & 0 & 0 & 0 & 0 & 1 \end{array} \right] \tag{3.93}
$$

Now, the conditions (3.75) can be expressed in the following form:

$$
ER_A = \begin{bmatrix} 1 & 0 & 0 & 0 & 0 & 0 \\ 0 & 1 & 0 & 0 & 0 & 0 \\ 0 & 0 & 0 & 1 & 0 & 0 \\ 0 & 0 & 0 & 0 & 1 & 0 \\ 0 & 0 & 0 & 0 & 0 & 1 \end{bmatrix} \begin{bmatrix} F_{Ax} \\ F_{Ay} \\ F_{Az} \\ M_{Ax} \\ M_{Ay} \\ M_{Az} \end{bmatrix} = \begin{bmatrix} F_{Ax} \\ F_{Ay} \\ M_{Ax} \\ M_{Ay} \\ M_{Az} \end{bmatrix} = 0 \tag{3.94}
$$

where F_{Ax}, F_{Ay}, F_{Az} are reaction force projections and M_{Ax}, M_{Ay}, M_{Az} are reaction moment projections on the Cartesian axes x, y, z. It is evident that two force components and three moment components equal zero. The third force component F_{Az} is equal to the radial component F_r. The complete mathematical dynamics model of the total system which has been used for unknown motion and component F_r calculations is presented in [2].

The manipulator geometric parameters are given in Table 3.5, dynamic parameters in Table 3.6, and actuator parameters in Table 3.7.

The internal manipulator coordinates in the initial instant (point A_0) are given in Table 3.8. The initial and final points on the nominal trajectory, and movement time are given in Table 3.9.

Adopting a trapezoidal velocity profile with an acceleration and deceleration time of 0.2 s, and using Eq. (3.23) for motion without constraints, and Eq. (3.68) for constrained motion, the nominal driving torques in the mechanism joints are calculated. On the trajectory path from A_1 to A_2 we adopted the

Table 3.5. Manipulator geometric parameters

Joint	Type of joint	Unit vectors of joint axes $\vec{e}_i^{\dagger}$	Vectors $\vec{r}_{ii}^{\dagger}$[m]	Vectors $\vec{r}_{i,i+1}^{\dagger}$[m]
1	$R^{\ddagger}$	0, 0, 1	0, 0, 0.4	0, 0, −0.4
2	R	1, 0, 0	0, 0.4, 0	0, −0.4, 0
3	R	1, 0, 0	0, 0, −0.4	0, 0, 0.4
4	R	1, 0, 0	0, 0.075, 0	0, −0.075, 0
5	R	0, 0, 1	0, 0.075, 0	0, −0.075, 0
6	R	0, 1, 0	0, 0.15, 0	0, −0.15, 0

† All vectors are shown in an internal (joint) coordinate system.
‡ R = rotational

Table 3.6. Dynamic parameters

Segment	Mass m_i[kg]	Moment of inertia [kg m²]		
		J_{xi}	J_{yi}	J_{zi}
1	0	0.	0.	0.2
2	5	0.25	0.01	0.25
3	5	0.25	0.25	0.01
4	1	0.002	0.002	0.002
5	1	0.002	0.002	0.002
6	2	0.01	0.002	0.01

Table 3.7. Actuator parameters

Actuator	C_M $\left[\dfrac{\text{Nm}}{\text{A}}\right]$	C_E $\left[\dfrac{\text{V}}{\text{rad/s}}\right]$	R_r [Ω]	B_C $\left[\dfrac{\text{Nm}}{\text{rad/s}}\right]$	N_V [−]	N_M [−]	η [−]	J_M [kg m²]
1	1.5	1.43	1.6	0.0058	31.17	24.94	0.8	0.00003
2	22.32	27.90	1.8	3.15	150	120	0.8	0.00079
3	14.88	18.6	1.8	1.4	100	80	0.8	0.00079
4	3.52	4.4	0.85	0.24	100	80	0.8	0.00001
5	3.52	4.4	0.85	0.24	100	80	0.8	0.00001
6	3.52	4.4	0.85	0.24	100	80	0.8	0.00001

Table 3.8. Internal coordinates

Point	Internal coordinates q_i[rad]					
A_0	q_1	q_2	q_3	q_4	q_5	q_6
	−1.57080	−0.52360	−2.09439	−0.52360	0	0

Table 3.9. Initial and final points on the nominal trajectory[†]

Point on trajectory	x [m]	y [m]	z [m]	q_4 [rad]	q_5 [rad]	q_6 [rad]	Movement	Movement time[s]
A_0	0.8	0	0.2	-0.52360	0	0	$A_0 \rightarrow A_1$	1
A_1	0.8	0.15	0	-0.52360	0	0		
A_2	0.8	0.75	0	-0.52360	0	0	$A_1 \rightarrow A_2$	1

[†] x, y, z – denote manipulator tip coordinates with respect to external immobile Cartesian coordinate system

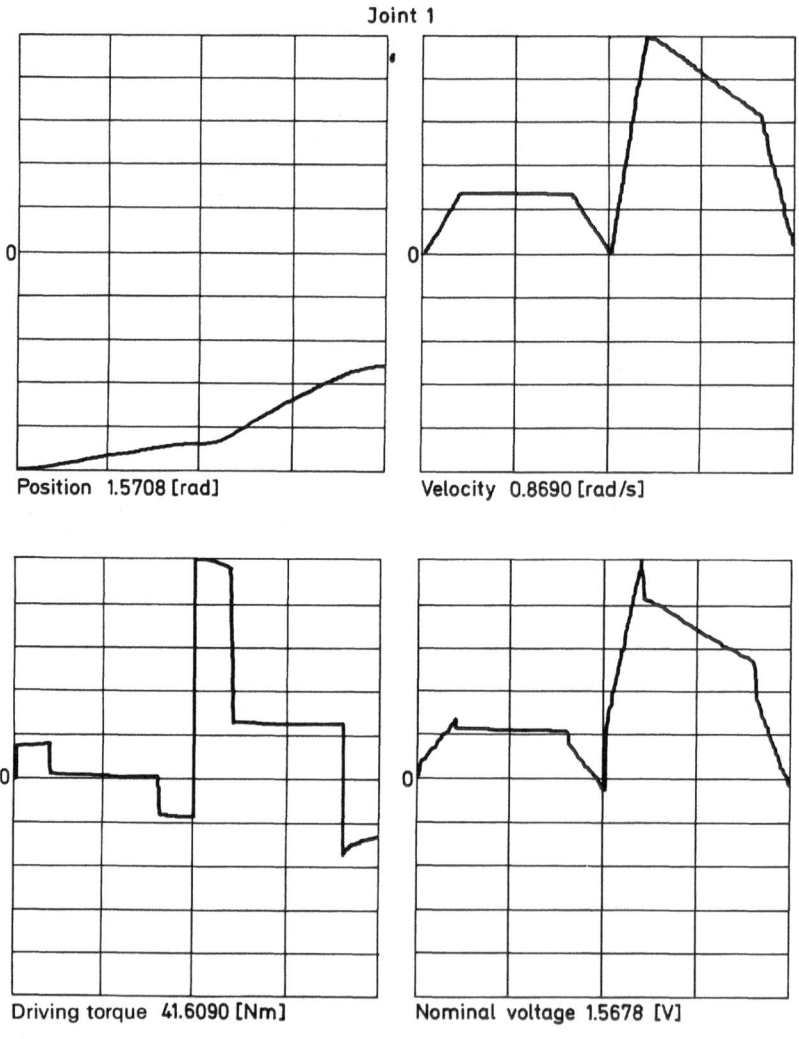

Position 1.5708 [rad]   Velocity 0.8690 [rad/s]

Driving torque 41.6090 [Nm]   Nominal voltage 1.5678 [V]

Fig. 3.33. Nominal values: position, velocity, driving torque and nominal voltage

nominal force $F_A = 30N$ which we want to realize during grinding. We assumed that the force F_A is acting upon the gravity centre of the last mechanism segment.
Using the following relation

$$u = \text{diag}\left(\frac{J_M^i R_r^i}{C_m^i} N_v^i\right)\ddot{q} + \left[\text{diag}\left(\frac{R_r^i B_c^i}{C_m^i} N_v^i\right) + \text{diag}(C_e^i N_v^i)\right]\dot{q}$$

$$+ \text{diag}\left(\frac{R_r^i}{C_m^i N_v^i \eta_i}\right) P \;,$$ (3.95)

where $B_c = B_c N_v N_M$, $C_M = C_m N_M$, $C_E = C_e N_v$, $N_M = N_v \eta$,

the nominal voltages (input to the corresponding actuators) are calculated. Equation (3.95) is obtained combining Eqs. (3.31) and (3.28).

The nominal dynamic results are shown in Fig. 3.33 which represents the position, velocity, driving torque and control signal time histories normalized on their maximal values.

Appendix A.3.1

Connection Between the Moving and Fixed System

The need for transformation of the values from the internal (moving) into the fixed (absolute) system lies in the requirement to present the kinematics and dynamics of robotic mechanisms in a unique, fixed system of coordinates (Fig. A.3.1.1).

Therefore let us consider two Cartesian coordinate systems: $Oxyz$ and $O'x'y'z'$ (Fig. A.3.1.2). Vector $\vec{r}$ is expressed via its projections onto the axes of both coordinate systems as:

$$\vec{r} = r_x\vec{i} + r_y\vec{j} + r_z\vec{k} \qquad (Oxyz) \tag{A.3.1.1}$$

$$\vec{r} = r_{x'}\vec{i'} + r_{y'}\vec{j'} + r_{z'}\vec{k'} \qquad (O'x'y'z') \tag{A.3.1.2}$$

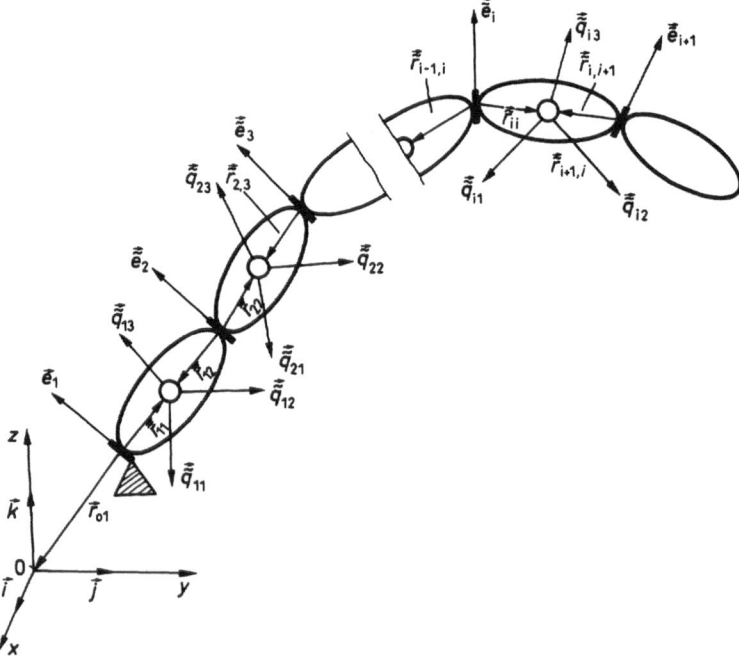

Fig. A.3.1.1. Representing the mechanism in the fixed coordinate system

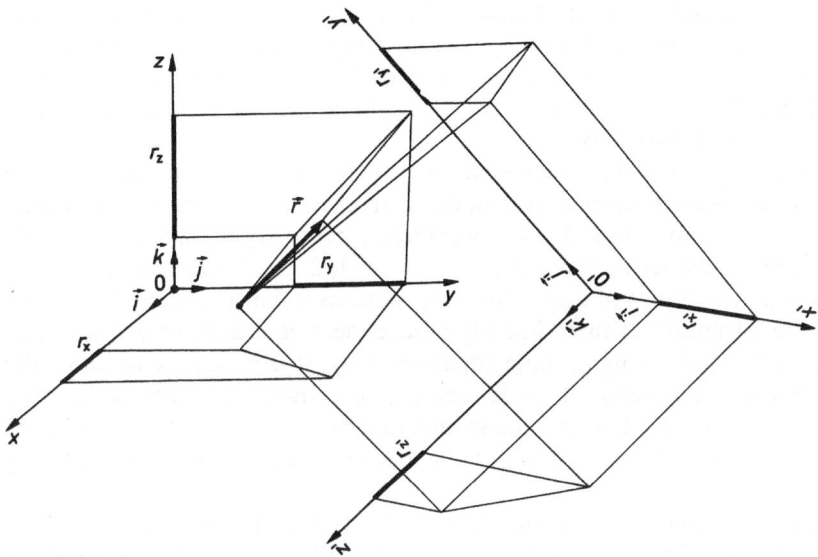

Fig. A.3.1.2. Representing vector $\vec{r}$ in two coordinate systems

Vector $\vec{r}$ is designated by $\vec{\tilde{r}}$, as seen by an observer in the $O'x'y'z'$ coordinate system. The vectors $\vec{i}, \vec{j}, \vec{k};\ \vec{i}', \vec{j}', \vec{k}'$ are unit vectors, associated to the corresponding coordinate axes.

Let us express the projections r_x, r_y, r_z via the data of vector $\vec{\tilde{r}}$ in $O'x'y'z'$. By scalar multiplication of Eq. (A.3.1.1) and Eq. (A.3.1.2) by $\vec{i}$, we obtain:

$$r_x \vec{i} \cdot \vec{i} + r_y \vec{j} \cdot \vec{i} + r_z \vec{k} \cdot \vec{i} = r_{x'} \vec{i}' \cdot \vec{i} + r_{y'} \vec{j}' \cdot \vec{i} + r_{z'} \vec{k}' \cdot \vec{i} \qquad (A.3.1.3)$$

Using the definition of the dot product of two vectors, and taking into account that $\vec{i} \cdot \vec{i} = 1$, $\vec{j} \cdot \vec{i} = 0$ and $\vec{k} \cdot \vec{i} = 0$, expression (A.3.1.3) becomes:

$$r_x = r_{x'} \vec{i}' \cdot \vec{i} + r_{y'} \vec{j}' \cdot \vec{i} + r_{z'} \vec{k}' \cdot \vec{i} \qquad (A.3.1.3a)$$

By analogous multiplication by $\vec{j}$ and $\vec{k}$, we obtain for r_y and r_z:

$$r_y = r_{x'} \vec{i}' \cdot \vec{j} + r_{y'} \vec{j}' \cdot \vec{j} + r_{z'} \vec{k}' \cdot \vec{j} \qquad (A.3.1.3b)$$

$$r_z = r_{x'} \vec{i}' \cdot \vec{k} + r_{y'} \vec{j}' \cdot \vec{k} + r_{z'} \vec{k}' \cdot \vec{k} \qquad (A.3.1.3c)$$

Using matrix notation, expressions for r_x, r_y, and r_z can be written in the form:

$$\begin{bmatrix} r_x \\ r_y \\ r_z \end{bmatrix} = \begin{bmatrix} \cos(\vec{i}'\,\vec{i}) & \cos(\vec{j}'\,\vec{i}) & \cos(\vec{k}'\,\vec{i}) \\ \cos(\vec{i}'\,\vec{j}) & \cos(\vec{j}'\,\vec{j}) & \cos(\vec{k}'\,\vec{j}) \\ \cos(\vec{i}'\,\vec{k}) & \cos(\vec{j}'\,\vec{k}) & \cos(\vec{k}'\,\vec{k}) \end{bmatrix} \begin{bmatrix} r_{x'} \\ r_{y'} \\ r_{z'} \end{bmatrix} \qquad (A.3.1.3d)$$

or $\vec{r} = Q \vec{\tilde{r}}$

where Q is the transformation matrix of the cosine, by which the observer from the coordinate system $Oxyz$ has to multiply the projections of vector $\vec{r}$ in the system $O'x'y'z'$ in order to obtain the values of the vector $\vec{r}$ projections in $Oxyz$. The stated relations connect the data by which one vector is described in two different coordinate systems.

The columns of matrix Q represent the projections of the unit vectors of the $O'x'y'z'$ system onto the axes of system $Oxyz$. In order to determine the elements of this matrix of dimensions 3×3, it is evidently necessary to have nine linearly independent equations with respect to q_{ij}; $i, j \in \{1, 2, 3\}$. We have already seen that knowing the values of one vector's projections in both coordinate systems enables the formation of three linearly independent equations of q_{ij} (A.3.1.3d). Consequently, for forming all nine equations of q_{ij} it is necessary to know the values of the projections of three non-colinear, non-zero vectors in both coordinate systems. In that case the system determinant of linear q_{ij} is different from zero, which enables calculation to be made of the transformation matrix Q (see Eqs. (3.1)–(3.4)).

Let us return to our task of determining the transformation matrices of the robot mechanism. First consider Figs. 3.4 and A.3.1.1. The resting two vectors which are not mutual and related to the unit vectors of the joint axes $\vec{e}_i$, colinear, are $\vec{a}_i$ and $\vec{b}_i = (\vec{e}_i \times \vec{a}_i)$. From Fig. 3.4 is evident that only for the case $q_i = 0$ does the identity $\vec{a}_i = \tilde{a}_i$ hold when it is possible to perform the transformation of the i-th segment coordinate system into the fixed system. The vector trihedron $\{\vec{e}_i, \vec{a}_i, \vec{b}_i\}$ and $\{\tilde{e}_i, \tilde{a}_i, \tilde{b}_i\}$ then conduct to coincide in space, so there exist only three non-colinear vectors, the position of which is known in both coordinate systems.

The system of Eq. (3.1) can be represented in an expanded form as:

$$q_{11}^0 \tilde{e}_{i,1} + q_{12}^0 \tilde{e}_{i,2} + q_{13}^0 \tilde{e}_{i,3} = e_{i,x}$$

$$q_{12}^0 \tilde{e}_{i,1} + q_{22}^0 \tilde{e}_{i,2} + q_{23}^0 \tilde{e}_{i,3} = e_{i,y}$$

$$q_{31}^0 \tilde{e}_{i,1} + q_{32}^0 \tilde{e}_{i,2} + q_{33}^0 \tilde{e}_{i,3} = e_{i,z} \qquad \text{(A.3.1.4)}$$

$$q_{11}^0 \tilde{a}_{i,1} + q_{12}^0 \tilde{a}_{i,2} + q_{13}^0 \tilde{a}_{i,3} = a_{i,x}$$

$$- - - - - - - - - - - - - - - - - - - -$$

$$q_{31}^0 \tilde{b}_{i,1} + q_{32}^0 \tilde{b}_{i,2} + q_{33} \tilde{b}_{i,3} = b_{i,z}$$

where $q_{k,1}^0$, $k, l \in \{1, 2, 3\}$ are the elements of matrix Q_i^0; $e_{i,(1,2,3)}$, $a_{i,(1,2,3)}$, and $b_{i(1,2,3)}$ are the projections of vectors $\tilde{e}_i$, $\tilde{a}_i$ and $\tilde{b}_i$ onto the axes of the coordinate system of the i-th segment, $e_{i,(x,y,z)}$, $a_{i,(x,y,z)}$ and $b_{i,(x,y,z)}$ are the projections of vectors $\vec{e}_i$, $\vec{a}_i$, and $\vec{b}_i$ onto the axes of the fixed coordinate system. By grouping the equations with respect to the unknowns $q_{k,1}^0$, $q_{k,2}^0$, $q_{k,3}^0$, $k \in \{1, 2, 3\}$, we obtain three systems of linear algebraic equations with three unknowns. By applying

Kramer's rule we obtain for instance the following value for q_{11}^0:

$$q_{11}^0 = \frac{\begin{bmatrix} e_{i,x} & \tilde{e}_{i,2} & \tilde{e}_{i,3} \\ a_{i,x} & \tilde{a}_{i,2} & \tilde{a}_{i,3} \\ b_{i,x} & \tilde{b}_{i,2} & \tilde{b}_{i,3} \end{bmatrix}}{\Delta} \quad ; \quad \Delta = \begin{bmatrix} \tilde{e}_{i,1} & \tilde{e}_{i,2} & \tilde{e}_{i,3} \\ \tilde{a}_{i,1} & \tilde{a}_{i,2} & \tilde{a}_{i,3} \\ \tilde{b}_{i,1} & \tilde{b}_{i,2} & \tilde{b}_{i,3} \end{bmatrix}$$

By generalizing the solution of system (A.3.1.4), the expanded matrix of the equations system over q_{kl}^0 is of the form:

$$B = \begin{bmatrix} \tilde{e}_{i,1}\tilde{e}_{i,2}\tilde{e}_{i,3} & 0 & 0 & e_{i,x} \\ 0 & \tilde{e}_{i,1}\tilde{e}_{i,2}\tilde{e}_{i,3} & 0 & e_{i,y} \\ 0 & 0 & \tilde{e}_{i,1}\tilde{e}_{i,2}\tilde{e}_{i,3} & e_{i,z} \\ \tilde{a}_{i,1}\tilde{a}_{i,2}\tilde{a}_{i,3} & 0 & 0 & a_{i,x} \\ 0 & \tilde{a}_{i,1}\tilde{a}_{i,2}\tilde{a}_{i,3} & 0 & a_{i,y} \\ 0 & 0 & \tilde{a}_{i,1}\tilde{a}_{i,2}\tilde{a}_{i,3} & a_{i,z} \\ \tilde{b}_{i,1}\tilde{b}_{i,2}\tilde{b}_{i,3} & 0 & 0 & b_{i,x} \\ 0 & \tilde{b}_{i,1}\tilde{b}_{i,2}\tilde{b}_{i,3} & 0 & b_{i,y} \\ 0 & 0 & \tilde{b}_{i,1}\tilde{b}_{i,2}b_{i,3} & b_{i,z} \end{bmatrix}$$

Let us note that from matrix B the equations of type (A.3.1.4) can be written directly by selecting the corresponding column.

In order to start the procedure of forming the transformation matrix, it is necessary to know the projections of the fixed vector $\vec{r}_{01}$ in the fixed system (Fig. A.3.1.1), connecting the coordinate origin to the first mechanism joint. It is also necessary to know the projections of vector $\vec{e}$ in the fixed system and the system connected to the mass centre of the first segment. It should be mentioned that the upper index "zero" in the transformation matrix designates the case of a mechanism assembly, when all the relative angular coordinates (q_i) are equal to zero (starting-initial configuration of the robot mechanism).

The transformation matrices are calculated in such a way that in each iteration to the robot chain the following segment is added and the corresponding transformation matrix is calculated recurrently. Thus Q_i^0 is calculated when adding the i-th segment to the chain, Q_{i-1}^0 having already been calculated. With known Q_{i-1}^0, $\vec{r}_{i-1,i} = Q_{i-1}^0 \vec{r}_{i-1,i}$ and $\vec{e}_i = Q_{i-1}^0 \vec{e}_i$ are also known.

Appendix A.3.2

Determining Velocities and Accelerations

The recurrent kinematic relations, by which the angular and linear velocities and angular and linear accelerations of the mechanism segments are determined, are performed applying the basic relations of the rigid body mechanics.

The absolute body velocity equals the sum of its transfer and relative velocity:

$$\vec{\omega} = \vec{\omega}_p + \vec{\omega}_r \qquad (A.3.2.1)$$

For two points A and B of the body it follows that:

$$\vec{v}_B = \vec{v}_A + \vec{\omega} \times \vec{r}_{AB} \qquad (A.3.2.2)$$

where $\vec{\omega}$ is the angular body velocity, $\vec{r}_{AB}$ is the vector originating in point A and ending in point B.

Let us consider the i-th kinematic pair, consisting of the i-th and $(i-1)$st mechanism segment (Fig. A.3.2.1). Motion of the $(i-1)$st segment will be considered as a transferring one, and motion in the i-th joint as relative. The following relations then hold: for the case of linear (translatory) kinematic pair

$$\vec{\omega}_i = \vec{\omega}_{i-1} \ , \quad \vec{v}_i = \vec{v}_{i-1} - \vec{\omega}_{i-1} \times \vec{r}_{i-1,i} + \dot{q}_i \vec{e}_i \qquad (A.3.2.3)$$

and for the case of rotational kinematic pair

$$\vec{\omega}_i = \vec{\omega}_{i-1} + \dot{q}_i \vec{e}_i \ , \quad \vec{v}_i = \vec{v}_{i-1} - \vec{\omega}_{i-1} \times \vec{r}_{i-1,i} + \vec{\omega}_i \times \vec{r}_{ii} \qquad (A.3.2.4)$$

Expressions (A.3.2.3, A.3.2.4) can be considered as recurrent; if gradually all the mechanism segments are circled, starting from the initial one, the velocities of all segments can be calculated using these expressions.

For the acceleration of points A and B in a rigid body, moving with angular velocity $\vec{\omega}$ and angular acceleration $\vec{\varepsilon}$:

$$\vec{\varepsilon} = \vec{\varepsilon}_p + \vec{\varepsilon}_r + \vec{\omega}_p \times \vec{\omega}_r \ , \quad \vec{w}_B = \vec{w}_A + \vec{\varepsilon} \times \vec{r}_{AB} + \vec{\omega} \times (\vec{\omega} \times \vec{r}_{AB}) \qquad (A.3.2.5)$$

Considering the coordinate system $Q_{i-1}(q_{i-1,j}, j=1, 2, 3)$, connected to the $(i-1)$st mechanism segment and $Q_i (q_{ij}, j=1, 2, 3)$, connected to the i-th mechanism segment, the relation is easily established via relations (A.3.2.1–

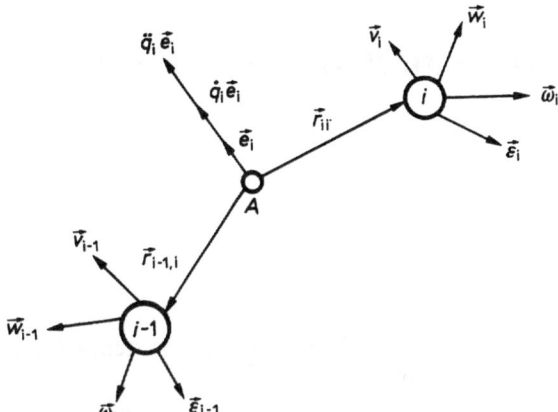

Fig. A.3.2.1. Velocities and accelerations of kinematic pair

A.3.2.3), considering motion of each previous $(i-1)$st segment as transferring and motion of i-th segment as relative:

$$\vec{v}=\vec{v}_i \ , \quad \vec{v}_p=\vec{v}_{i-1} \ , \quad \vec{\omega}=\vec{\omega}_i \ , \quad \vec{\omega}_p=\vec{\omega}_{i-1}$$

$$\vec{\varepsilon}=\vec{\varepsilon}_i \ , \quad \vec{\varepsilon}_p=\vec{\varepsilon}_{i-1} \ , \quad \vec{w}=\vec{w}_i \ , \quad \vec{w}_p=\vec{w}_{i-1} \qquad\qquad (A.3.2.6)$$

Now the following relations can be written: for the linear kinematic pair $\vec{\varepsilon}_i=\vec{\varepsilon}_{i-1}$ and for the rotational kinematic pair $\vec{\varepsilon}_i=\vec{\varepsilon}_{i-1}+\ddot{q}_i\vec{e}_i+\dot{q}_i\vec{\omega}_i\times\vec{e}_i$.

Let us pass on to determining the linear accelerations. Let A be the point of the momentary i-th joint centre. Acceleration of the point of the $(i-1)$st segment, coinciding with point A, can be $\vec{w}_{A,i-1}$. Then, as shown in Fig. A.3.2.1, this can be written as:

$$\vec{w}_{A,i-1}=\vec{w}_{i-1}-\vec{\varepsilon}_{i-1}\times\vec{r}_{i-1,i}-\vec{\omega}_{i-1}\times(\vec{\omega}_{i-1}\times\vec{r}_{i-1,i}) \qquad (A.3.2.7)$$

For the i-th segment point, coinciding momentarily with point A, can be written in the case of linear kinematic pair as $\vec{w}_{A,i}=\vec{w}_{A,i-1}+\ddot{q}_i\vec{e}_i+2\dot{q}_i\vec{\omega}_{i-1}\times\vec{e}_i$ and for the rotational pair $\vec{w}_{A,i}=\vec{w}_{A,i-1}$, which can now be written as $\vec{w}_i=\vec{w}_{A,i}+\vec{\varepsilon}_i\times\vec{r}_{ii}+\vec{\omega}_i\times(\vec{\omega}_i\times\vec{r}_{ii})$.

It should be mentioned that the relative movement q_i in the above expression comprises both linear and angular displacement.

Appendix A.3.3

Momentum of Rigid Body with Respect to a Fixed Pole

Consider the rigid body shown in Fig. A.3.3.1. The coordinate system $0\xi\eta\zeta$ is firmly connected to the body, and is moving with it.

Momentum of rigid body with respect to fixed pole 0 is:

$$\vec{L}_0 = \int_M \vec{\rho} \times \vec{v} \, dM = \int_M \vec{\rho} \times (\vec{\omega} \times \vec{\rho}) \, dM \qquad (A.3.3.1)$$

where vector $\vec{\rho}$ is the position vector of the elementary mass dM with respect to pole 0, $\vec{v}$ is the velocity vector of elementary mass and $\vec{\omega}$ is the angular velocity vector of the rigid body. After expanding the double cross product in expression (A.3.3.1) one obtains:

$$\vec{L}_0 = \vec{\omega} \int_M \rho^2 \, dM - \int_M \vec{\rho}(\vec{\omega} \cdot \vec{\rho}) \, dM \qquad (A.3.3.2)$$

or

$$\vec{L}_0 = J_0 \vec{\omega} - \int_M (\omega_\xi \xi + \omega_\eta \eta + \omega_\zeta \zeta)\vec{\rho} \, dM \qquad (A.3.3.3)$$

where J_0 is the polar moment of inertia for pole 0. As the double polar moment of inertia equals the sum of the axial moments of inertia for the coordinate axes, $2J_0 = J_\xi + J_\eta + J_\zeta$, the projections of the momentum onto the coordinate axes are:

$$
\begin{aligned}
L_\xi &= J_\xi \omega_\xi - J_{\xi\eta} \omega_\eta - J_{\xi\zeta} \omega_\zeta \\
L_\eta &= -J_{\xi\eta} \omega_\xi + J_\eta \omega_\eta - J_{\eta\zeta} \omega_\zeta \\
L_\zeta &= -J_{\xi\zeta} \omega_\xi - J_{\eta\zeta} \omega_\eta + J_\zeta \omega_\zeta
\end{aligned}
\qquad (A.3.3.4)
$$

If the coordinate axes are the main axes of inertia, the components of the momentum for those axes are:

$$L_1 = J_1 \omega_1 \ , \qquad L_2 = J_2 \omega_2 \ , \qquad L_3 = J_3 \omega_3 \qquad (A.3.3.5)$$

Euler's Dynamic Equations

Between the rigid body momentum and external forces the relation holds:

$$\frac{d\vec{L}_0}{dt} = \vec{M}_0 \qquad (A.3.3.6)$$

that the time derivative of the momentum for pole 0 equals the moment of external forces for the same pole as the moment point.

If this moment of external forces equals zero, i.e. if external forces are not acting on the body (inertial rotation), the momentum is constant. In every other case the momentum varies with time.

Let us designate by $\vec{L}_0$ the position vector of the terminal point of the momentum (Fig. A.3.3.2); the time derivative of this vector then represents the absolute velocity of point N.

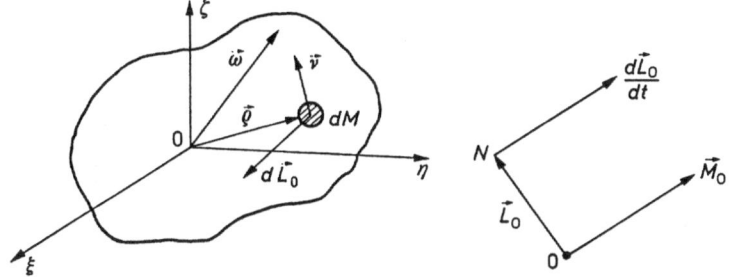

Fig. A.3.3.1. Momentum of rigid body elemen-   **Fig. A.3.3.2.** Moment of momentum
tary mass

As the absolute velocity consists of the relative and transfer velocity, the moment of momentum consists of the relative and transfer derivative part and can be written in the form:

$$\frac{d\vec{L}_0}{dt} = \dot{\vec{L}}_0 + \vec{\omega} \times \vec{L}_0 = \vec{M}_0. \qquad (A.3.3.7)$$

where $\dot{\vec{L}}_0 = \dot{L}_\xi \vec{\xi}_0 + \dot{L}_\eta \vec{\eta}_0 + \dot{L}_\zeta \vec{\zeta}_0$ is the relative time derivative of the position vector $\vec{L}_0$, and $\vec{\xi}_0, \vec{\eta}_0, \vec{\zeta}_0$ are unit vectors of the ξ, η, ζ axes, respectively.

Equation (A.3.3.7) represents Euler's dynamic equation in vector form. This equation corresponds with three scalar equations for the coordinate axes of the moving trihedron:

$$\begin{aligned}
\dot{L}_\xi + \omega_\eta L_\zeta - \omega_\zeta L_\eta &= M_\xi \\
\dot{L}_\eta + \omega_\zeta L_\xi - \omega_\xi L_\zeta &= M_\eta \\
\dot{L}_\zeta - \omega_\xi L_\eta - \omega_\eta L_\xi &= M_\zeta
\end{aligned} \qquad (A.3.3.8)$$

where the momentum components for the coordinate axes are determined by equation (A.3.3.4). If the coordinate axes are the main axes of inertia, Euler's equations obtain a simpler form:

$$J_1\dot{\omega}_1 - (J_2 - J_3)\omega_2\omega_3 = M_1$$
$$J_2\dot{\omega}_2 - (J_3 - J_1)\omega_3\omega_1 = M_2 \qquad\qquad (A.3.3.9)$$
$$J_3\dot{\omega}_3 - (J_1 - J_2)\omega_1\omega_2 = M_3$$

where J_1, J_2, J_3 are the main moments of inertia, $\omega_1, \omega_2, \omega_3$ are the projections of the angular velocity vector onto the main axes of inertia and moments M_1, M_2, M_3 are the main moments of all external forces for these axes.

Appendix A.3.4

Example of Mathematical Model Derivation

Using the algorithm presented in Chapter 3, the mathematical model of the SCARA (basic configuration) robot dynamics "by hand" is derived.

First Segment

After assembling the mechanism in Fig. A.3.4.1, the positions of all the vectors are obtained in the fixed system (xyz), when the internal coordinates equal zero, i.e. $q_i = \psi = \varphi = 0$ (Fig. A.3.4.2). Note that in this position the local coordinate systems coincide with the xyz system. Hence (expression (3.2), (A.3.1.4)):

$$\vec{q}_{11}^{\,0} = [1 \quad 0 \quad 0]^T , \quad \vec{q}_{12}^{\,0} = [0 \quad 1 \quad 0]^T , \quad \vec{q}_{13}^{\,0} = [0 \quad 0 \quad 1]^T$$

i.e.

$$Q_1^0 = [\vec{q}_{11}^{\,0} \ \vec{q}_{12}^{\,0} \ \vec{q}_{13}^{\,0}] = \begin{bmatrix} 1 & 0 & 0 \\ 0 & 1 & 0 \\ 0 & 0 & 1 \end{bmatrix}$$

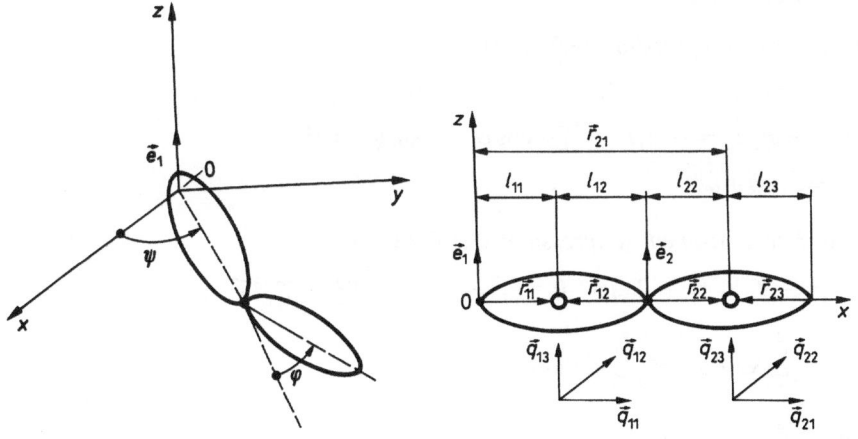

Fig. A.3.4.1. Mechanical robot configuration   **Fig. A.3.4.2.** Initial mechanism position

By rotating the first segment for angle ψ and by applying the formula of Rodrigues (expression (3.4)), we obtain

$$\vec{q}_{11} = [\cos\psi \quad \sin\psi \quad 0]^{\mathrm{T}}$$
$$\vec{q}_{12} = [-\sin\psi \quad \cos\psi \quad 0]^{\mathrm{T}}$$
$$\vec{q}_{13} = [0 \quad 0 \quad 1]^{\mathrm{T}}$$

Hence the transformation matrix is

$$Q_1 = [\vec{q}_{11} \quad \vec{q}_{12} \quad \vec{q}_{13}] = \begin{bmatrix} \cos\psi & -\sin\psi & 0 \\ \sin\psi & \cos\psi & 0 \\ 0 & 0 & 1 \end{bmatrix}$$

$$\vec{r}_{11} = Q_1\vec{r}_{11} = Q_1[l_{11} \quad 0 \quad 0]^{\mathrm{T}} = l_{11}[\cos\psi \quad \sin\psi \quad 0]^{\mathrm{T}}$$
$$\vec{r}_{12} = Q_2\vec{r}_{12} = -l_{12}[\cos\psi \quad \sin\psi \quad 0]^{\mathrm{T}}$$
$$\vec{e}_2 = [0 \quad 0 \quad 1]^{\mathrm{T}}$$

Velocities and Accelerations of First Segment (Expressions (3.6)–(3.8))

$$\vec{\omega}_1 = \dot{\psi}\vec{e}_1 = [0 \quad 0 \quad \dot{\psi}]^{\mathrm{T}}, \quad \vec{\omega}_0 = 0$$
$$\vec{\varepsilon}_1 = \dot{\vec{\omega}}_1 = [0 \quad 0 \quad \ddot{\psi}]^{\mathrm{T}}$$
$$\vec{\alpha}_{11} = \vec{e}_1 = [0 \quad 0 \quad 1]^{\mathrm{T}}, \quad \vec{\alpha}_1^0 = 0$$
$$\vec{\beta}_{11} = \vec{e}_1 \times \vec{r}_{11} = l_{11}[-\sin\psi \quad \cos\psi \quad 0]^{\mathrm{T}}$$
$$\vec{\beta}_1^0 = \dot{\vec{\omega}}_1 \times (\vec{\omega}_1 \times \vec{r}_{11}) = \dot{\psi}^2\vec{e}_1 \times (\vec{e}_1 \times \vec{r}_{11}) =$$
$$= l_{11}\dot{\psi}^2[-\cos\psi \quad -\sin\psi \quad 0]^{\mathrm{T}}$$
$$\vec{w}_1 = \vec{\beta}_{11}\ddot{\psi} + \vec{\beta}_1^0$$

Inertial Force (expressions (3.9), (3,10))

$$\vec{a}_{11} = -m_1\vec{\beta}_{11} = -m_1 l_{11}[-\sin\psi \quad \cos\psi \quad 0]^{\mathrm{T}}$$
$$\vec{a}_1^0 = -m_1\vec{\beta}_1^0 = -m_1 l_{11}\dot{\psi}^2[-\cos\psi \quad -\sin\psi \quad 0]^{\mathrm{T}}$$
$$\vec{F}_1 = \vec{a}_{11}\ddot{\psi} + \vec{a}_1^0$$

Inertial Force Moment (expressions (3.13)–(3.17))

$$\vec{b}_{11} = -T_1\vec{\alpha}_{11} = -T_1[0 \quad 0 \quad 1]^{\mathrm{T}} = -[T_1^{13} \quad T_1^{23} \quad T_1^{33}]^{\mathrm{T}}$$

$$T_i^{jk} = \sum_{l=1}^{3} Q_i^{jl} J_{il} q_{il}^k = \sum_{l=1}^{3} q_{il}^j q_{il}^k J_{il}$$

$$T_1^{13} = \sum_{l=1}^{3} Q_1^{1l} J_{1l} q_{1l}^3 = 0$$

$$T_1^{23} = \sum_{l=1}^{3} Q_1^{21} J_{1l} q_{1l}^{3} = 0$$

$$T_1^{33} = \sum_{l=1}^{3} Q_1^{31} J_{1l} q_{1l}^{3} = \sum_{l=1}^{3} q_{1l}^{3} q_{1l}^{3} J_{1l} = J_{13}$$

$$\vec{b}_{11} = [0 \quad 0 \quad -J_{13}]^T$$

$$\vec{b}_1^0 = -T_1 \vec{\alpha}_1^0 + \vec{\lambda}_1 = \vec{\lambda}_1 =$$

$$= Q_1 \begin{bmatrix} (J_{12} - J_{13}') (\vec{\omega}_1 \cdot \vec{q}_{12}) (\vec{\omega}_1 \cdot \vec{q}_{13}) \\ (J_{13} - J_{11}) (\vec{\omega}_1 \cdot \vec{q}_{13}) (\vec{\omega}_1 \cdot \vec{q}_{11}) \\ (J_{11} - J_{12}) (\vec{\omega}_1 \cdot \vec{q}_{11}) (\vec{\omega}_1 \cdot \vec{q}_{12}) \end{bmatrix} = 0$$

$$\vec{M}_1 = \vec{b}_{11} \ddot{\psi} + \vec{b}_1^0$$

Second Segment

As $\vec{q}_{21}^0 = \vec{q}_{11}$, $\vec{q}_{22}^0 = \vec{q}_{12}$ and $\vec{q}_{23}^0 = \vec{q}_{13}$, applying Rodrigues' formula, we obtain

$$Q_2 = [\vec{q}_{21} \quad \vec{q}_{22} \quad \vec{q}_{23}] = \begin{bmatrix} \cos(\psi + \varphi) & -\sin(\psi + \varphi) & 0 \\ \sin(\psi + \varphi) & \cos(\psi + \varphi) & 0 \\ 0 & 0 & 1 \end{bmatrix}$$

In a further text, for simplicity, we will use substitution $\sigma = \psi + \varphi$

$$\vec{r}_{22} = l_{22} [\cos \sigma \quad \sin \sigma \quad 0]^T$$

$$\vec{r}_{23} = -l_{23} [\cos \sigma \quad \sin \sigma \quad 0]^T$$

$$\vec{e}_3 = [0 \quad 0 \quad 1]^T$$

Velocities and Accelerations of Second Segment

$$\vec{\omega}_2 = \vec{\omega}_1 + \dot{\varphi} \vec{e}_2 = [0 \quad 0 \quad \dot{\psi} + \dot{\varphi}]^T = [0 \quad 0 \quad \dot{\sigma}]^T$$

$$\vec{\alpha}_{21} = \vec{e}_1 = [0 \quad 0 \quad 1]^T$$

$$\vec{\alpha}_{22} = \vec{e}_2 = [0 \quad 0 \quad 1]^T, \quad \vec{\alpha}_2^0 = 0$$

$$\vec{\beta}_{21} = \vec{\beta}_{11} + \vec{e}_1 \times (\vec{r}_{22} - \vec{r}_{12})$$

$$= [-l_1 \sin \psi - l_{22} \sin \sigma \quad l_1 \cos \psi + l_{22} \cos \sigma \quad 0]^T$$

$$l_1 = l_{11} + l_{12}$$

$$\vec{\beta}_{22} = \vec{e}_2 \times \vec{r}_{22} = l_{22} [-\sin \sigma \quad \cos \sigma \quad 0]^T$$

$$\vec{\beta}_2^0 = \vec{\beta}_1^0 + \vec{\alpha}_1^0 \times (\vec{r}_{22} - \vec{r}_{12}) + \dot{\varphi}(\vec{\omega}_1 \times \vec{e}_2) \times \vec{r}_{22} + \vec{\gamma}_{22} - \vec{\gamma}_{12}$$

$$\vec{\gamma}_{ij} = \vec{\omega}_i \times (\vec{\omega}_i \times \vec{r}_{ij})$$

$$\vec{\gamma}_{22} - \vec{\gamma}_{12} = \vec{\omega}_2 \times (\vec{\omega}_2 \times \vec{r}_{22}) - \vec{\omega}_1 \times (\vec{\omega}_1 \times \vec{r}_{12})$$
$$= [-\dot{\sigma}^2 l_{22} \cos \sigma - \dot{\psi}^2 l_{12} \cos \psi$$
$$- \dot{\sigma}^2 l_{22} \sin \sigma - \dot{\psi}^2 l_{12} \sin \psi \quad 0]^T$$
$$\vec{\beta}_2^0 = \dot{\psi}^2 l_{11} [-\cos \psi \quad -\sin \psi \quad 0]^T +$$
$$+ [-\dot{\sigma}^2 l_{22} \cos \sigma - \dot{\psi}^2 l_{12} \cos \psi \quad -\dot{\sigma}^2 l_{22} \sin \sigma - \dot{\psi}^2 l_{12} \sin \psi \quad 0]^T =$$
$$= [-\dot{\sigma}^2 l_{22} \cos \sigma - \dot{\psi}^2 l_1 \cos \psi \quad -\dot{\sigma}^2 l_{22} \sin \sigma - \dot{\psi}^2 l_1 \sin \psi \quad 0]^T$$
$$\vec{w}_2 = \vec{\beta}_{21} \ddot{\psi} + \vec{\beta}_{22} \ddot{\varphi} + \vec{\beta}_2^0$$

Inertial Force

$$\vec{a}_{21} = -m_2 \vec{\beta}_{21} = [m_2 l_1 \sin \psi + m_2 l_{22} \sin \sigma \quad -m_2 l_1 \cos \psi - m_2 l_{22} \cos \sigma \quad 0]^T$$
$$\vec{a}_{22} = -m_2 \vec{\beta}_{22} = [m_2 l_{22} \sin \sigma \quad -m_2 l_{22} \cos \sigma \quad 0]^T$$
$$\vec{a}_2^0 = -m_2 \vec{\beta}_2^0 = [m_2 \dot{\sigma}^2 l_{22} \cos \sigma + m_2 \dot{\psi}^2 l_1 \cos \psi \quad m_2 \dot{\sigma}^2 l_{22} \sin \sigma +$$
$$+ m_2 \dot{\psi}^2 l_1 \sin \psi \quad 0]^T$$
$$\vec{F}_2 = \vec{a}_{21} \ddot{\psi} + \vec{a}_{22} \ddot{\varphi} + \vec{a}_2^0$$

Inertial Force Moment

$$\vec{b}_{21} = -T_2 \vec{a}_{21} = -T_2 [0 \quad 0 \quad 1]^T = [-T_2^{13} \quad -T_2^{23} \quad -T_2^{33}]^T =$$
$$= [0 \quad 0 \quad -J_{23}]^T$$
$$\vec{b}_{22} = -T_2 \vec{a}_{22} = \vec{b}_{21} = [0 \quad 0 \quad -J_{23}]^T$$
$$\vec{b}_2^0 = -T_2 \vec{a}_2^0 + \vec{\lambda}_2 = 0$$
$$\vec{M}_2 = \vec{b}_{21} \ddot{\psi} + \vec{b}_{22} \ddot{\varphi} + \vec{b}_2^0$$

Inertial Matrix H (expression (3.24))

$$H_{ik} = -\vec{e}_i \cdot \sum_{j=\max(i,k)}^{n} (\vec{b}_{jk} + \vec{r}_{ji} \times \vec{a}_{jk})$$

$$H_{11} = -\vec{e}_1 \cdot \sum_{j=1}^{2} (\vec{b}_{j1} + \vec{r}_{j1} \times \vec{a}_{j1}) = -\vec{e}_1 \cdot (\vec{b}_{11} + \vec{r}_{11} \times \vec{a}_{11} +$$
$$+ \vec{b}_{21} + \vec{r}_{21} \times \vec{a}_{21})$$
$$\vec{r}_{21} = (\vec{r}_{11} - \vec{r}_{12}) + \vec{r}_{22} = [l_1 \cos \psi + l_{22} \cos \sigma \quad l_1 \sin \psi + l_{22} \sin \sigma \quad 0]^T$$
$$H_{11} = m_1 l_{11}^2 + m_2 (l_1^2 + l_{22}^2 + 2 l_1 l_{22} \cos \varphi) + J_{13} + J_{23}$$
$$H_{12} = -\vec{e}_1 \cdot (\vec{b}_{22} + \vec{r}_{21} \times \vec{a}_{22}) = m_2 l_{22} (l_{22} + l_1 \cos \varphi) + J_{23}$$
$$H_{22} = -\vec{e}_2 \cdot (\vec{b}_{22} + \vec{r}_{22} \times \vec{a}_{22}) = m_2 l_{22}^2 + J_{23}$$

Column Matrix h (expression (3.24))

$$h_i = -\vec{e}_i \cdot \sum_{j=i}^{n} [\vec{r}_{ji} \times (\vec{a}_j^0 + \vec{G}_j) + \vec{b}_j^0]$$

$$h_1 = -\vec{e}_1 \cdot \sum_{j=1}^{2} [\vec{r}_{j1} \times (\vec{a}_j^0 + \vec{G}_j) + \vec{b}_j^0] =$$

$$= -\vec{e}_1 \cdot [\vec{r}_{11} \times (\vec{a}_1^0 + \vec{G}_1) + \vec{b}_1^0 + \vec{r}_{21} \times (\vec{a}_2^0 + \vec{G}_2) + \vec{b}_2^0] =$$

$$= -\vec{e}_1 \cdot (\vec{r}_{11} \times \vec{a}_1^0) - \vec{e}_1 \cdot (\vec{r}_{21} \times \vec{a}_2^0) =$$

$$= -m_2(l_1 l_{22} \dot{\sigma}^2 \sin\varphi - l_1 l_{22} \dot{\psi}^2 \sin\varphi) =$$

$$= -m_2 l_1 l_{22}(\dot{\sigma}^2 - \dot{\psi}^2)\sin\varphi = -m_2 l_1 l_{22}(2\dot{\psi}\dot{\varphi} + \dot{\varphi}^2)\sin\varphi$$

$$h_2 = -\vec{e}_2 \cdot (\vec{r}_{22} \times \vec{a}_2^0 + \vec{b}_2^0) = m_2 l_1 l_{22} \dot{\psi}^2 \sin\varphi$$

Differential Equations of Motion (expression (3.23))

$$\begin{bmatrix} H_{11} & H_{12} \\ H_{21} & H_{22} \end{bmatrix} \begin{bmatrix} \ddot{\psi} \\ \ddot{\varphi} \end{bmatrix} + \begin{bmatrix} h_1 \\ h_2 \end{bmatrix} = \begin{bmatrix} P_1 \\ P_2 \end{bmatrix}$$

or

$$\begin{bmatrix} m_1 l_{11}^2 + m_2(l_1^2 + l_{22}^2 + 2l_1 l_{22}\cos\varphi) + J_{13} + J_{23} & m_2 l_{22}(l_{22} + l_1 \cos\varphi) + J_{23} \\ m_2 l_{22}(l_{22} + l_1 \cos\varphi) + J_{23} & m_2 l_{22}^2 + J_{23} \end{bmatrix} \cdot$$

$$\begin{bmatrix} \ddot{\psi} \\ \ddot{\varphi} \end{bmatrix} + m_2 l_1 l_{22}\sin\varphi \begin{bmatrix} -\dot{\varphi}^2 - 2\dot{\psi}\dot{\varphi} \\ \dot{\psi}^2 \end{bmatrix} = \begin{bmatrix} P_1 \\ P_2 \end{bmatrix}$$

References

[1] Vukobratović, M., Vujić, D., "Nominal Tracking Simulation in Conditions of Mechanical Vibrations Impact on the Manipulation Robots", Mech. Mach. Theory, Vol. 22, No. 5, pp. 441–451, 1987.

[2] Vukobratović, M., Vujić, D., "Contribution to Solving Dynamic Robot Control in a Machining Process", Mech. Mach. Theory, Vol. 22, No. 5, pp. 421–429, 1987.

[3] Vukobratović, M., Stokić, D., Control of Manipulation Robots: Theory and Application, Vol. 2, Springer-Verlag, Berlin, 1982.

[4] Vukobratović, M., Potkonjak, V., Applied Dynamics and CAD of Manipulation Robots, Vol. 6, Springer-Verlag, Berlin, 1985.

[5] Vukobratović, M., Kirćanski, N., Real-Time Dynamics of Manipulation Robots, Vol. 4, Springer-Verlag, Berlin, 1985.

[6] Vukobratović, M., Potkonjak, V., Dynamics of Manipulation Robots: Theory and Application, Vol. 1, Springer-Verlag, 1982.

[7] Vukobratović, M., Applied Dynamics of Manipulation Robots, Vol. 1 Textbook Series, Springer-Verlag, 1988.

[8] Wittenburg, J., Dynamics of Systems of Rigid Bodies, B.G. Teubner, Stuttgart, 1977.

Chapter 4

Control of Robots

In this chapter we shall consider the problems concerning control synthesis for manipulation robots. The control system represents a very important part of the robotic system. Application of the robots in industry and other fields depends on efficiency, reliability and the capabilities of the control system which has to ensure successful application of robots in various tasks.

The control system of robots can be realized in different ways, with varying degrees of complexity depending on the tasks imposed upon a specific robot. The simplest control task which can be imposed upon robotic systems is their positioning in various points in a working space, for example driving the robots in various postures in space (robots might be positioned "empty", without any payload or with a payload so that their task is to move a payload from one position to another). Positioning of the robot can be realized in various ways. The task is to put a gripper of the robot (or a payload) in some specified point with a corresponding orientation of the gripper. If we deal with simple structures of the robots (e.g. a robot with three linear joints, or a robot with a cylindrical structure (Fig. 1.6)), determination of the joints coordinates corresponding to the desired position of the gripper is simple. Here we can directly specify the positions of the joints of the robot. The control system has to ensure the desired positions of each joint of the robot. The simplest control systems are implemented for robots with fixed programmes. These are control systems without feedback loops (i.e. the so-called open-loop control systems). Actuators driving the joints of such robots are rated to maximum until the desired joint positions are reached. The desired positions of the joints are imposed by mechanical limiters or by the positions of the switches (sensors). When the joint reaches the given limiter or sensor (the desired position) the actuator has to be stopped. Thus, the control system has no permanent information on any position (angles or linear displacements) of the joints, i.e. there is no feedback loop from the manipulation system. This solution of the control system is implemented for robots of "zero generation" which represent mechanical hands, mostly driven by pneumatic actuators. The capabilities of these manipulators are very constrained; since the specification of any desired positions of the joints is performed by limiters, the capabilities of precise specification and changes of the desired positions of the robot are rather poor.

The positions of the joints of such "robots" can be set in other ways. If we introduce a microcomputer into the control system, the positions of the robot can be specified by the computer and memorized. The position of the joint obtained by sensor (potentiometer, shaft-encoder etc.) can then be supervised and compared with specified positions in the computer. When the joint position is close to the desired one, the actuator is stopped. In this case the control system gets information on the joint position: the actuators are controlled (on-off) and the information from the sensors is used just to stop the actuator (i.e. the feedback loop is not permanently utilized). The advantage of this solution lies in its simplicity, but it is unconvenient from the actuator overloading standpoint (since the actuator is rated to maximum, and then deaccelerated regardless how far the desired position of the joint is from the real temporary position). Programming of complex movements for these robots are very limited.

In order to ensure uniform work of the actuators and more flexible specifications of the desired positions of the joints, the "servosystem" control of the robot joints has been introduced (i.e. control based on feedback loops which supply information on positions, speeds and accelerations of the joints). Such control is applied with most modern robots. In practically all robots (of the first, second and third generations) control by closed feedback loops is unavoidable. In the following we shall deal exclusively with this type of robot control.

In closed-loop control systems the desired positions of the joints are usually imposed by the computer. The computer sends the desired positions to the control subsystems which control the joints. On this basis of feedback loops, the control signals are generated driving the actuators and joints to the desired positions.

We shall consider the synthesis of servosystems for joints control. Servosystems enable much more flexible specifications of desired joint positions, and more uniform work of the actuators.

However, imposing desired positions by direct specification of the joint positions has several drawbacks. As previously explained, in order to accomplish various tasks the gripper of the robot (or the payload) has to be placed in the desired positions in the working space. If the operator wants to place the gripper of the robot in a desired position by specifying the joints positions, he has to determine iteratively the corresponding positions of the joints. This can be a very tedious and time-consuming job (unless the structure of the robot is simple). Thus, we have to enable the user to specify directly any desired positions of the robot hand, either by computer (programming), or by a teaching-box (see Chap. 6). In both cases the operator of the robot has to specify directly the desired position and orientation of the gripper (or of the payload), and the control system has to compute automatically the corresponding positions of the joints; that is, the control system has to compute internal (joint) coordinates on the basis of the desired "values" of the external coordinates. This calculation can be performed by digital computers in various ways (see Chap. 2). The majority of modern robots are equipped with control systems which enable direct specifications of the external coordinates.

Today more complex tasks are assigned to robots in industry. Apart from the simplest tasks considered above, which can be reduced to free movement of the robot (and the payload) from one position to another, modern robots have to ensure movement along prespecified paths in the working space (for example, arc welding, movement of the robot in a working space with many obstacles, etc.). In this, the operator of the robot has to specify the desired path of the robot hand and the control system has to calculate the corresponding trajectories of the robot joints and to carry them out.

Sometimes robotic tasks can be so complex that the operator requires longer time to specify the positions through which the robot has to move, or to specify hand trajectories in order to perform given tasks. For example, if the robot has to move near various machines and equipment in its working space (i.e. if the robot has to move near various obstacles), the operator has to plan all intermediate positions through which the robot has to pass, or to plan trajectories along which the robot hand has to move in order to prevent collision of the gripper or any links of the robot with the obstacles. Obviously, such trajectory planning can be very tedious and difficult, which is why it is necessary to develop a control system capable of solving such problems automatically, and so release the operator from the trajectory planning tasks. There have been several robots on the market which enable automatic planning of trajectories. The user has to specify the general task (e.g. replace the object from one machine to another); the control system then plans automatically all the movements of the robot (approaching the object, orientation of the hand, grasping the object, lifting the object, replacement to another machine with obstacle avoidance, putting the object into the machine, etc.). This automatic planning of the trajectories of the robot is of great significance when introducing robots into the *flexible technological systems*, a prime requirement in modern robotics.

4.1 Hierarchical Control of Robots

The control system which can accommodate the above is very complex. In order to simplify the synthesis and implementation of the control system, control is usually realized *hierarchically* in several levels, so that each control level solves its specific task. The hierarchical structure shown in Fig. 4.1 is the one usually adopted for robotic control systems. The control is realized in three levels [1].

1. *Strategical control level* has to plan the trajectories of the robot. This level receives its task from the operator by communicating with the control system by a special programming language. The strategical control level has to plan each motion of the robot. The operator specifies the tasks which

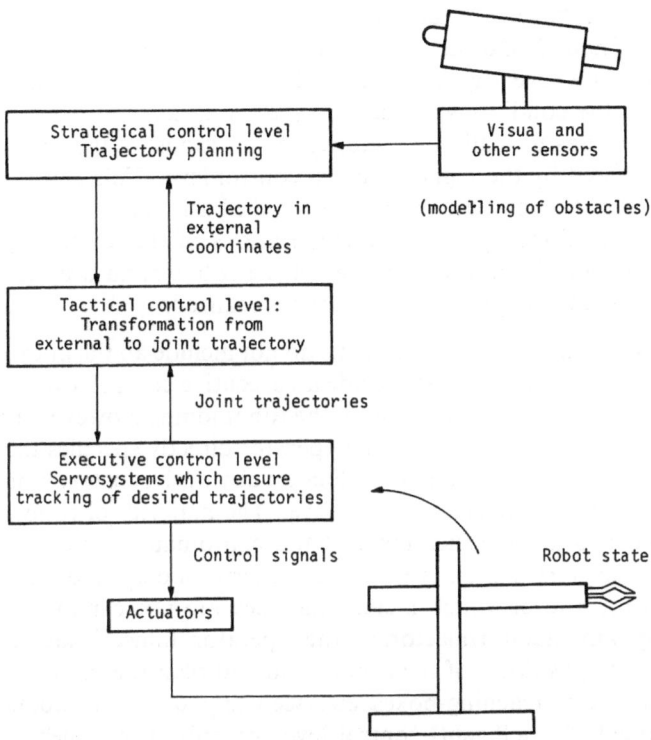

Fig. 4.1. Hierarchical structure of the robot control

have to be accomplished by the robot, and the strategical control level defines the paths of the robot hand which have to be realized. If the working space of the robot is pre-defined, the strategical control level can plan the paths in the space without additional information from any sensors. However, if the positions of all machines and objects in the working space is not (precisely) defined in advance or they change during the operation of the robot, the path planning must be performed using the sensor information (information from cameras, proximity sensors, etc.). In this case the strategical control level must solve the path planning problem in real time, i.e. during the process execution, which is a much more complex problem than if this can be done off-line (before task execution). In both cases the strategical control level defines the trajectories of the robot hand, i.e. it defines trajectories of the external coordinates of the robot.

2. *Tactical control level* has to map the trajectories from external into internal (joint) coordinates of the robot. That is, the strategical control level defines the trajectories of the robot hand coordinates. The tactical control level has to compute corresponding trajectories of the robot joints which have to be realized in order to carry out the imposed hand trajectories. This problem is

solved using a so-called "inverse kinematic model of the robot", which has been considered in Chap. 2. Output of the tactical control level are joint trajectories. This control level can operate in either off-line or in on-line mode, depending on the conditions imposed in specific tasks.

3. *Executive control level* has to realize trajectories (or positions) of the robot joints which are imposed by the higher, tactical control level. This control level must ensure realization of these trajectories on the basis of information on the robot state (positions, speeds, and accelerations of the joints). By tracking the joint trajectories, the trajectories of the robot hand are also accomplished, and the task imposed by operator is realized.

It should be emphasized that all control systems do not include all the above listed control levels. All control levels must include an executive control level in order to realize various positions or trajectories of the robot joints. However, as we have already explained, with some robots the operator directly specifies the desired joint coordinates, but not the hand coordinates: obviously, such robots do not include a tactical (nor strategical) control level. The majority of robots now on the market incorporate specification of hand coordinates, which includes a tactical control level. However, most of the modern robots still do not include a strategical control level, which means that they are not capable of planning automatically any hand trajectories; the operator himself has to impose any trajectories (or positions) of the robot hand and plan the paths by programming languages, or by teaching-boxes, etc. (see Chap. 6). These robots have no strategical (so-called intelligent) control level, or only in rudimental form. Even so, given the tasks demanded of the modern robot, robots today or in the near future must include very elaborate and complex strategical control levels. This is specially true if we consider inclusion of the robots into flexible manufacturing systems.

In Chap. 2 we considered the kinematic model of the robot and the problems concerning tactical control level. In this chapter we shall consider problems concerning the synthesis of executive control level. This means that we shall consider control of actuators which drive the joints of the robot to maintain positions and trajectories imposed either by a higher (tactical) control level, or by the operator directly. In this we shall observe both problems: if the robot moves "point-to-point" (from one position to another), and if it should realize certain continual trajectories. It should be mentioned that synthesis of the executive control level is relevant for all three generations of automatic robots, and for remote and manual robot control.

4.2   Control of a Single Joint of the Robot

First we shall consider a simple case when a single joint of the robot is moving (e.g. the i-th joint), while all other joints are fixed. The joints of the robot are

driven by actuators. Assume the synthesis of control for an actuator driving the
i-th joint where all other joints of the robot are fixed.

4.2.1   Model of Actuator and Joint Dynamics

Actuators driving the joints of the robot might be D.C. and A.C. electro-motors,
hydraulic or pneumatic actuators. Robots with a closed-loop control are usually
driven by D.C. electro-motors. Thus, in the following we shall consider these
actuators. However, these considerations might be easily extended to hydraulic
actuators.

The model of the dynamics of a D.C. electro-motor with a permanent
magnet driving the i-th joint can be written in the following form [2]. The
equation of moments equilibrium around the motor axis is given in Eqn (3.50)

$$N_M^i N_V^i J_M^i \ddot{q}_i + P_i = N_M^i C_M^i i_r^i - B_C^i \dot{q}_i \tag{4.1}$$

where J_M^i is the moment of inertia of the rotor of the motor, q_i is the angle of
rotation of the motor (joint), P_i is the load acting around the motor axis, C_M^i is
the mechanical constant of the motor (the coefficient of proportionality between
the moment which is developed by the motor and the current of the rotor coil),
i_M^i is the current in the rotor coil, B_C^i is the coefficient of the viscous friction of the
motor, N_M^i is the moment reduction ratio on the motor axis (ratio between the
moments behind and in front of any reductor), N_V^i is the speed reduction ratio of
the reductor (ratio between the speed of the input and output shaft of the
reductor). The equation describing the equilibrium in the electric circuit of the
rotor coil can be written in the form (if the inductivity of the coil is neglected):

$$R_r^i i_r^i + C_e^i N_V^i \dot{q}_i = u_i \tag{4.2}$$

where u_i is the input voltage on the rotor circuit, C_e^i is the coefficient of
proportionality between the contra-electromotor force of the rotor and the
rotation velocity of the motor (this force is the voltage developing due to
rotation of the rotor coil in the magnetic field), and R_r^i is the rotor coil resistance.
Based on Eqs. (4.1) and (4.2) we can write:

$$N_M^i N_V^i J_M^i \ddot{q}_i + P_i = C_M^{i'} u_i - (B_C^i + C_E^{i'}) \dot{q}_i \tag{4.3}$$

where the following notations are introduced:

$$C_M^{i'} = N_M^i C_M^i / R_r^i \quad \text{and} \quad C_E^{i'} = N_V^i N_M^i C_M^i C_e^i / R_r^i$$

In order to write the model of actuators in the state space, let us introduce
the state vector in the form:

$$x_i = (q_i, \dot{q}_i)^T \tag{4.4}$$

Now, instead of Eq. (4.3) we may write:

$$\dot{x}_i = A_i x_i + b_i u_i + f_i P_i \tag{4.5}$$

where A_i is matrix of dimensions 2×2, b_i and f_i are vectors of dimensions 2×1 given by:

$$A_i = \begin{bmatrix} 0 & 1. \\ 0 & -(B_C^i + C_E^{i'})/(J_M^i N_V^i N_M^i) \end{bmatrix}, \quad b_i = \begin{bmatrix} 0 \\ C_M^{i'}/(J_M^i N_V^i N_M^i) \end{bmatrix},$$

$$f_i = \begin{bmatrix} 0 \\ -1./(J_M^i N_V^i N_M^i) \end{bmatrix} \tag{4.6}$$

The actuator is driving the i-th joint while all other joints are in some fixed positions $q_j = q_j^0, j = 1, 2, \ldots, n, j \neq i$. The i-th actuator is driving the mechanical part of the robot around the i-th joint. In the given fixed positions of the joints $q_j^0, j > i$, the mechanical part of the robot has a constant moment of inertia around the i-th joint $H_{ii}(q_j^0)$ (Fig. 4.2). The actuator practically drives the set of links which all together have a moment of inertia around the i-th joint $H_{ii}(q_j^0)$. These links also produce gravitational moment around the axis of the i-th joint; this moment depends on the fixed positions of all joints ($j > i$) and the instant (variable) angle of the i-th joint, i.e. $G_i(q_j^0, q_i)$. Thus, the moment which the mechanism produces around the shaft of the i-th motor might be written as:

$$P_i = H_{ii}(q_j^0) \ddot{q}_i + G_i(q_j^0, q_i) \tag{4.7}$$

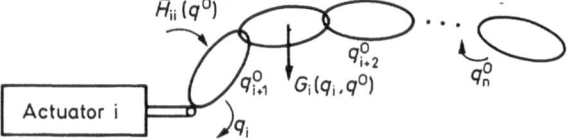

Fig. 4.2. Actuator in the i-th joint of robot (the remaining joints are fixed)

If we introduce the dynamic model of the mechanism rotation around the i-th joint in the model of actuator (Eq. (4.5)), we obtain the model of dynamics of the actuator and the mechanism driven by the actuator in the following form (for simplicity we shall write $H_{ii} = H_{ii}(q_j^0)$ and $G_i = G_i(q_j^0, q_i)$):

$$\dot{x}_i = \bar{A}_i x_i + \bar{b}_i u_i + \bar{f}_i G_i \tag{4.8}$$

where

$$\bar{A}_i = \begin{bmatrix} 0 & 1 \\ 0 & -(B_C^i + C_E^{i'})/(J_M^i N_V^i N_M^i + H_{ii}) \end{bmatrix},$$

$$\bar{b}_i = \begin{bmatrix} 0 \\ C_M^{i'}/(J_M^i N_V^i N_M^i + H_{ii}) \end{bmatrix}, \quad \bar{f}_i = \begin{bmatrix} 0 \\ -1./(J_M^i N_V^i N_M^i + H_{ii}) \end{bmatrix}$$

The model (4.8) represents the object of control (actuator and the mechanism of the robot which has to be controlled).

4.2.2 Synthesis of Servosystem

We have to synthesize control of the actuator and the joint which should ensure that when the desired position of the joint is set at q_i^0, the joint will be driven to this position in an adequate way. The control accomplishing this task is usually a servosystem, the scheme of which is presented in Fig. 4.3.

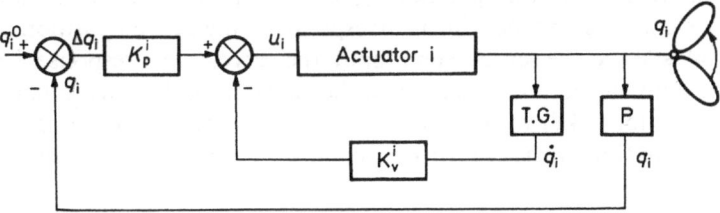

Fig. 4.3. Positional servosystem

The servosystem for control of the i-th actuator and joint consists of the following elements: sensor of position which gives information on the temporary position of the i-th joint and of the shaft of actuator q_i (potentiometer, op-to-encoder, etc. – see Chap. 7); sensor of rotational (or displacement) velocity of the joint and motor $\dot{q}_i$ (usually tachogenerators are used); differentiator which gives the difference between the set (desired) position of the i-th joint q_i^0 and real position q_i^0 which is obtained by the position sensor; amplifier of the position error which amplifies the position error signal $\Delta q_i = q_i^0 - q_i$ for k_p^i times, where k_p^i represents the so-called position gain; amplifier of information on velocity which amplifies the signal from the sensor of rotational velocity of the joint $\dot{q}_i$ for k_v^i times, where k_v^i represents the velocity feedback gain (in the following we shall call it velocity gain).

The mode of operation of the servosystem is obvious from Fig. 4.3. Information on the joint position q_i is returned as feedback and the difference between the desired and real position is amplified by k_p^i times. This represents the input voltage signal for the actuator. If $q_i^0 > q_i$ this produces a positive voltage signal which drives the motor so that q_i is increased until it reaches q_i^0; if $q_i^0 < q_i$ the negative signal appears which drives the actuator towards decrement of the angle q_i until it reaches q_i^0; when q_i reaches q_i^0 the error Δq_i falls to zero, and the signal on the actuator input also falls to zero and the actuator is stopped. However, due to inertia of the rotor of the motor and inertia of the mechanism $J_M^i N_M^i N_V^i + H_{ii}(q_j^0)$, the motor cannot be stopped instantly, as it could incur "over-shooting", i.e. the real position might overshoot the desired position q_i^0 before the motor stops. In order to ensure adequate positioning of the joint (without "overshoot") we introduce velocity feedback: information from the speed sensor is amplified k_v^i times and brought to the actuator input in order to dampen sudden changes in the actuator motion which might be caused by position feedback.

Thus, the servosystem gives the following voltage input signal for the actuator:

$$u_i = -k_p^i(q_i - q_i^0) - k_v^i \dot{q}_i = k_p^i \Delta q_i - k_v^i \dot{q}_i \tag{4.9}$$

In synthesis of the servosystem we have to select the position and velocity gains so as to achieve a satisfactory positioning of the joint in the given position q_i^0. That is, if the signal of step type which corresponds to the desired position of the joint (Fig. 4.4) is fed to the servosystem input, the servosystem response, i.e. movement of the joint, depends on selection of the gains. It can be shown that if the differential equations (4.8) with input (4.9) are solved, the response of the servosystem can be in three various forms depending on the choice of k_p^i and k_v^i (Fig. 4.4):

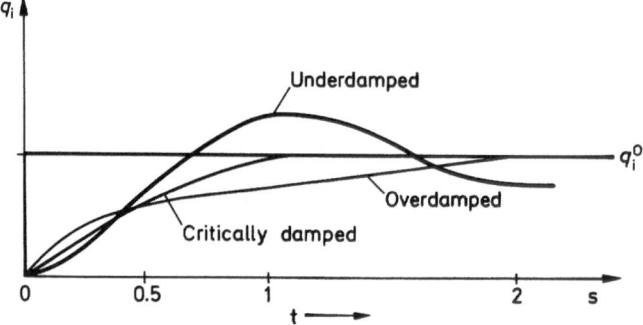

Fig. 4.4. Responses of servosystem to step input

(a)   the servosystem could be *underdamped*, in which case the joint rapidly moves from its initial position, and reaches the desired position q_i^0 but "overshoots" it, i.e. q_i gets values which are greater than q_i^0, and then it oscillates around the desired position before setting;

(b)   the servosystem could be *critically damped*; in this case the joint reaches the desired position relatively quickly, but there is no "overshoot" or oscillations, and the joint quickly settles on the given q_i^0;

(c)   the servosystem could be *overdamped*, in which case the joint slowly moves towards the desired position, there is no "overshoot" nor oscillations, but the settling period is considerably longer than in the case of a critically damped servosystem [2].

These three types of servosystem responses can be described by the following functions (as the solutions of differential equations (4.8), (4.9) depending on k_p^i and k_v^i):

(a)   the underdamped servosystem:

$$q_i - q_i^0 \sim C_1 e^{-\xi_i \omega_i t} \sin(\omega_i \sqrt{1-\xi_i^2}\, t) + C_2 e^{-\xi_i \omega_i t} \cos(\omega_i \sqrt{1-\xi_i^2}\, t) \qquad (4.10)$$

(b)   the critically damped servosystem:

$$q_i - q_i^0 \sim C_3 e^{-\omega_i t} \qquad (4.11)$$

(c)   overdamped servosystem:

$$q_i - q_i^0 \sim C_4 e^{-(\xi_i \omega_i + \omega_i \sqrt{\xi_i^2-1})t} + C_5 e^{-(\xi_i \omega_i - \omega_i \sqrt{\xi_i^2-1})t} \qquad (4.12)$$

In the functions (4.10)–(4.12), C_1–C_5 represent constants (which depend on q_i^0), while ξ_i represents the damping factor and ω_i is the characteristic frequency of the servosystem. The damping factor ξ_i defines whether the servosystem is critically damped, underdamped or overdamped. If:

$$\begin{aligned}
\xi_i &< 1 \quad \text{the servosystem is underdamped,} \\
\xi_i &= 1 \quad \text{the servosystem is critically damped,} \qquad (4.13) \\
\xi_i &> 1 \quad \text{the servosystem is overdamped}
\end{aligned}$$

The damping factor ξ_i and characteristic frequency ω_i of the system are the features of the servosystem which are direct functions of the selection of the gains k_p^i and k_v^i and the parameters of the actuator and mechanism. It can be shown that [2]:

$$\xi_i = (C_M^{i\prime} k_v^i + C_E^{i\prime} + B_C^i)/[2\sqrt{C_M^{i\prime} k_p^i (J_M^i N_V^i N_M^i + H_{ii})}] \qquad (4.14)$$

$$\omega_i = [\sqrt{C_M^{i\prime} k_p^i}/\sqrt{J_M^i N_V^i N_M^i + H_{ii}}] \qquad (4.15)$$

It is quite obvious that by selection of the gains k_p^i and k_v^i we may directly involve ξ_i and ω_i and by this we may change the character of the servosystem response, i.e. the way in which the joint is driven to the imposed position. In selecting the gains k_p^i and k_v^i several requirements have to be satisfied:

1.   The servosystem which controls the joint of the robot must not be under-damped in any way. If the servosystem is underdamped, "overshoot" of the joint position occurs and oscillations appear. This is not acceptable with robots for if the desired position is near some obstacle in the working space, and overshooting occurs, the robot could hit or collide with some obstacle. The servosystem has therefore to be overdamped ($\xi_i > 1$) or critically ($\xi_i = 1$) damped. As the response of the servosystem is significantly slower if it is overdamped, in order to achieve as fast a response as possible (but without "overshoot" and oscillations), it is more suitable that the servosystem is critically damped.

2.   Up to now we have ignored the influence of the gravitational moment around the actuator axis G_i. All the above considerations are valid as-

suming that the external moments are not acting upon the actuator (except the inertia moment $H_{ii}\ddot{q}_i$). Let us consider the influence of the gravitational moment. When the joint comes near the desired position q_i^0, the gravitational moment of the mechanism $G_i(q_j^0, q_i^0)$ is acting around the axis of the actuator. Since the error between the desired and real position Δq_i drops to zero and as the actuator is stopped the velocity $\dot{q}_i$ also falls to zero, the signal on the actuator input has also to drop to zero in accordance with Eq. (4.9), as does the moment produced by the actuator. However, the actuator must produce the moment which should compensate the gravitational moment G_i (if not, the gravitational moment causes movement of the joint). To obtain this actuator moment which will compensate the external load G_i, some voltage signal u_i must appear on the actuator input. It is obvious from Eq. (4.9) that this signal can be produced only if there is some error between the real and desired position when the joint motion is finished. The error in positioning of the joint which appears in a steady state due to external load G_i is called the *steady state error*, $\Delta q_i(\infty)$. From Eqs. (4.8) and (4.9) it is obvious that:

$$\Delta q_i(\infty) = [G_i(q_j^0, q_i^0)/C_M^{i'}k_p^i] \tag{4.16}$$

i.e. the steady state error is inversely proportional to the position gain. Since our aim is to reduce the error in the robot positioning to minimum, it is obvious that it is necessary to increase the position gain k_p^i as much as possible.

3. The structure of the robot itself has its own frequency on which the resonant oscillations of the whole structure appear. This frequency is called the *structural frequency*, ω_0. According to requirement 1 the gains have to be selected in such a way as to ensure that the servosystems are always (over)critically damped. However, since the damping factor ξ_i depends on other parameters of the actuators and the mechanism, it is possible that oscillations of the servosystem with frequency ω_i yet might appear. If the characteristic frequency of the servosystem ω_i is close (equal) to the structural frequency ω_0 resonant oscillations of the whole structure could appear. Since these oscillations must not under any circumstances be allowed, it should be ensured that the characteristic frequency of the servosystem is sufficiently below the range of any possible structural frequency; that is, the characteristic frequency must satisfy [3]:

$$\omega_i \leqslant 0.5\omega_0 \tag{4.17}$$

If condition (4.17) is met, the characteristic frequency is sufficiently low so that the structural frequency cannot be excited and the undesired oscillations cannot appear. The problem lies in the fact that structural frequency ω_0 is hard to determine theoretically and usually it is defined experimentally. Since according to Eq. (4.15) the characteristic frequency of the servosystem is proportional to the position gain k_p^i, condition (4.17) means that the position gain must not be too high in order to prevent the characteristic frequency of the servosystem becoming too high and reaching the range of the structural frequency.

4. The electrical signals in the servosystem on Fig. 4.3 are never ideally "clear", but always include some "noise" as well as useful information. For example, apart from useful information, signals from sensors (potentiometers, tachogenerators, etc), always include some noise which originates from various sources (signals from the voltage sources are never exact; some oscillatory modes always appear, etc.). The noise is usually by an order of magnitude a lower signal than the useful signal. These signals are amplified by the amplifiers k_p^i and k_v^i. If these gains are too high, they amplify not only useful signal but also the noises, and so their influence on the servosystem performance can become significant, which is why in selecting these gains their values must be restricted.

Based on the above requirements it is possible to select gains k_p^i and k_v^i. Requirements 3 and 4 are in essence the same and they both demand the gains to be restricted (i.e. the gain must not be too high). Usually if requirement 3 is satisfied, requirement 4 is also met. However, requirement 2 is opposite to these two, as it demands that the position gain should be as high as possible. This is why we have to select the gains in the following way.

First the maximum allowed position gain k_p^i is selected for requirement 3; based on Eqns (4.15) and (4.17) we obtain:

$$k_p^i = \frac{\omega_0^2}{4C_M^{i'}} (J_M^i N_V^i N_M^i + H_{ii}) \tag{4.18}$$

We must check whether the gain k_p^i calculated by Eq. (4.18) also satisfies requirement 4. Since we have selected the maximum allowed k_p^i we have also satisfied requirement 2 (to the highest possible degree).

When the servosystem has to be critically damped $\xi_i = 1$ the velocity gain has to be determined:

$$k_v^i = [2\sqrt{C_M^{i'} k_p^i} \sqrt{(J_M^i N_V^i N_M^i + H_{ii})} - C_E^{i'} - B_C^i] / C_M^{i'} \tag{4.19}$$

Thus we obtain the gains which satisfy all necessary requirements to the maximum possible degree.

It should be noted that, since the linear servosystems are applied not only in robotics but also for control in various other systems, it is possible to synthesize the feedback gains using various other methods developed in the automatic control theory. These methods can be easily found in the relevant references [2, 4] (method in frequency domain, pole-placement, linear optimal regulator etc.).

Example

For the first joint of the manipulator in Fig. 4.5 (the photograph of this robot is given in Fig. 1.6), synthesis of the servosystem gains should be performed. The joint is driven by a D.C. electromotor of the type IG2315-P20, the parameters of which are given in Table 4.1. The data on masses, moments of inertia, lengths and positions of centres of masses of the links are given in Table 4.2. It is easy to show that the moment of intertia of the mechanism around the axis of the first joint is given by

$$H_{11}(q_j^0) = J_{z1} + J_{z2} + J_{z3} + m_3(l_3 + q_3)^2 \tag{4.20}$$

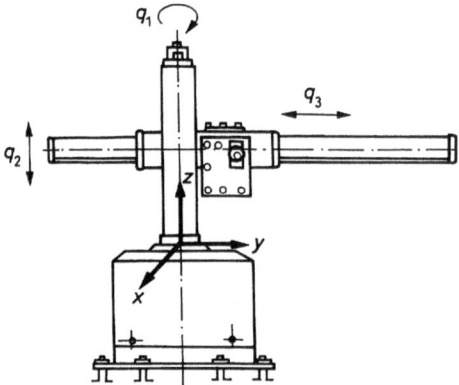

Fig. 4.5. Robot UMS-2

Table 4.1. Data on actuators for robot UMS-2

Actuator	1	2	3
$C_e\left[\dfrac{V}{\text{rad/s}}\right]$	0.0459	0.0459	0.0459
$C_m\left[\dfrac{M}{A}\right]$	0.0481	0.0480	0.0480
$J_M[\text{kg m}^2]$	0.00003	0.00003	0.00003
$N_V[-]$	31.17	2616	1570
$N_M[-]$	31.17	2616	1570
$R_r[\Omega]$	1.6	1.6	1.6
$B_C\left[\dfrac{\text{Nm}}{\text{rad/s}}\right]$	0.0058	0.0154	0.000923

Table 4.2. Data on robot UMS-2

Link	1	2	3
Mass [kg]	10.0	7.	4.15
Length [m]	0.213	0.026	0.035
$J_x[\text{kg m}^2]$	—	—	—
$J_y[\text{kg m}^2]$	—	—	—
$J_z[\text{kg m}^2]$	0.0294	0.055	0.318

if the third link is fixed in the position $q_3^0 = 0$, the moment of inertia of the mechanism around the axis of the first joint is $H_{ii} = 0.403$ kg ($l_3 = 0.035$ m). Using the values of the actuator parameters in Table 4.1 we get the model of actuator and the joint dynamics in the form (4.8) where the matrices are given by:

$$\bar{A}_1 = \begin{bmatrix} 0 & 1 \\ 0 & -3.117 \end{bmatrix}, \quad \bar{b}_1 = \begin{bmatrix} 0 \\ 2.17 \end{bmatrix}, \quad \bar{f}_1 = \begin{bmatrix} 0 \\ -2.31 \end{bmatrix} \tag{4.21}$$

The structure of the servosystem to be synthesized is given in Fig. 4.3. The gains of the servosystem are selected according to the above described approach. Let us assume that the structural frequency is determined (experimentally) to be $\omega_0 \approx 24$ Hz. Based on Eq. (4.18) we obtain the position gain as:

$$k_p^1 = 62.2 \left[\frac{V}{\text{rad}} \right]$$

Assuming that the noises in the sensor for measuring the position do not exceed 1% of the useful signal and assuming that the total angle of rotation of this joint is $\pm 180°$, we can determine that the signal on the amplifier output due to noises is 0.3 V, which could be considered as negligible. The velocity feedback gain is obtained based on Eq. (4.19):

$$k_v^1 = 9.62 \left[\frac{V}{\text{rad/s}} \right]$$

This gain is also relatively low so it will not cause significant influence of the noise.

4.2.3   The Influence of the Variable Moment of Inertia of the Mechanism

The described synthesis of a servosystem is in essence the usual synthesis of a servosystem for mechanical systems. However, the robotic systems, being different from some other mechanical systems, have a variable moment of inertia of the mechanism. We have assumed that only the i-th joint can move while all the other joints are fixed in given positions q_j^0. The moment of inertia of the mechanism around the axis of the i-th joint $H_{ii}(q_j^0)$ depends on the angles (positions) q_j^0 in which the joints which are "behind" the i-th joint in the kinematic chain are fixed (see Chap. 3). If the position of any joint which is "behind" the i-th joint is changed, the moment of inertia of the mechanism around the axis of the i-th joint will also change. Let us consider how the variation of the moment of inertia of the mechanism influences the performance of the servosystem in the i-th joint.

Let us assume that the gains k_p^i and k_v^i are calculated for such a position of the joints of the robot q_j^0 for which the moment of inertia around the axis of the i-th joint has the value of $\bar{H}_{ii}$. In this case the gains are given by:

$$k_p^i = \frac{1}{4C_M^{i'}} \omega_0^2(\bar{H}_{ii})(J_M^i N_V^i N_M^i + \bar{H}_{ii})$$

$$k_v^i = [2\sqrt{C_M^{i'} k_p^i} \sqrt{(J_M^i N_V^i N_M^i + \bar{H}_{ii})} - C_E^{i'} - B_C^i]/C_M^{i'} \tag{4.22}$$

whereby $\omega_0(\bar{H}_{ii})$ we have denoted the structural frequency of the robot for the moment of inertia $\bar{H}_{ii}$. It has been shown [3] that the structural frequency is inversely proportional to the square root of the moment of inertia of the mechanism, i.e.:

$$\omega_0(H_{ii}) = \frac{k}{\sqrt{(J_M^i N_V^i N_M^i + H_{ii})}} = \frac{\omega_0(\bar{H}_{ii})\sqrt{(J_M^i N_V^i N_M^i + \bar{H}_{ii})}}{\sqrt{(J_M^i N_V^i N_M^i + H_{ii})}} \tag{4.23}$$

(k is the proportionality factor).

If any of the joints in the kinematic chain of the robot (which is "behind" the i-th joint) change their position $q_j \neq q_j^0$ the moment of inertia around the axis of the i-th joint will also change and become $H_{ii}(q_j) \neq \bar{H}_{ii}$. In this case the characteristic frequency of the servosystem in the i-th joint can be obtained if we introduce the expression (4.22) for the position gain in the expression (4.15):

$$\omega_i(H_{ii}) = \frac{\omega_0(\bar{H}_{ii})\sqrt{(J_M^i N_V^i N_M^i + \bar{H}_{ii})}}{2\sqrt{(J_M^i N_V^i N_M^i + H_{ii})}} \tag{4.24}$$

Obviously the characteristic frequency $\omega_i(H_{ii})$ must satisfy the following inequality:

$$\omega_{ii}(H_{ii}) = \frac{\omega_0(\bar{H}_{ii})\sqrt{(J_M^i N_V^i N_M^i + \bar{H}_{ii})}}{2\sqrt{(J_M^i N_V^i N_M^i + H_{ii})}} \leqslant \frac{1}{2} \omega_0(H_{ii}) \tag{4.25}$$

which can be easily checked if the expression (4.23) for $\omega_0(H_{ii})$ is introduced in Eq. (4.25). This means that regardless of the moment of inertia of the mechanism, requirement 3 (given by Eq. (4.17)), which stipulates that the characteristic frequency of the servosystem has to be sufficiently beyond the structural frequency, is always satisfied (if the position gain is selected according to Eq. (4.22)).

However, the damping factor of the servosystem in the i-th joint if the moment of inertia of the mechanism changes to the new value H_{ii}, becomes (according to Eqs. (4.14) and (4.22)):

$$\xi_i = \frac{C_M^{i'} k_v^i + C_E^{i'} + B_C^i}{2\sqrt{C_M^{i'} k_p^i (\sqrt{J_M^i N_V^i N_M^i + H_{ii}})}} = \frac{\sqrt{(J_M^i N_V^i N_M^i + \bar{H}_{ii})}}{\sqrt{(J_M^i N_V^i N_M^i + H_{ii})}} \qquad (4.26)$$

If the j-th joint changes its position to that moment of inertia of the mechanism around the i-th joint H_{ii} is less than $\bar{H}_{ii}$ for which the gains have been computed, i.e. if $\bar{H}_{ii} > H_{ii}$, the servosystem is obviously overdamped in the new position of the mechanism, i.e. $\xi_i > 1$. However, if the mechanism comes in the position in which the moment of inertia of the mechanism around the i-th joint H_{ii} is greater than the moment of inertia $\bar{H}_{ii}$ for which the gains have been computed, i.e. if $\bar{H}_{ii} < H_{ii}$ it is obviously $\xi_i < 1$. This means that the servosystem would be underdamped. As we have explained above (requirement 1), the servosystems for robots must not be underdamped under any circumstances. In order to ensure that the servosystem is always (over)critically damped ($\xi_i > 1$) we must not allow $\bar{H}_{ii} < H_{ii}$. This leads to the following conclusion: that in order to ensure that the servosystem is always (over)critically damped the gains have to be selected for such position of the mechanism for which the moment of inertia of the mechanism around the i-th joint is maximal. As can be seen from Eq. (4.26), the damping factor does not depend on the selection of the position gain (if the velocity gain is selected according to Eq. (4.22)). Thus, we must select the velocity gain for the position of the mechanism for which the moment of inertia of the mechanism around the axis of the i-th joint $\bar{H}_{ii}$ is at the maximum possible.

All possible positions of the mechanism should be examined (by varying the joints angles q_j) and the maximum moment of inertia of the mechanism $\bar{H}_{ii} = \max H_{ii}(q_j)$ should be determined. For the defined moment of inertia we have to compute the velocity gain k_v^i according to Eq. (4.22). In all positions of the mechanism for which $H_{ii} \neq \bar{H}_{ii}$ the servosystem must be overdamped (according to Eq. (4.26) since $H_{ii} < \bar{H}_{ii}$). However, if the moment of inertia varies so much that in some positions $\bar{H}_{ii} \gg H_{ii}$, the damping factor can become too high $\xi_i \gg 1$, which means that the servosystem is very overcritically damped and the positioning is very slow, and the performance of the servosystem then becomes very un-uniform depending on the mechanism position which is unacceptable for any robot application. In order to ensure that the robot performance is nearly uniform in all positions of the mechanism, we have to ensure that the damping factor ξ_i is approximately constant ($\xi_i \approx 1$). To achieve this we must introduce the *variable* velocity gains k_v^i (since the damping factor does not depend on any position gain). For each position of the mechanism we compute the moment of inertia $H_{ii}(q_j)$ and determine the gain k_v^i so as to achieve $\xi_i = 1$. The implementation of a variable gain is significantly more complex than implementation of constant gains, specially if we take into account that it requires on-line computation of the moment of inertia of the mechanism (see Chap. 5). Another way to compensate for the influence of the variable moment of inertia of the mechanism is by introduction of global control (see Sect. 4.4.2).

However, if the variation of the moment of inertia of the mechanism is not too large satisfactory performance of the servosystem can be obtained even with the constant velocity gain (computed for max $H_{ii}(q_j)$). If we consider expression (4.26) for the damping factor it is obvious that the moment of inertia of the motor rotor and the reduction ratio of the reductor has an effect upon the variation of the damping factor with the variation of H_{ii}. If $J_M^i N_V^i N_M^i \gg (\bar{H}_{ii} - H_{ii})$ it is obvious that the damping factor will not change significantly regardless of the variation of the moment of inertia of the mechanism. In other words, if the equivalent moment of inertia of the rotor is large with regard to the variation of the moment of inertia of the mechanism we may expect that the performance of the servosystem will be uniform (and approximately critically damped) for all positions of the mechanism even if we keep the velocity gain constant. Thus, by selecting the large motor and reductor we may eliminate the influence of the variable moment of inertia of the mechanism. This approach is often applied in robot construction. However, it is clear that such a solution has some drawbacks from the point of view of energy consumption, unnecessary loading of joints, as well as use of unnecessary powerful actuators and large reductors.

The large reducers are specially inconvenient due to large backlash and dry friction that they might introduce in the system. In the last few years large efforts have been made towards introducing the so-called direct-drive actuators (i.e. motors without reducers). By this the problems with backlash and friction are reduced, but on the other hand variation of the moment of inertia of the mechanism might affect the servosystem performance, and more complex control law (with variable velocity gain, for example) has to be introduced.

Example

For the servosystem in the first joint of the robot in Fig. 4.5 in the previous example we have computed the gains if the third joint is in position $q^0 = 0$. Considering the expression (4.20) for the moment of inertia of the mechanism around the axis of the first joint, it is obvious that if the third joint is set in the position $q^0 > 0$ the moment of inertia of the mechanism H_{ii} will be greater and the damping factor will be less than 1. Using Eq. (4.26) the damping factor for the position of the third joint $q_3^0 = 0.3$ [m] can be calculated as:

$$\xi_1 = \frac{\sqrt{J_M^i N_V^i N_M^i + J_{z1} + J_{z2} + J_{z3} + m_3 l_3^2}}{\sqrt{J_M^i N_V^i N_M^i + J_{z1} + J_{z2} + J_{z3} + m_3 (l_3 + q_3^0)^2}} = \frac{\sqrt{0.435}}{\sqrt{0.895}} < 1$$

Thus, the gains selected in the previous example will not be satisfactory for all positions of the mechanism. In Fig. 4.6 the responses of the servosystem for various positions of the third joint q_3^0 are presented. This is why the gains must be selected for the position of the mechanism for which $\bar{H}_{ii} = \max(H_{ii}(q))$. In this case H_{ii} is maximum if the q_3^0 is maximum, i.e. for $q_3^0 = 0.8$ [m]. We obtain $\bar{H}_{ii}(q_3^0 = 0.8\text{ m}) = 3.323$ [kg m^2], and the gains are obtained as:

$$k_p^1 = 62.212 \left[\frac{V}{\text{rad}}\right], \quad k_v^1 = 27.5 \left[\frac{V}{\text{rad/s}}\right]$$

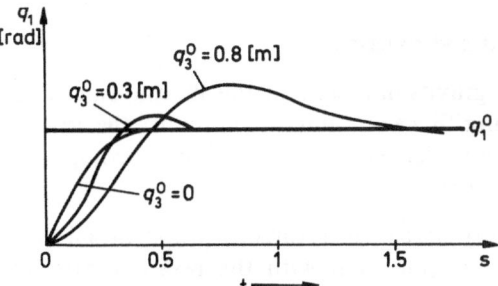

Fig. 4.6. Responses of the servosystem in the first joint of the robot UMS-2 for various positions of the third joint

If we compute the gains in this way the servosystem will be overdamped for all positions of the mechanism. According to Eq. (4.26) the damping factor changes with variation of q_3^0 as presented in Fig. 4.7. It can be seen that for $q_3^0 \approx 0$ the servosystem is strongly overdamped which causes slow positioning of the first joint. In Fig. 4.8 the response of the first joint for various positions of the third joint is presented. In order to achieve more uniform positioning of the joint it is necessary to introduce either (a) variable gain, or (b) global control, or (c) to apply a *larger* actuator and reductor with a *larger* moment of inertia.

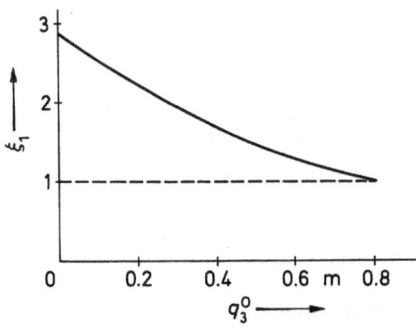

Fig. 4.7. The variation of the damping factor of the servosystem in the first joint of the robot UMS-2 for various positions of the third joint

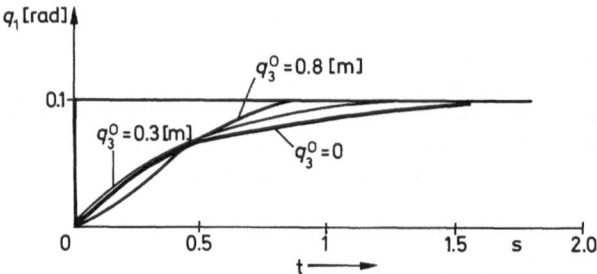

Fig. 4.8. Responses of the servosystem in the first joint of the robot UMS-2 for various positions of the third joint

4.2.4 Influence of Gravity Moment and Friction

We have already shown that the gravity moment of the mechanism causes steady state error in robot positioning. Since our aim is to minimize the errors in the robot positioning, we have to consider various possibilities to compensate for the influence of the gravity moments.

1. We have shown above that steady state error is directly proportional to the gravity moment and inversely proportional with the position gain and moment coefficient of the actuator. We have shown that if we select a higher position gain k_p^i the steady state error will be less. However, the position gain is limited by the resonant structural frequency and noises, so the steady state error cannot be eliminated beyond a certain limit by increasing the position gain. Obviously, by selecting the more powerful actuator (with greater coefficient C_m^i) and a larger reductor (with higher reduction ratio N_M^i) we may decrease the steady state error but this solution has certain drawbacks, as we have pointed out.

2. Gravity moments can be compensated by introducing an additional signal at the actuator input which is proportional to the gravity moment (Fig. 4.9). In this case the control system has to compute the gravity moment in the i-th joint $G_i(q_j^0, q_i)$ as a function of the coordinates (positions) of the robot joints, and to generate at the actuator input an additional signal which will

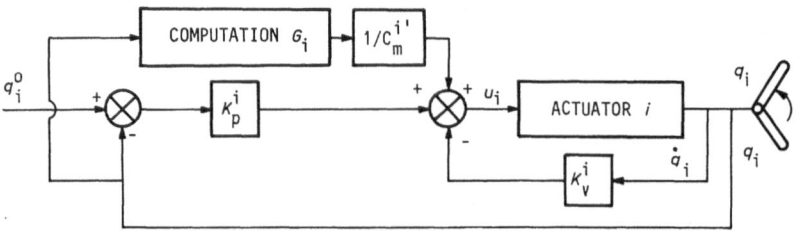

Fig. 4.9. Positional servosystem with gravity moment compensation

produce the compensating moment. Thus, the input signal for the actuator is given by:

$$u_i = -k_p^i(q_i - q_i^0) - k_v^i \dot{q}_i + \frac{G_i}{C_M^{i'}} \qquad (4.27)$$

In this way we can eliminate the steady state error due to the gravity moment. However, the drawback of this solution lies in the fact that it requires the control system to compute gravity moments which might be relatively complex functions of instant joint coordinates.

3. Steady state error can be eliminated by the introduction of integral feedback, i.e. feedback of the integral of positional error (Fig. 4.10). Thus, we obtain the so-called PID regulator (P – proportional, I – integral action, D – differential) which is applied with many various systems. In this case the actuator input is given by:

$$u_i = -k_p^i(q_i - q_i^0) - k_v^i \dot{q}_i + k_I^i \int_0^t (q_i^0(t) - q_i) dt \qquad (4.28)$$

where k_I^i is the integral feedback gain. The integral feedback has the role of producing the signal proportional to the integral of the position error when the servosystem approaches the set position. This signal will compensate the external load and eliminates the steady state error. Synthesis of the gains for the PID servosystem will not be considered here [2].

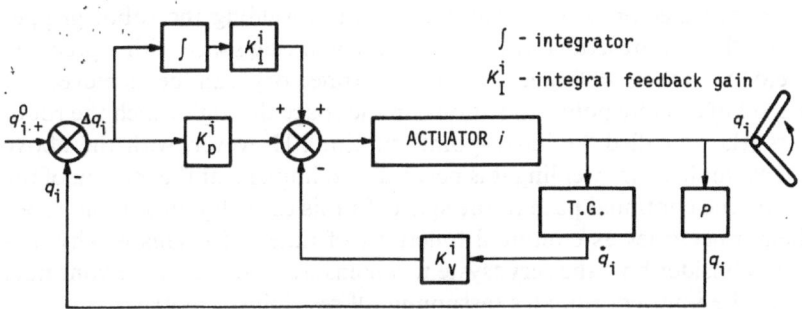

Fig. 4.10. PID regulator in the i-th joint of the robot

4. Finally, steady state error due to gravity moments can be reduced by introducing brakes in the joint, which should hold the joint in the desired position. This solution is simple regarding control but often technically it cannot be applied and is inconvenient for eliminating errors due to the gravity moments if the trajectory tracking have to be realized.

Besides the gravity moments friction forces also affect the servosystem performance. These forces in the joints can also cause errors in the servosystem positioning and operation. In this, special problems arise due to static friction forces which appear when the joint starts to move from the still position which differs from the "dynamic" friction forces during the motion. Compensation of these forces can be realized by one of above listed methods for compensations of the gravity moments. However, the model and parameters of these forces are often not known, so any computation and introduction of additional compensation signals (analogous to the solution in Fig. 4.9) cannot be easily implemented. Actually this compensation signal can be defined experimentally. Backlash in the reducer, elastic modes and other nonlinear effects, the models of which are not simple, also affect performance of the servosystem. However, we must be careful about these effects during synthesis and implementation of the servosystems for robots. It should be mentioned that the input signal to the actuators is constrained by amplitude, which limits the speed of positioning of the servosystem if the given (desired) position is "far" from the initial position of the joint.

4.2.5   Synthesis of the Servosystem for Trajectory Tracking

Up to now we have considered the problem of positioning of the joint in the set position q_i^0. At the input of the servosystem we put the constant signal which corresponds to the desired position of the joint q_i^0 (Fig. 4.11) and the joint is positioned according to the above described procedure. However, as we have already underlined, modern industry requires robots which can be not only precisely positioned in various points in the working space, but which can also track continual trajectories. For example, with arc welding the robot gripper should move along a prescribed trajectory in the working space with a precisely defined velocity. Often defining the desired trajectory can be achieved by imposing a set of discrete points (positions) in the space through which the robot has to pass (the so-called "point-to-point" motion). However, with the above mentioned example of arc welding it is necessary to implement the motion of the gripper along the continual path in the space. In this case all joints of the robot realize their trajectories as continual functions of time $q_i^0(t)$. This is why it is necessary to consider how the servosystem can ensure tracking of the continual trajectory of the joint coordinates (assuming all rest joints are fixed).

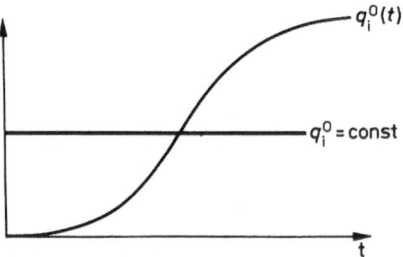

Fig. 4.11. Continual and time variable input for the servosystem

Let us assume that at the servosystem input (Fig. 4.3) there is a signal $q_i^0(t)$ which is a continual function of time (see Fig. 4.11). This signal corresponds to the desired nominal trajectory of the i-th joint, i.e. to the desired change of the joint angle. This means that the joint angle $q_i(t)$ has to track trajectory $q_i^0(t)$; the servosystem has to ensure that the real joint $q_i(t)$ is as close as possible to $q_i^0(t)$ at each moment of time. Even more important, it should be ensured that the rotational velocity of the joint $\dot{q}_i(t)$ is as close as possible to the desired velocity $\dot{q}_i^0(t)$ at each moment of time. However, if we just feed the desired trajectory $q_i^0(t)$, (i.e. corresponding signal) at the input of the servosystem (Fig. 4.3), the servosystem output – joint angle $q_i(t)$ – will undoubtedly have a delay with the given input $q_i^0(t)$. This delay is due to dynamic characteristics of the actuator and the mechanism driven by the actuator (i.e. inertia of the actuator and mechanism, friction and contra-electromotive force which appears in the actuator). Here we shall not analyse mathematically this phenomenon, but it is clear that it is necessary to compensate for this delay in order that the joint can implement precisely the desired trajectory. In order to compensate for this delay caused by servosystem dynamics we can introduce the so-called *feedforward* signal (*precompensation* signal).

The feedforward term has to compensate for delay of the servosystem along the given nominal trajectory and can be synthesized in various ways. Here we shall briefly present one simple procedure for synthesis of feedforward for the robot servosystem. The model of the actuator in the i-th joint is given by Eq. (4.8) in the state space. The nominal trajectory of the joint $q_i^0(t)$ has to be realized. Since the trajectory of the joint is set $q_i^0(t)$, by differentiating we can obtain the desired change (trajectory) of the joint velocity $\dot{q}_i^0(t)$. The state vector of the servosystem and actuator (Eq. (4.8)) is given by $x_i = (q_i, \dot{q}_i)$. This means that the trajectory of the state vector of the actuator $x_i^0(t) = (q_i^0(t), \dot{q}_i^0(t))^\mathsf{T}$ is given. It should be ensured that at each moment t the difference between the real state vector $x_i(t)$ and nominal trajectory $x_i^0(t)$ is as small as possible. The feedforward term represents the signal at the actuator input $\bar{u}_i^0(t)$ which satisfies the equation [5]:

$$\dot{x}_i^0(t) = \bar{A}_i x_i^0(t) + \bar{b}_i \bar{u}_i^0(t) + \bar{f}_i G_i(q_j^0, q_i^0) \tag{4.29}$$

i.e. the signal $\bar{u}_i^0(t)$ satisfies the model of the actuator and joint (Eq. (4.8)) along the specified trajectory $x_i^0(t)$. The signal $\bar{u}_i^0(t)$ represents the *programmed* (voltage) signal as the function of time and is called the *local nominal programmed control*. The name local originates from the fact that this signal is computed for one local actuator and joint of the robot ignoring the other joints (which are assumed to be fixed). The name programmed originates from the fact that this control is a function of time only and not of the temporary state of the joint and actuator (i.e. it is not a function of any temporary position and speed of the joint), so that it represents the "programmed" input for the actuator corresponding to the programmed trajectory $x_i^0(t)$. If we remember the form of the matrix $\bar{A}_i$ and vectors $\bar{b}_i, \bar{f}_i$ in Eq. (4.8) it can be easily shown that the signal $\bar{u}_i^0(t)$ satisfying Eq. (4.29) might be computed according to the expression:

$$\bar{u}_i^0(t) = [(J_M^i N_V^i N_M^i + \bar{H}_{ii})\ddot{q}_i^0(t) + (B_C^i + C_E^{i'})\dot{q}_i^0(t)$$
$$+ G_i(q_j^0, q_i^0(t))]/C_M^{i'} \tag{4.30}$$

where $\ddot{q}_i^0(t)$ represents the desired variation of the joint acceleration along the specified trajectory $q_i^0(t)$, which is obtained by differentiating the nominal trajectory of the velocity $\dot{q}_i^0(t)$. Based on Eq. (4.30) we obtain the local nominal control using the specified trajectory $q_i^0(t)$; the local nominal control, obviously, is not function of the real position and speed of the joint. If the local nominal control is fed into the input of the actuator (as a programmed voltage signal), and if no perturbation is acting upon the joint, the actuator and joint will move along the specified trajectory $q_i^0(t)$. However, some perturbations always must act upon the servosystem. In the initial moment $t = 0$ the joint angle $q_i(0)$ need not correspond to the nominal angle $q_i^0(0)$. Because of this the motion of the joint always deviates from the nominal trajectory if we feed the actuator with the programmed nominal control $\bar{u}_i^0(t)$ only. The behaviour of the actuator and the joint in that case is described by:

$$\dot{x}_i(t) = \bar{A}_i x_i(t) + \bar{b}_i \bar{u}_i^0 + \bar{f}_i G_i(q_j^0, q_i) \tag{4.31}$$

Obviously if $x_i(0) \neq x_i^0(0)$ the real state vector $x_i(t)$ will not coincide with the nominal trajectory $x_i^0(t)$. Due to this the additional signal Δu_i must be fed into the input of the actuator so as to ensure that the state vector $x_i(t)$ is as close to $x_i^0(t)$ as possible, if the perturbations are acting upon the actuator and if $x_i(0) \neq x_i^0(0)$. Let us introduce the vector of deviation of the state from the nominal trajectory $x_i^0(t)$ as the difference between the real state and nominal state $\Delta x_i(t) = x_i(t) - x_i^0(t)$. Model (4.31) can then be written in the following form:

$$\Delta \dot{x}_i(t) = \bar{A}_i \Delta x_i(t) + \bar{b}_i \Delta u_i(t) + \bar{f}_i [G_i(q_j^0, q_i) - G(q_j^0, q_i^0)] \tag{4.32}$$

Model (4.32) is called the *model of deviation of the actuator state from the nominal trajectory*. For this model of the actuator and the joint we must ensure that the deviation of the state $\Delta x_i(t)$ approaches zero. Since the elements of the vector Δx_i are $(-\Delta q_i, -\Delta \dot{q}_i)$, where $\Delta q_i = q_i^0 - q_i$ and $\Delta \dot{q}_i = \dot{q}_i^0 - \dot{q}_i$ deviation of the joint angle Δq_i and velocity $\Delta \dot{q}_i$ should also approach zero. The additional signal at input Δu_i has to ensure that the deviation vector Δx_i is near zero. Model (4.32) is analogous to the basic model of the actuator and joint (4.8). It is therefore quite obvious that the additional control signal Δu_i for the deviation model (4.32) can be generated quite analogously as for the actuator and joint positioning. Namely, the problem of reducing the state vector Δx_i of the deviation model (4.32) to zero is analogous to the problem of positioning the actuator and joint (4.8) into the position $x_i = (0, 0)^T$. Thus, the additional control signal might be generated as:

$$\Delta u_i = k_p^i \Delta q_i + k_v^i \Delta \dot{q}_i = -k_i^T \Delta x_i \tag{4.33}$$

where by k_i is denoted the vector given by $k_i = (k_p^i, k_v^i)^T$. The total signal which has to be fed to the actuator input is:

$$u_i = \bar{u}_i^0(t) + \Delta u_i(t) = \bar{u}_i^0(t) - k_i^T \Delta x_i \tag{4.34}$$

Figure 4.12 shows the control scheme which ensures tracking of the trajectory. The servosystem for trajectory tracking has a similar structure as the servosystem for positioning (Fig. 4.3). The only difference is in the pre-compensator which represents the computation of the local nominal control according to Eq. (4.30), and in the fact that instead of the feedback by velocity $\dot{q}_i$ here we introduce the difference between the real speed $\dot{q}_i(t)$ and the nominal velocity $\dot{q}_i^0(t)$. This difference is amplified by k_v^i. Since the deviation model (4.32) is similar to model (4.8), synthesis of the gains k_p^i and k_v^i for the servosystem with a precompensator is analogous to synthesis of the gains for positioning of the servosystem.

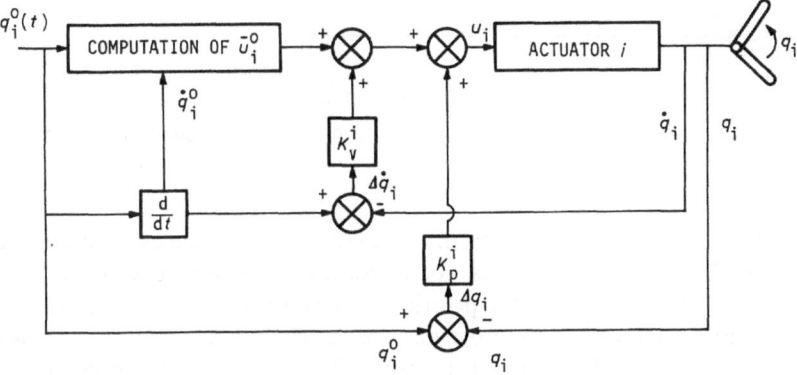

Fig. 4.12. The servosystem with local nominal control

It should be mentioned that for computation of local nominal control according to Eq. (4.30), for the moment of inertia of the mechanism H_{ii}, the least possible value should be taken (which this moment might have, depending on the position of the rest of the joints q_j). The reason for this is to avoid overshoots. Thus, the procedure for choice of the moment of inertia of the mechanism for calculating the local nominal control is opposite to that for computation of the speed feedback gain.

Example

Let us assume that at input of the servosystem for the first joint of the manipulator in Fig. 4.5 (the servosystem has been synthesized in the example in Sect. 4.2.2), instead the position of the signal corresponding to joint trajectory $q_1^0(t)$ is fed. This trajectory is presented in Fig. 4.13 and can be described by:

$$q_1^0(t) = \begin{cases} \dfrac{a_1 t^2}{2}, & 0 < t \leqslant \dfrac{\tau}{2} \\[3mm] a_1\left(\tau \cdot t - \dfrac{t^2}{2} - \dfrac{\tau^2}{4}\right), & \dfrac{\tau}{2} < t \leqslant \tau \end{cases} \qquad (4.35)$$

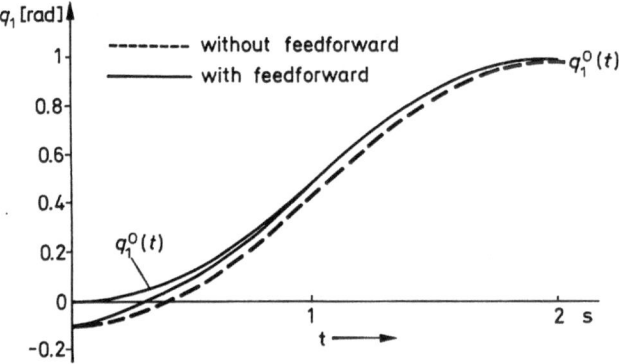

Fig. 4.13. Trajectory tracking with and without precompensator (the first joint of the robot UMS-2)

where $a_1 = 1$ [rad/s^2] is the acceleration, and $\tau = 2$ [s] is the time duration of the movement. If the precompensator is not introduced, but we directly feed the desired trajectory to the input of the positioning servosystem, the real trajectory of the joint will be delayed to the nominal one, as can be seen in Fig. 4.13 (for initial error $\Delta q_1(0) = q_1^0(0) - q_1(0) = 0.1$ [rad]). This is why it is necessary to introduce feedforward for compensating delay. Based on data from Table 4.1 and assuming that the third joint is fixed in the position $q_3^0 = 0$, so that the moment of inertia of the mechanism around the first joint is given by $H_{ii} = 0.405$ [kg m^2], we can obtain the local nominal control according to Eq. (4.30):

$$\bar{u}_1^0(t) = \begin{cases} [0.435 \cdot a_1 + 1.346 \cdot a_1 t]/0.9375 & 0 \leqslant t \leqslant \dfrac{\tau}{2} \\[3mm] [-0.435 \cdot a_1 + 1.346 \cdot a_1(\tau - t)]/0.9375 & \dfrac{\tau}{2} < t \leqslant \tau \end{cases}$$

4.3   Control of Simultaneous Motion of Several Robot Joints

Up to now we have considered the control of one joint of the robot while the remaining joints are fixed. However, in order to execute some control task by the robot the gripper of the robot has to be positioned in some position in space. To do this it is necessary to drive all joints of the robot into certain positions (angles) which correspond to the desired position of the gripper. In principle, it is possible to drive the joints of the robot to certain positions successively, one by one, so that while each joint is moving the remaining joints are fixed. In this case the control which we have observed up to now will be satisfactory. However, it is clear that such positioning of the robot, joint by joint, is not efficient from the point of view of the time then required for the task to be accomplished. Let us denote by τ_i the time required to drive the i-th joint to a certain given position. If the robot is driven to the desired position so that its joints are positioned successively the time required for positioning of the robot is given by

$$\tau = \sum_{i=1}^{n} \tau_i \tag{4.36}$$

where n is the number of joints of the robot. If all the joints of the robot are moving simultaneously towards the given positions, the time required for robot positioning is given by

$$\tau = \max_i \tau_i \tag{4.37}$$

Obviously the time required for positioning of the gripper into the desired position if all joints are moving simultaneously (Eq. (4.37)) is considerably less than if the joints are moving successively (Eq. (4.36)). Since one of the main goals in robot design is to achieve as high a working speed as possible (so that the time required for task execution is as low as possible and the efficiency of the robot is thereby increased), it is clear that with modern robots the simultaneous positioning of all joints must be ensured. Even more, if the tracking of a given path of the gripper is required, it is obvious that all the joints of the robot must move simultaneously (i.e. they have to track their trajectories simultaneously). Thus, the simultaneous movement of all robot joints is unavoidable for all modern robots, specially for new generation robots which include those capable for automatic path planning.

If several joints of the robot are moving simultaneously dynamic coupling between the joints must appear. This dynamic coupling must affect performance of the servosystems around the robot joints. In this section we shall consider the influence of the dynamic forces, and synthesis of control which ensures satisfactory performance of the robot if its joints are moving simultaneously.

4.3.1 Analysis of the Influence of Dynamic Forces in Simultaneous Motion of Several Joints

In Chapter 3 we presented the model of robot dynamics. We have seen that the driving torques P in the robot joints might be expressed as functions of angles q, velocities $\dot{q}$, and accelerations $\ddot{q}$ of the joints:

$$P = H(q)\ddot{q} + h(q, \dot{q}) \tag{4.38}$$

where $H(q)$ is the inertia matrix $(n \times n)$, $h(q, \dot{q})$ is the vector of centrifugal, Coriolis and gravity moments $(n \times 1)$. If several joints are moving simultaneously the dynamic moment P_i is acting around the i-th joint. This dynamic moment is given by Eq. (4.38). This means that an external load is acting upon the i-th actuator and this load P_i is the function of the angles, velocities, and accelerations of all the joints. If only the i-th joint is moving (and all the remaining joints are fixed) upon the actuator is acting the load given by Eq. (4.7) which we have taken into account in the synthesis of the servosystem around the i-th joint. The dynamic moment which is caused by simultaneous motion of several joints loads the i-th servosystem and affects the servosystem performance. In the following we shall consider qualitatively how certain dynamic forces (moments) in simultaneous motion of several joints affect the performance of the servosystems around the robot joints. In this we shall consider both the positioning problem (by simultaneous positioning of all the robot joints) and the problem of tracking of the robot gripper trajectory (by simultaneous tracking of all the joint trajectories). Let us assume that the desired positions of the joints q_i^0, or the trajectories $q_i^0(t)$ are fed simultaneously at the inputs of all the servosystems.

1. *Variable moments of inertia.* If several joints are moving simultaneously the moment of inertia of the mechanism around the i-th joint is varying during the motion, since H_{ii} depends on the positions of all the joints which are in the kinematic chain "behind" the i-th joint. In Sect. 4.2.3 we have already considered the influence of the variation of the moment of inertia of the mechanism upon the servosystem performance. We have shown that it is necessary to compute the feedback gains for the maximum possible value of the moment of inertia of the mechanism in order to prevent the system being underdamped. However, we have also seen that if the moment of inertia is significantly varied, performance of the robot can be uneven. This is specially inconvenient with simultaneous motion of several joints, as the moment of inertia is varied during the motion, which can cause oscillatory tracking of trajectories or positioning. However, this problem can be solved in one of the ways previously mentioned.

2. *Cross-inertia members.* Acceleration of the j-th joint causes the moment in the i-th joint through so-called cross-inertia moments, which themselves represent the elements of the inertia matrix H off the main diagonal $H_{ij}(i \neq j)$. Thus, due to acceleration in the j-th joint $\ddot{q}_j$ upon the i-th

servosystem acts as an external load $H_{ij}(q)\ddot{q}_j$. As we have explained above, the external load upon the shaft of the actuator causes errors in the positioning of the joint, or in the tracking of trajectories (since the servosystem must overcome this external load through the position error). However, this moment only appears if the accelerations are relatively high. When the robot starts to move or stops, the accelerations drop to zero, so they do not influence the positioning of the joints, i.e. they do not cause steady state errors. These moments can cause errors in tracking trajectories if these are with high accelerations $\ddot{q}_j$. If for implementation of a given task it is not essential to achieve the precise tracking of fast trajectories, the effects of these moments are then no longer significant and can be ignored. However, if precise tracking of fast trajectories is essential (which means that in each moment the difference between the real position of the joint and the nominal trajectory must be minimal), then the moments due to cross-inertia members must be compensated for.

3. *Gravity moments.* The effects of gravity moments have already been considered in Sect. 4.2.4. In simultaneous motion of several joints the gravity moments vary during the movement, causing errors both in positioning and in tracking of trajectories. Compensations for these moments can be realized in one of the previously mentioned ways, but it should be kept in mind that the gravity moments vary during the tracking of trajectories so they cannot be completely compensated by PID regulators.

4. *Centrifugal and Coriolis moments.* These moments are produced by the velocities $\dot{q}_j$ in the robot joints. They also act as external loads upon the servosystems around the joints. However, since the centrifugal and Coriolis forces are directly proportional to rotational (or translational) velocities of the joints $\dot{q}_j$, these forces are significant only if the joints are moving at high speeds. When the robot is starting to move or is stopping these forces are negligible, which means they do not influence the positioning of the robot in any desired positions and do not cause steady state errors. These forces only cause errors in the tracking of fast trajectories. Similarly, as in the case of cross-inertia moments, here we can also conclude that if the precise tracking of fast trajectories is not required the influence of these forces can be ignored. If the tracking of fast trajectories is essential, we must take into account centrifugal and Coriolis forces in the synthesis of control.

As can be seen from these considerations, in simultaneous movement of several joints dynamic coupling between the joints appears which affects the positioning and tracking of trajectories of the robot. Also the influence of these dynamic forces upon the positioning and tracking of slow trajectories is not significant, so the servosystems synthesized for isolated joints of the robot can overcome them. In a previous section we have shown how at the level of local servosystems we can compensate for the influence of variable inertia and external loads. These compensations are often quite sufficient to ensure positioning and tracking of slow trajectories even if several joints are moving simultaneously. For such tasks it is quite acceptable to control the robot by local servosystems synthesized for isolated joints. However, if we have to ensure tracking of fast trajectories the influence of dynamic forces cannot be ignored. Since these forces act as external loads, if the servosystem feedback gains are high, the errors caused by these forces might be small, so that even in the case of fast trajectories we may accept the servosystems synthesized for isolated joints (Sect. 4.3.5). However, since high gains are limited, if we must ensure precise tracking of fast trajectories we cannot only apply local servosystems, but the dynamic forces must be compensated for.

Example

For the first joint of the robot in Figure 4.5 we have synthesized the servosystem as in the example in Sect. 4.2.3, assuming that all the remaining joints are fixed. The positioning of this joint is shown in Fig. 4.8. If simultaneously with the first joint the third joint is also moving, the Coriolis moment acts around the first joint given by $m_3(l_3+q_3)^2 \dot{q}_1 \dot{q}_3$ which acts as an external load upon the servosystem. If we just consider positioning of the first joint (as in the Example in Sect. 4.2.3) but assuming that the third joint is moving with the velocity $\dot{q}_3^0 = 0.4$ [m/s], it can be seen that the positioning is quite satisfactory and the influence of the dynamic moments can be ignored (Fig. 4.14), which also means that the local servosystem satisfies the simultaneous positioning of several joints.

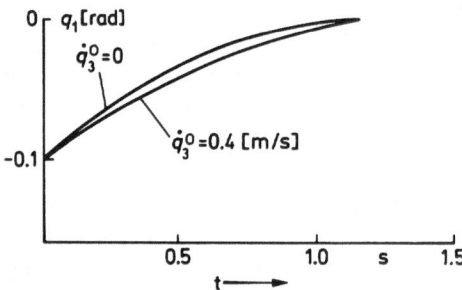

Fig. 4.14. Positioning of the first joint of the robot UMS-2 when the third joint is moving with the velocity $\dot{q}_3^0$

For tracking of the trajectory from Fig. 4.13 we have synthesized the servosystem with local nominal control as in the example in Sect. 4.2.5. If together with motion of the first joint, the third joint is moving with the velocity $\dot{q}_3^0 = 0.2$ [m/s], the tracking of the trajectory is as presented in Fig. 4.15. It can be seen that the Coriolis moment causes the error during the tracking of the trajectory (but not in final position). If we increase the velocity of the third joint up to $\dot{q}_3^0 = 0.4$ [m/s] (the tracking is also presented in Fig. 4.15), it can be seen that the tracking error is increased. If this error is unacceptable, we should not apply just the local servosystem (with a precompensator) but also introduce an additional control signal which has to compensate for the influence of the dynamic moment.

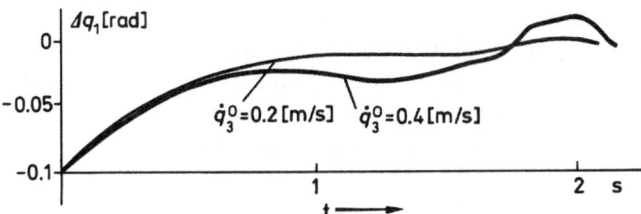

Fig. 4.15. The tracking of the trajectory in the first joint of the robot UMS-2 when the third joint is moving with the velocity $\dot{q}_3^0$

It should be kept in mind that the movement of the third joint causes variation of the moment of inertia around the first joint which also affects the tracking of the trajectory of the first joint.

4.3.2 Dynamic Control of Robots

Most of the robots on the market today are not capable of ensuring precise tracking of fast trajectories, so for them it has been sufficient just to apply local servosystems without introducing the compensation for dynamic forces. However, since the requirements imposed on robots concerning the speed and quality of operations (precision of tracking) are increasing, the control at the executive level must take into account the dynamics of the robot. The control which concerns all (or some) dynamic forces in the robotic system is called the *dynamic control of robots*.

The basic problem when applying dynamic control lies in the fact that the dynamic forces acting within the robotic mechanism are generally very complex functions comprising coordinates, speeds and accelerations of the joints. Thus, if we want the control to compensate the effects of these forces, this usually leads to complex control laws. As we have already emphasized, the control of robots now is implemented solely by microprocessors. If we want to compensate the dynamic forces it is necessary for the microprocessor to compute these forces as the functions of state, i.e. the coordinates, velocities and accelerations of the joints during the motion of the robot. This computation requires a large number of additions and multiplications which the microprocessor has to process for a relatively short period of time (see Chap. 5). This is why the problem of minimizing of number of additions and multiplications arises in the synthesis of the dynamic control.

Various approaches have been developed for synthesis of the dynamic control of robots. (A survey of these approaches can be found in [6]). Here we shall briefly consider just two approaches: by nominal programmed control and by global control [5, 6].

In Sect. 4.2.5 we have shown that by applying nominal local programmed control $\bar{u}_i^0(t)$ delay in the servosystem along the trajectory is compensated with satisfactory tracking of the nominal trajectory of one joint, assuming that all the rest of the joints are fixed. If several joints are moving simultaneously the local nominal control $\bar{u}_i^0(t)$ cannot compensate dynamic moments which act upon the i-th joint, which is why instead of the local nominal control we may apply *nominal programmed control computed on the basis of the complete model of the system*. This control can be computed in the following way. Let the nominal trajectories of all joints of the robot $q_i^0(t)$, $i = 1, 2, \ldots, n$ be given. By differentiating we get nominal trajectories of the velocities $\dot{q}_i^0(t)$ and accelerations $\ddot{q}_i^0(t)$ of the joints. Now, based on the model of the mechanism dynamics (Eq. (4.38)) we can compute the *nominal driving torques* $P_i^0(t)$ in the robot joints:

$$P^0(t) = H(q^0(t))\ddot{q}^0 + h(q^0, \dot{q}^0) \tag{4.39}$$

where by P^0 we have denoted the vector $P^0 = (P_1^0, P_2^0, \ldots, P_n^0)^T$, and by $q^0(t)$ we have denoted the vector $q^0 = (q_1^0, q_2^0, \ldots, q_n^0)^T$. The nominal driving torques $P_i^0(t)$ represent the moments which have to be implemented in the joints of the robot in order to ensure that the robot moves along the desired nominal trajectories $q^0(t)$.

Models of the actuators in the joints are given by Eq. (4.5) where P_i represents the external load by which the mechanism acts upon the actuators, i.e. P_i is the driving torque realized by the actuator. The nominal voltage signal in the actuator input $u_i^0(t)$ which has to implement the nominal driving torque $P_i^0(t)$ must satisfy the equations:

$$\dot{x}_i^0(t) = A_i x_i^0(t) + b_i u_i^0(t) + f_i P_i^0(t) , \qquad i = 1, 2, \ldots, n \tag{4.40}$$

where $x_i^0(t) = (q_i^0(t), \dot{q}_i^0(t))^T$.

The nominal programmed control $u_i^0(t)$ which satisfies Eq. (4.40) can be computed by (on the basis of Eq. (4.6)):

$$u_i^0(t) = [J_M^i N_V^i N_M^i \ddot{q}_i^0(t) + (B_C^i + C_E^{i\prime})\dot{q}_i^0(t) + P_i^0(t)]/C_M^{i\prime} \tag{4.41}$$

Obviously the nominal control $u_i^0(t)$ differs from the local nominal control $\bar{u}_i^0(t)$, given by Eq. (4.30), since the former includes the total nominal driving torque $P_i^0(t)$ which represents the dynamic moment due to the movement of all joints of the robot. Thus, the nominal programmed control $u_i^0(t)$ compensates not only the dynamics of the actuator and a single joint, but also it compensates the dynamics of the complete mechanism (but only along the nominal trajectory). If we feed the nominal programmed signals $u_i^0(t)$ at the inputs of all actuators, and if the models of the actuators and the mechanism are exact, and the robot in the initial moment is in such a position that $q_i(0) = q_i^0(0)$ for all joints, and if no perturbation is acting upon the robot, the joints of the robot would move along the imposed trajectories $q_i^0(t)$. However, all the above assumptions seldom hold. This is the reason why deviations of the joint coordinates from the nominal trajectories always appear if we just apply the nominal programmed control $u_i^0(t)$. The motion of the joints is described by the model

$$\dot{x}_i(t) = A_i x_i(t) + b_i u_i^0(t) + f_i(P_i^0(t) + \Delta P_i) \tag{4.42}$$

where ΔP_i represents deviation of the real load in the joint of the robot $P_i(t)$ from its nominal value $P_i^0(t)$, $\Delta P_i = P_i - P_i^0$. If $x_i(0) \neq x_i^0(0)$ then $x_i(t)$ will not coincide with $x_i^0(t)$, so we must introduce the additional control signal u at the actuator input; this additional signal has to ensure that the $x(t)$ is as close to the $x^0(t)$ as possible. The model (4.42) can be written as (based upon Eq. (4.40)):

$$\Delta \dot{x}_i(t) = A_i \Delta x_i(t) + b_i \Delta u_i + f_i \Delta P_i , \quad i = 1, 2, \ldots, n \tag{4.43}$$

where $\Delta x_i = x_i(t) - x_i^0(t)$. The model (4.43) represents the model of deviation of the actuator state from its nominal trajectory if all joints of the robot deviate from their nominal trajectories. This model is similar to the basic model of the actuator (4.5) except for the fact that instead the total external moment P_i here acts just as a deviation of the dynamic moment from the nominal driving torque. The nominal programmed control $u_i^0(t)$ compensates the nominal dynamic moments $P_i^0(t)$ and reduces the effects of the dynamic forces upon the actuator. If we apply the servosystem synthesized in Sect. 4.2.5 the reduced dynamic moment ΔP_i (as an external load) acts upon it. The servosystems can then overcome this external load more efficiently if we apply the nominal programmed control (Fig. 4.16).

However, application of the nominal programmed control involves numerous difficulties. To compute $u_i^0(t)$ on the basis of Eq. (4.41), it is necessary to first compute the nominal moments based on Eq. (4.39). This means that the complete dynamic model of the mechanism which, as we have mentioned above is very complex, has to be computed, making on-line computation of the nominal moments during motion of the robot not easy. Since the nominal moments and control are functions just of the nominal trajectories of the robot $q_i^0(t)$ (and not of the real coordinates and velocities of the robot) it is possible to compute these functions in advance before the process starts and store them in the memory of the microcomputer. If the robot task is known in advance (which is commonly the case in robotics) it is possible to compute $P_i^0(t)$ and $u_i^0(t)$ in advance (off-line) and store them. When the execution of the task starts we extract from the microcomputer the corresponding values of $u_i^0(t)$ (at each sampling period) which are sent to the actuator's inputs. The requirements regarding the speed necessary for computing the driving torques and the selection of microprocessors are then relaxed. However, there are certain drawbacks. In many robotic tasks the trajectories are not known in advance but must be generated during execution (for example, if the robot has to plan its trajectories to avoid collisions with obstacles which are not known in advance and which are not fixed). This is often encountered in modern robotics. Also storage of the nominal programmed control can require an extremely large

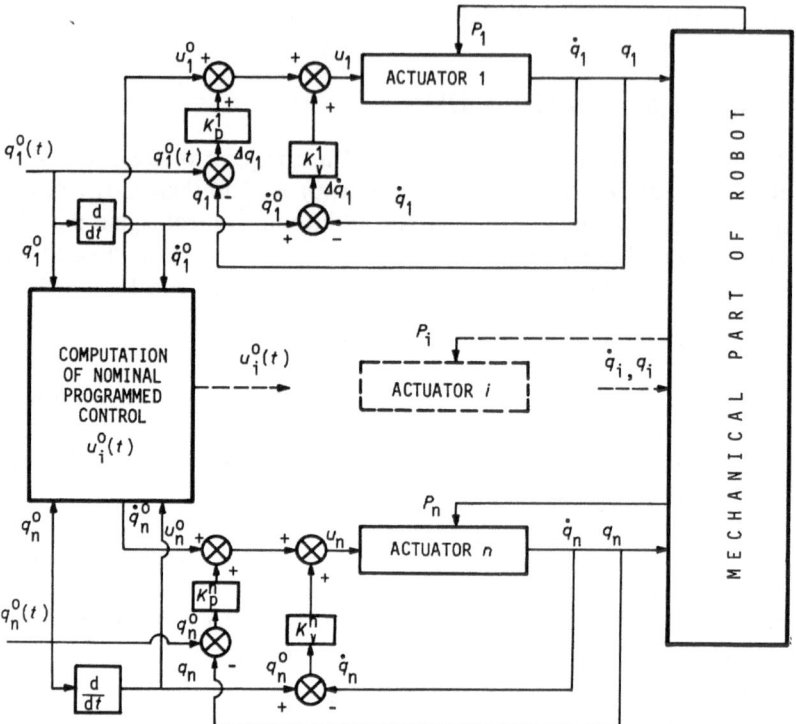

Fig. 4.16. The control scheme of the robot if the nominal programmed control is applied

memory capacity of the microcomputer. Finally, the main drawback of the nominal programmed control lies in the fact that it assumes that the model of the dynamics of the mechanism and actuators is known (including all the relevant parameters), which is not always the case.

The second possibility to compensate the dynamic moments in the trajectory tracking is by *global control*. If we keep local servosystems with a local nominal control (Sect. 4.2.5), the external load from dynamic moments then acts; this external load is given by Eq. (4.38). To compensate the effects of these dynamic moments we can introduce an additional compensating signal at the actuator input; this additional signal has to be proportional to the dynamic moment P_i acting upon the i-th joint. This additional signal Δu_i^G might be introduced in the form:

$$\Delta u_i^G = -k_i^G \tilde{P}_i \tag{4.44}$$

where k_i^G represents the global gain, and $\tilde{P}_i$ represents the value which is proportional (or equal) to the dynamic moment P_i in the i-th joint. This additional control signal is called the global control since it represents the

feedback between the joints of the robot. The *local* servosystem includes only the
feedback by *local* coordinates and speeds of the joint which is controlled by
the servosystem so they just have local information, compared with the global
control which represents an interchange of the information between the servosystems (so-called *cross-feedback (global)* loops – see Fig. 4.17).

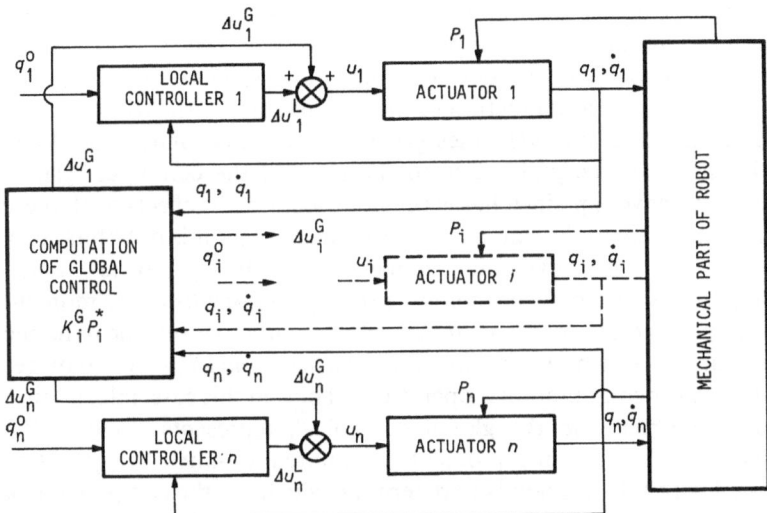

Fig. 4.17. The robot control scheme with global control

Global control has to compensate for the effects of the dynamic moment P_i
by generating the signal at the actuator input which is proportional to this
moment. The basic problem regarding the global control is how to get information on the dynamic moment, P_i.

Two versions for implementing global control are suggested.

1. The dynamic force which acts upon the servosystem in the joint might be
    directly measured by a force transducer. By placing the force transducer in
    the shaft of the actuator (and joint) the dynamic force P_i is measured
    directly. The signal from the force transducer is (through the micropro-
    cessor) fed back to the actuator input (multiplied by the gain k_i^G), so that in
    this case $\tilde{P}_i = P_i$. By this, the signal is generated at the actuator input which is
    proportional to the dynamic force and which produces the actuator torque
    to compensate for the dynamic moment. Global control, in this case, is
    realized by *force feedback*. It can therefore be understood that this realiz-
    ation of global control is very simple and does not require any additional
    computation in the microprocessor (except for Eq. (4.44)). However, the
    force transducers are not cheap and require additional technical solutions
    when being installed in the actuator shaft.

2. The second way to realize global control is by *on-line computation of the moment*. Based on the information of the instant coordinates q_i and velocities $\dot{q}_i$ of the joints which are obtained by corresponding sensors, the microcomputer computes the dynamic moments P_i according to the model (4.38). Thus, the microcomputer generates the signal at the actuator input which is proportional to computed value of the dynamic moment. However, as we have already explained, the model of the mechanism dynamics (Eq. (4.38)) can be extremely complex, requiring very fast (and expensive) microcomputer capable of computing on-line (during the robot movement) the dynamic moments as a function of the instant state of the robot (i.e. the instant coordinates q_i and velocities $\dot{q}_i$). Therefore, this solution of global control is often not acceptable due to the price of the computer system. However, as we have explained in the previous section, the effects of all these dynamic forces need not be significant, for in many cases it is not obligatory to compute all the dynamic moments of the complete model (Eq. (4.38)). Instead, we may just compute some components of the moment (gravity moments, cross-inertia members, etc.). In other words, the dynamic moments can be computed by an *approximative model* which requires significantly fewer computational operations. Thus, in this case, information on dynamic moments in the global control $\tilde{P}_i$ represents some of the components of the dynamic moments computed on-line. The problem encountered with this solution is to determine which are the components of the dynamic moments which must be compensated by global control, i.e. which components have to be computed on-line. To answer this question it is necessary to *analyse the stability of the nonlinear model* of *the robot dynamics*. The stability analysis gives an answer as to whether the robot is stable around the nominal trajectory with the selected servosystems and with the global control (computed by the approximative model of the robot). Generally, this analysis requires application by the computer. Selection of adequate dynamic control of robots is therefore best performed by *computer-aided control synthesis*.

It should be mentioned that various combinations of the control laws are possible. It is possible to apply a nominal programmed control (which compensates the nominal dynamic moments), and the local servosystems and global control (which in this case has to compensate for the deviation of the real moments from the nominal ones, ΔP_i).

Example

For the first joint of the robot in Fig. 4.5 the nominal programmed control can be obtained in the following way, assuming that the nominal trajectories of the first and the third joint are given by Eq. (4.35) where for the first joint $a_1 = 1$ [rad/s^2], and for the third joint $a_3 = 0.8$ [m/s^2], while $\tau = 2$[s]. The nominal driving torque is given by:

$$P_1^0 = (J_{z1} + J_{z2} + J_{z3} + m_3(l_3 + q_3^0(t)^2)\ddot{q}_1^0(t)$$
$$+ 2m_3(l_3 + q_3^0(t))\dot{q}_3^0(t)\dot{q}_1^0(t))$$

or (taking into account Eq. (4.35) and the parameters given in Table 4.2):

$$P_1^0 = \begin{cases} \left[0.402 + 4.15 \cdot \left(0.035 + a_1 \cdot \frac{t^2}{2} \right)^2 \right] a_1 + \\ \quad + 2 \cdot 4.15 \left(0.035 + a_1 \frac{t^2}{2} \right) a_1 \cdot a_3 t^2, \quad 0 \leqslant t \leqslant \frac{\tau}{2} \\ - \left\{ 0.402 + 4.15 \cdot \left[0.035 + a_1 \left(\tau t - \frac{t^2}{2} - \frac{\tau^2}{4} \right) \right]^2 \right\} a_1 + \\ \quad + 2 \cdot 4.15 \left[0.035 + a_1 \left(\tau t - \frac{t^2}{2} - \frac{\tau^2}{4} \right) \right] \cdot a_1 a_3 (\tau - t)^2, \quad \frac{\tau}{2} < t \leqslant \tau \end{cases}$$

Based on Eq. (4.41) and on the parameters of the actuators from Table 4.1, we can obtain the nominal programmed control as:

$$u_1^0(t) = \begin{cases} \left\{ 0.0291a_1 + 1.346a_1 t + \left[0.402 + 4.15 \left(0.035 + a_1 \frac{t^2}{2} \right)^2 \right] a_1 + \right. \\ \quad \left. + 2 \cdot 4.15 \cdot \left(0.035 + \frac{a_1 t^2}{2} \right) a_1 a_3 t^2 \right\} \Big/ 0.9375, \quad 0 \leqslant t \leqslant \frac{\tau}{2} \\ \left\{ -0.0291a_1 + 1.346a_1(\tau - t) - \left[0.402 + 4.15 \cdot \left(0.035 + \right. \right. \right. \\ \quad \left. + a_1 \left(\tau t - \frac{t^2}{2} - \frac{\tau^2}{4} \right) \right)^2 \right] a_1 + 2 \cdot 4.15 \cdot \left[0.035 + \right. \\ \quad \left. \left. + a_1 \left(\tau t - \frac{t^2}{2} - \frac{\tau^2}{4} \right) \right] a_1 a_3 (\tau - t)^2 \right\} \Big/ 0.9375, \quad \frac{\tau}{2} < t \leqslant \tau \, . \end{cases}$$

If this nominal programmed control and the servosystem feedback gains computed in the example in Sect. 4.2.3 are applied, we obtain the ideal tracking of the trajectory from Fig. 4.13, assuming that in the third joint the nominal programmed control and the corresponding servosystem feedback loops are also applied, again assuming $q_1(0) = q_1^0(0)$. If $q_1(0) \neq q_1^0(0)$ but $\Delta q_1(0) = q_1^0(0) - q_1(0) = 0.1$ [rad] the tracking of the nominal trajectory is as presented in Fig. 4.18, which is quite acceptable. This shows that the nominal programmed control and the local servosystems, in this case, can compensate the dynamic moments acting in the joints of the robot.

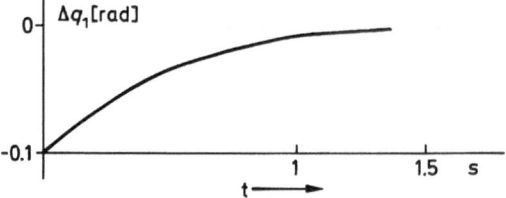

Fig. 4.18. Tracking of the trajectory for the first joint of the robot UMS-2 with nominal programmed control

4.3.3   Computer-aided Synthesis of Robot Control

As previously explained, there are various possibilities for synthesis of the dynamic and non-dynamic control of robots. The designer of the robot has to select adequate control laws. The choice depends on the robot structure and the dynamic characteristics of the robot mechanism, on the selection of actuators, and the requirements imposed upon the robot being designed. The choice also depends on the tasks of the particular robot. Does the robot just have to ensure positioning or must it track trajectories of the gripper as required in particular tasks. To take into account all these requirements and problems the designer has to analyse carefully the dynamic performance of the robot. Due to the high non-linearity of the dynamic model of the robot, this analysis is very complex; it is not possible to implement "by hand". However, there are computer systems available today which can significantly help the designer select the control for his robot.

Application of computers for control synthesis of large-scale dynamic systems is now broadly encountered. In robotic systems computers enable fast and efficient analysis of robot dynamics with the various control laws. Dynamic models of the robot can be set efficiently by digital computers, so that the user has only to impose data on the particular robot (the number of the joints, orientations of the joint axes, lengths, masses and moments of inertia of the links, etc.), and the computer automatically computes the model of the robot. Using these models the computer can analyse the dynamic performance of the robot. In principle, it is possible to develop algorithms for automatic synthesis of control for robots. However, these procedures are often very time consuming and inefficient, so that it is more appropriate for the designer of the robot to select the control himself with the computer just serving as an assisting device to help the designer. This *interactive control synthesis* enables the user to use his experience in the control system design, and to utilize the digital computer for the complex analysis of the robot dynamics and for the computation of adequate parameters of the control (the feedback gains, approximate models, etc.).

One such *software package* for computer-aided synthesis of control for manipulation robots has been developed by the Mihailo Pupin Institute, in Beograd. In order to illustrate the computer-aided control synthesis, we shall briefly describe the basic structure (flow-chart) of this software package, where we shall emphasize the interaction between the user and the computer.

The package consists of several programme blocks which can be easily combined by the user. However, the algorithm takes care of the logical sequence of the blocks (for example, it is impossible to start synthesis of the servosystems if data on actuators and mechanism have not been previously imposed). The package enables the control synthesis for robots of arbitrary structure with various numbers of joints and actuators. The flow-chart of the package is shown in Fig. 4.19. Here, we shall briefly describe the modules which are included in this package.

1. *Module for input data on robot mechanism.* This module enables the user to impose data about geometry and structure of the mechanism, as well as dynamic parameters of the robot (lengths of the links, masses, moments of inertia). Based on these data, the software package automatically sets kinematic and dynamic models of the robot, which are used in the synthesis of control.
2. *Module for input data on actuators.* The user has to select the type of actuator which he wants to apply and the basic data on actuators (moments of inertia of the rotor, moment and electromotor constants of the motor, electrical resistance of the rotor circuit, etc., if the D.C. motors are applied as actuators). Based on these data the package automatically sets the models of actuators.
3. *Module for imposing the desired trajectory.* The user has to specify the task which the robot has to implement, i.e. the user has to choose whether the robot just has to be positioned in various positions in its working space, or whether it should move "point-to-point", or if it should track the desired trajectory. The user has to impose data on the desired positions or trajectories of the gripper of the robot. Based on these data, the algorithm, using the kinematic model of the robot, determines the corresponding positions or trajectories, i.e. it solves the so-called inverse kinematic problem (see Chap. 2).
4. *Module for synthesis of the local servosystem.* The package can synthesize the local servosystems using various methods in accordance with the user option. The package can select servosystem feedback gains for all joints of the robot.
5. *Module for nominal control synthesis.* The package enables automatic computation of the nominal driving torques which correspond to the imposed trajectories of the robot gripper, and synthesis of local nominal control or nominal programmed control which is computed using the complete dynamic model of the robot.

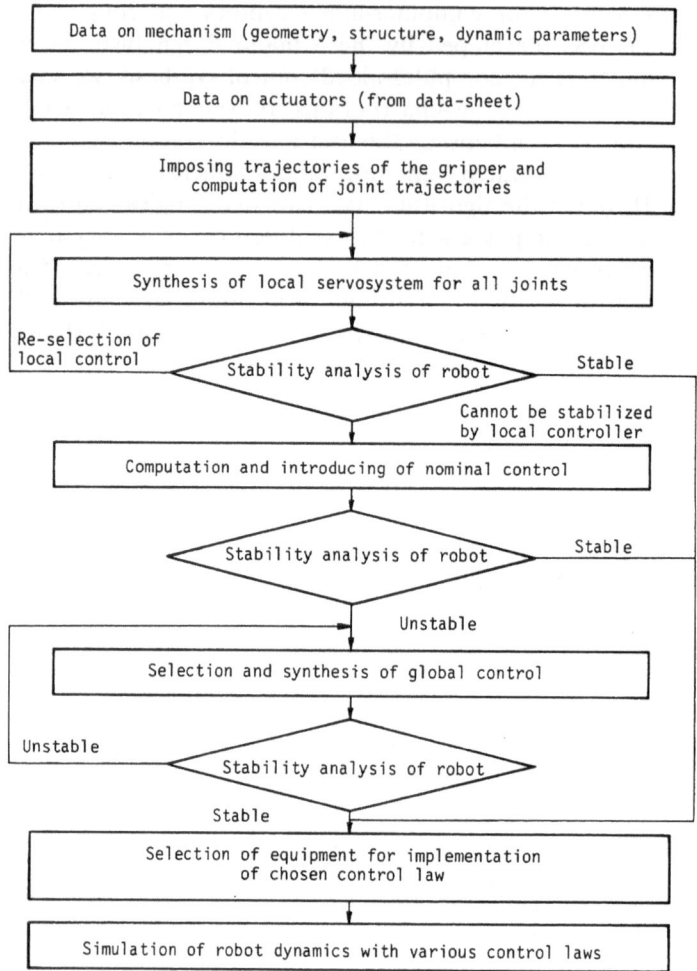

Fig. 4.19. Flow-chart of the software package for computer-aided control synthesis for manipulation robots

6. *Module for synthesis of global control.* The user is able to synthesize the global control using the software package, i.e. he can introduce global feedback loops which compensate for the influence of dynamic coupling between the joints. Global control can be introduced either as force feedback, or as on-line computation of dynamic moments; in the latter case the package helps the user to select the approximative model of the robot dynamics which can be used for computation of global control.

7. *Module for analysis of robot performance.* This module analyses the stability of the complete non-linear model of the robot when various control laws are

applied: local servosystems around the robot joints, with or without local nominal control, with or without nominal control (feedforward), computed on the basis of the complete robot dynamic model, with or without global control. The stability analysis of the robot around the nominal trajectory (position) is performed using various methods for stability analysis of large-scale non-linear dynamic systems. The package gives the answer to the question whether the robot is stable or not with the selected control law and calculated feedback gains.

8. *Module for the simulation of robot dynamics.* The module for *simulation* of robot dynamics enables verification of robot behaviour with various control laws. Using this module, the designer of the robot obtains precise insight into the performance, quality and speed of the robot which helps him to make a final decision on selecting the most appropriate control law.

The basic function of this package is to enable the designer of the robot to make a precise and valid decision on selecting the control for the executive control level; the designer can choose the simplest control which still *satisfies* the set requirements as to the quality and speed of task execution. The package also enables the user to select the appropriate equipment (microprocessors, sensors, etc.) which are required for implementation of the synthesized control (see Chap. 5).

4.4 Effects of Payload Variation and Notion of Adaptive Control of Robots

Up to now we have assumed that all parameters of the robot are constant and precisely known in advance. This assumption is valid for the majority of the robot parameters, such as lengths, masses and moments of inertia of the links, and for the majority of the actuator parameters. However, some parameters in robotic systems are not sufficiently precisely defined in advance and can vary during task execution. Such are the coefficients of viscous and dry friction in ethe actuators and joints, backlash and some actuator coefficients. Usually determination of these parameters has to be done experimentally, so their specification during the robot design is difficult. During the operation of the robot these parameters vary, but rather *slowly*. Often they do not affect the robot performance if the values of parameters stay within the limits allowed (which can be ensured by good design and precise details of all mechanical parts of the robot, etc.). Thus, it is often unnecessary to include in the control law any compensation for variations of these parameters.

However, there are some parameters of the robotic system which vary significantly and relatively fast, which have considerable effect upon robot performance. Such parameters are masses, dimensions and moments of inertia of the *payload* which is carried by the robot. It is clear that the robot must behave uniformly when it moves without a payload and when it carries the payload. When the robot carries the payload the equivalent mass and moments of inertia of the last link (gripper) of the robot change, which causes variation in the performance of the servosystems of the robot. The dynamics of the payload cause changes in the dynamic moments around the robot joints.

The presence of the payload causes the change of the moments of inertia of the mechanism around the axes of the robot joints. In Sect. 4.2 we have considered the influence of the variation of the moment of inertia of the mechanism upon the servosystem performance. We have seen that in selection of velocity feedback gain we have to take into account the maximum value of the moment of inertia of the mechanism around the joint in order to prevent overshoots. When the gripper of the robot takes the payload, the moments of inertia of the mechanism around the joint axis increase. If during synthesis of the servosystem we have not taken into account the mass and the moment of inertia of the payload the servosystems can become underdamped which causes oscillations in the robotic system. Due to this, it is necessary to synthesize the servosystem gains taking into account the maximum payload mass and the moments of inertia which can be carried by the robot. However, if a relatively large variation of the payload is assumed, this can cause uneven performance of the servosystems as for example when the gripper is empty, or when it carries a payload lighter than the maximum assumed payload, the servosystem can become very overdamped ($\xi_i \gg 1$) with a resulting slow response.

This problem can be solved by introducing the variable velocity feedback gain. Depending on the moment of inertia of the mechanism, we have to change the velocity feedback gain. Such *adaptive* control ensures that for each payload (and for the "empty" robot), the performance of the servosystems will be nearly equal. However, such control requires the parameters of the payload to be known. In some cases the parameters of the payload are known in advance and can be set through the language for robot programming or by some other means.

However, generally, the parameters of the payload are not known, so that if the control has to adapt to variation of the payload parameters it is necessary to ensure *identification* of these parameters. Because of this, the control system has to include some algorithm for identification of the payload and based on the identified parameters of the payload the servosystem gains are adapted. Identification of the payload parameters can be realized in various ways. The most efficient approach is by direct measuring of the forces in the contact points between the gripper and the payload. In these contact points force transducers are implemented which give direct information on the dynamics of the payload, so the payload parameters can be easily identified [5].

Adaptive control can be introduced in various ways. However, if variation of the payload with respect to the parameters of the robot links and actuators is relatively small (for example if the permitted payload masses are relatively small with respect to the mass of the links and moments of inertia of the actuators and reductors), then it is not often necessary to introduce adaptive control. It can be assumed that servosystems synthesized with constant feedback gains are sufficiently robust to withstand variations of the payload parameters. This assumption is valid for the majority of the robots on the market today for which *non-adaptive* (*robust*) control is applied. However, new robots are appearing which can carry the payload with much larger masses than the masses of the links. Obviously for such robots the influence of payload variations can cause uneven performance of the servosystems, making it necessary to apply adaptive control (or the control with the variable gains). However, with such robots *elastic effects* of the links usually appear (since the links are of relatively small masses with respect to payload mass). In previous considerations, we have assumed that all links of the robot are *rigid*. However, if the masses of the links are small with respect to mass of the payload the elastic effects can have significant effect upon the system performance. The control system in these cases must take care not only about variation of the payload parameters, but also about the elastic effects, which can complicate the control law and its implementation.

4.5 Control of Robots in Assembly Tasks

One of the most complex tasks in robotics is the so-called task of *assembling* mechanical parts by robots. In this task one or several robots have to assemble two or more elements. The specificity and complexity of this task lies in the phase of part mating, when the part carried by the robot has to come into contact with another part, which can be fixed or carried by another robot. As we have explained in Chap. 1 (see Figs. 1.21, 1.25), in this phase variation of the robot structure appears. The robot changes its structure from the open chain to the closed kinematic chain and vice versa depending on whether the parts are in contact, or not. In the moment when there is contact between the parts (or contact between the robot and its environment), the external forces are acting upon the robot. The control of robots has to take into account all these external forces (the reaction forces), which complicates the control law.

The assembly process by robots consists of several phases: first the robot has to approach the payload, the robot takes the payload, carries it towards the place where the contact is made, then the parts are assembled etc. (Figs. 1.23–1.25). As we have already emphasized, the most delicate phase is the one in which the parts come into contact, such as putting the peg into the hole. The control task in the "peg into hole" process can in principle be reduced to the precise positioning of the robot, i.e. to precise placing of the payload in the hole without the contact between the payload and the hole. However, this can be implemented only if all the dimensions and positions of all the elements and if the positioning (or tracking of the defined trajectories) of the robot are precise. In practice there always appear some contacts between the peg and the hole, and reaction forces appear. These reaction forces have to be minimized. This often happens if tolerances between the dimensions of the peg and the hole are small. The control system has to minimize the reaction forces between the payload and the hole.

This task is solved in several ways. One of the approaches most often applied in practice is by the so-called *passive compliance*. A special system of springs is installed on the gripper which enables both easy displacement of the payload in the plane normal to the axis of the hole and rotation around the corresponding axes (see Sect. 7.2.1, Fig. 7.12). In this way the position of the payload is "complied" with the axis of the hole, and the peg is easily introduced into the hole [7]. The advantage of this solution lies in the fact that complexity of the control system is not increased as no additional feedback loops nor additional computation are required. However, the drawback is that passive compliance cannot be applied for various payloads and holes (with various dimensions, shapes etc.), but only for each particular task can specific compliance be applied. Since the basic characteristic of robots is their *reprogrammability* (i.e. their ability to transfer from one task to another simply and quickly, or their ability to be moved from one working place to another for a very short period of time), any application of such passive compliance reduces the robots' capabilities making them similar to machines which can only be applied for one very narrow set of tasks. Usually the solution is to develop several different compliances for the same robot. The robot has to change compliance when the task (or the payload) is changed.

The second solution for robot control in an assembly task is by *force feedback*. The force transducers are installed on the robot gripper which give information on the reaction forces between the payload and hole. Based on this information the executive control level can generate signals for the actuators of the gripper so that the gripper moves towards reduction of the reaction forces. In this way by force feedback active compliance of the gripper (and the payload) is achieved.

Various solutions of the force feedback control laws are possible both from the point of view of the force transducers' positions on the gripper and the algorithm for control of the gripper on the basis of the measured forces (see Sect. 7.2). Force transducers can be installed in various places on the gripper. Usually they are placed between the last link of the robot and the gripper (Fig. 7.9), and from there all reaction forces and moments due to reaction forces are measured [8]. However, these transducers also measure the inertia forces due to the gripper motion, so we do not obtain "pure" information on the reaction forces. Thus, it is more convenient to place the force transducers directly in the contact points between the gripper and the payload. However, the drawback of this solution is that the measured reaction forces are significantly less than in the previous case, and this requires very precise and sensitive transducers [9].

Various solutions have been put forward concerning the algorithm for control of the gripper to achieve active compliance of the payload. It is possible to implement direct feedback from the force transducers to the actuator inputs, or to calculate the necessary displacement of the gripper depending on the measured forces. In this way the reprogrammable compliance of the gripper for various payloads and holes can be achieved (i.e. the same robot can easily be applied to the various assembly tasks). However, execution speed of the robot with force feedback is considerably less than if passive compliance is applied. Finally, passive compliances are much more precise and allow less tolerances (less difference between the dimensions of the payload and the hole).

It should be mentioned that the problem of assembling by robots can be solved in other ways too (by visual feedback, or by applying some other types of sensors, etc.). One of the problems concerning assembly by robots is path planning, which can be solved either automatically at the strategical control level, or by the operator of the robot (by programming the language of the robot, or by the teaching-box, etc. see Chap. 6).

References

[1] Popov, E.P., Veschagin, F.A., Zenkevich, S.L., Manipulation Robots: Dynamics and Algorithms (in Russian). In series Scientific Fundamentals of Robotics, Nauka, Moscow, 1978.

[2] Chestnut, H., Mayer, R.W., Servomechanisms and Regulating System Design. John Wiley and Sons, Inc., New York, 1963.

[3] Paul, P., Robot Manipulators: Mathematics, Programming, and Control, The MIT Press, 1981.

[4] Athans, M., Falb, P.I., Optimal Control: An Introduction to the Theory and its Application, McGraw-Hill, New York, 1966.

[5] Vukobratović, K.M., Stokic, M.D., Control of Manipulation Robots: Theory and Application, Monograph. Springer-Verlag, Berlin, 1982.

[6] Vukobratović, K.M., Stokic, M.D., Kircanski, M.N., Non-adaptive and Adaptive Control of Manipulation Robots. Monograph, Springer-Verlag, Berlin, 1985.

[7] Drake, S.H., "The Use of Compliance in a Robot Assembly System", IFAC Symp. on Information and Control Problems in Manufacturing Technology, Tokyo, 1977.

[8] Nevins, L.J., Whitney, D.F., "Assembly Research", Automatica, 16, pp. 595–613, 1980.

[9] Stokic, M.D., Vukobratović, K.M., Hristic, S.D., "Implementation of Force Feedback in Manipulation Robots", International Journal of Robotic Research, No. 4, 1985.

Chapter 5

Microprocessor Implementation of Control Algorithms

5.1 Introduction

This chapter is devoted to microcomputer systems suitable for controllers of industrial manipulators. As mentioned, a robotic controller represents a single or multi-processor system aimed at driving a robot arm to move in accordance with a user-written program. Robotic controllers are much more complex in their hardware and software structure than most industrial regulators and programmable automata for numerically controlled (NC) machines. The industrial regulators commonly represent PID single-input single-output regulators designed through the use of digital or analog integrated circuits or microcomputers. Some advanced industrial controllers represent self-tuning regulators [1] and automatically adjust parameters (proportional, differential and integral gains) to the parameters of the controlled process. Such a multi-channel PID regulator corresponds to the executive level of a robotic controller (see Chap. 4). However, the executive level of an advanced robotic controller can be much more complex than a set of PID regulators when designed to accomplish trajectory following control (in contrast to point-to-point control). The executive module must compensate for dynamic effects of the mechanical arm. This compensation is usually more complex numerically and imposes the need for fast microprocessors. Robot control systems are usually much more complex than programmable automata, which were used earlier to control simple pick-and-place manipulators. The difference is remarkable in numerical complexity both at a tactical and strategical level. Inverse kinematics and singularities-avoiding algorithms are solved at tactical control level. These problems are much more complicated than the corresponding ones with NC machines. The difference is even more noticeable at a strategical control level. Programmable industrial controllers use relatively simple languages while up-to-date robotic controllers use sophisticated high-level robot languages (see Chap. 6). And, trajectory planning, sensor data processing and other algorithms must be implemented at the strategic level of robot controllers.

Robot controllers communicate with other computers and various terminal devices such as video-display terminals and teaching units (Fig. 5.1). Communication between a robot controller and other computers is usually designed using

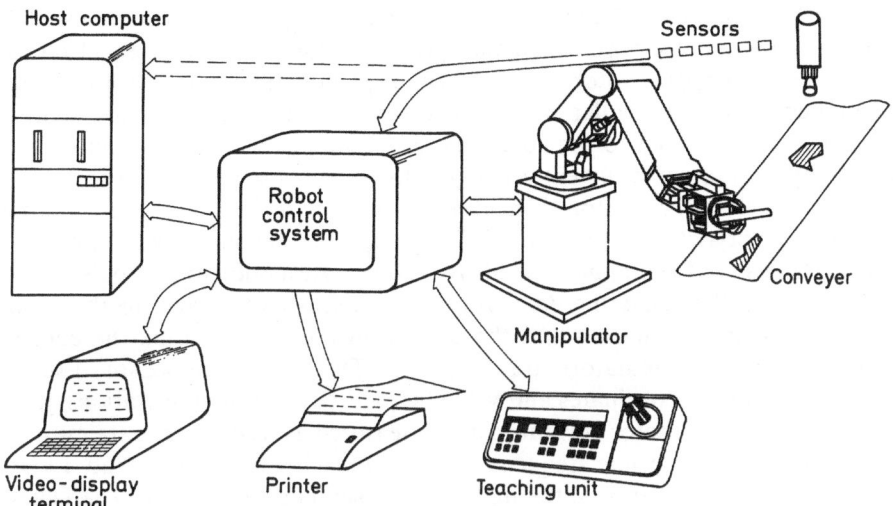

Fig. 5.1. Robot control system

standard serial or parallel interfaces. The purpose of these computers can be very different. For example, a host computer can be used for synchronizing several robots on a production line. A computer can also be used to load particular robot programs to corresponding local robot controllers, activate them or interrupt them according to information coming from the ambient. The host computer is also used for on-line coordination between robots and for optimization of production time. Nevertheless, a host computer is sometimes used even in the case of single robot system. Then it is used for processing visual or other complex information. From the hardware point of view host computers are mostly either special bit-slice processors or even vector (matrix) processors.

Except for classical general-purpose video-display terminals, dedicated "teaching units" are very often used. By combining the functions both of teaching units and robot languages (see Chapter 6); even very complex robot operations can be easily described. Thus, by pressing only one key, the manipulator can move along an operational space coordinate. These functional keys are very useful for describing simple robot operations. Teaching units are often provided with alpha-numeric displays that make various messages to be followed by the operator possible.

The "robot-pilot" is also used during the learning process. It has the same kinematic structure and joint sensors as the manipulator itself. It is not equipped with actuators. While the operator moves the pilot in accordance with a prespecified task, the controller reads the joint angles from the position sensors, optimizes their number and stores them into memory. Thus, the recorded motion of the pilot can be reproduced by the robot itself. Such a learning process is very useful for painting and arc-welding tasks.

The basic function of a controller is to generate appropriate driving signals for the actuators of a robot. As mentioned, most robot actuators are electrical (D.C., A.C. or stepper motors), hydraulical or pneumatical. The control signals, computed by the controller processor, must be converted to analog signals, amplified and adapted to the actuators. For example, hydraulic actuators are driven by a servo-valve current, usually in the range of several mA. The driving currents of electrical motors can exceed even 100 A. Thus powerful output transistors or tyristors must be used. The stepper motors are driven by current pulses. One step (fixed angle of shaft rotation) corresponds to one pulse.

The controller reads the data from various sensors attached to the robot arm. Combining these data and the data from the tactical level, the control signals driving the actuators are computed. The robot arm sensors usually measure the joint coordinates and velocities, although joint accelerations, joint or wrist forces and motor currents can be measured as well. These data must be sampled with appropriate frequencies, commonly between 50 Hz and 1 kHz. The faster the robot arm is, the higher is the sampling frequency. The fastest robots (direct-drive arms with hand velocity over 10 m/s and accelerations over 5 G ($1 G = 9.81$ m/s^2)) require sampling frequency up to 1 kHz. Industrial robots are 5 to 10 times slower and are sampled with 50 to 100 Hz. We see that data flow between the sensors and the controller is intensive. As processing these data can overburden the central processing unit (cpu) of the controller, the faster robots require an additional processor with the following functions: sampling the data from sensors, normalizations, sensors fault detection, etc. The processed data are stored in the memory which is in common with the executive-level-processor memory.

Finally, the controller accepts the data from a variety of external sensors which measure the states of various devices and machine tools in the environment. For example, sensors which measure velocity of a conveyor, positions of moving machine parts and others, are scanned by the controller, thus synchronizing the pre-programmed motion of the robot with the actual events occurring in its environment. An additional processor for the external sensor signal processing may be needed and for image-processing an additional high-performance processor must be used.

5.2 Basic Subsystems of Modern Robot Controllers

Research and development in the field of new solid-state technologies and computer-aided design of large-scale integration circuits represent the basis for obtaining a suitable hardware for robotic controllers. High quality, reliability and speed of operation are standard characteristics which must satisfy an up-to-date robotic controller.

In order to satisfy these requirements Very Large-Scale Integration (VLSI) technology is used. The number of cells (transistors) per square-inch increases from year to year. Figure 5.2 shows the memory-cell area of dynamic memories and EPROMs (Electrically Programmable Read-Only Memories) from 1970 TO 1990 [2]. The basic element of any modern robotic controller is the microprocessor in which processing power directly determines the performance of the entire system. The development of microcomputers started in the early

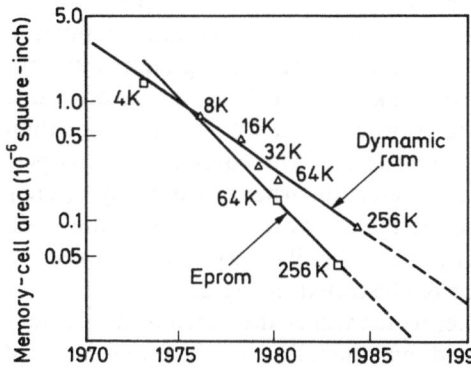

Fig. 5.2. Memory-cell area from 1970 to 1990

1970s with the occurrence of 4-bit microprocessors. Later, 8-bit processors appeared on the market. They are still used in robotic controllers, especially to drive simple industrial pick-and-place manipulators. The main deficiency of these processors is their low speed of operation (arithmetical and logical operations). As the data sampled from various transducers (see Chap. 7) are often 12- or 16-bit data, the corresponding arithmetics must be at least 16-bit which demonstrates that 16-bit microcomputers are more suitable for robotic controllers. These microprocessors include several 16-bit registers, 16-bit arithmetic-logic units, interface to memory and input/output (I/O) devices, interrupt controllers, programmable counters, direct-memory-access controllers for high speed data transfer between memory and I/O devices and memory to memory. An example of a microprocessor (INTEL 80186) is given in Fig. 5.3 [3]. The

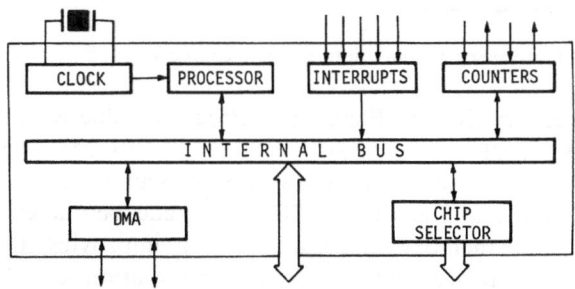

Fig. 5.3. Block-diagram of iAPX 186

processing power of one 16-bit microprocessor is not always enough for solving inverse kinematic problems (Chap. 2) in real time. The same holds for dynamic servoregulators (Chap. 4). These algorithms include several hundred arithmetic operations which must be repeated 10 to 1000 times per second. The fixed-point implementation of these algorithms is very tedious because of trigonometric functions, square roots and other operations that must be computed so often. Thus, modern robotic controllers use floating-point arithmetics that can not be realized by the use of 16-bit processors. In order to overcome this problem, a numeric data coprocessor is usually added to the main microprocessor. A coprocessor enhances the speed of computation from 10 to 100 times compared to the system with no coprocessor. For example, one floating-point multiplication takes 20 to 30 μs, one square root takes about 40 μs, a sine or cosine function about 100 μs, etc. Without the coprocessor it would take more than 10 ms on an average speed 16-bit microprocessor. The coprocessor also performs a series of control functions (fault detection when dividing by 0, etc.). A special advantage of the coprocessor is that it works parallel to the main microprocessor. This saves computing time significantly. The connection between the processor and coprocessor is illustrated in Fig. 5.4.

The next module of a robot controller which will be described is the memory unit. Various types of memories are used, but the following three are the most

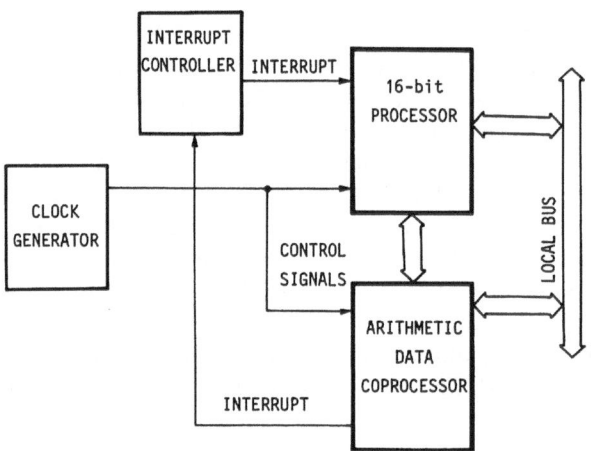

Fig. 5.4. Connection between a 16-bit processor and an arithmetic data coprocessor

common: Random Access Memory (RAM), Electrically Programmable Read-Only Memory (EPROM) and Electrically Erasable PROM (EEPROM). The development of dynamic RAM memories in Metal Oxide Semiconductor (MOS) LSI technology originated in the early 1970s and made permanent progress not only in decreasing the memory cell area, but also in improving the reliability, temperature range and robustness with regard to radiation, etc. A typical capacity of the RAM memories used within robot controllers is in the range of 64 Kbyte to 1 Mbyte.

Read-Only Memories (ROM) are not very often used within robotic controllers, because these memories are not reprogrammable. EPROMs are commonly used because their content can be erased after exposing them to ultraviolet radiation for 10–15 minutes. A new content can then be put into the memory. Even these memories are often replaced by EEPROMs, which are electrically erasable and reprogrammable. The content of these EEPROMs can be exchanged more than 10,000 times, without loss of reliability of data. These memories are developed by using a new transistor technology (floating-gate transistors) which are loaded by the Flower–Nordheim tunnel-effect through an 200 Å oxide layer [2]. Up-to-date EPROMs are produced as up to 64 Kbyte chips, and EEPROMs as up to 8 KByte chips. The connection between a memory module, I/O module and a processor module is shown in Fig. 5.5. The development of memory media is very important for robot controllers, especially for the robots intended for carrying out long spatial movements (painting, arc welding, etc.). For example, in the case when the joint coordinates are memorized with the frequency of 10 Hz, then a 10-minute task consumes about 100 KByte memory space. If we wish to store several memory-consuming tasks, we obviously need external magnetic memories, such as flexible and fixed disks.

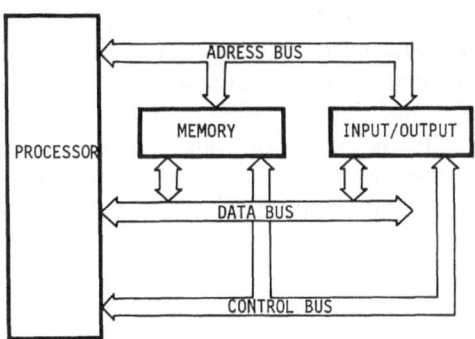

Fig. 5.5. Processor, memory and I/O module interconnection

Flexible disks (floppy disks) are circular plastic folia covered with a magnetic layer. As their capacity exceeds 1 MByte, they are very useful for storing large robot program libraries. The basic disadvantage of floppy disks is their sensitivity to industrial environment (dust). Fixed (Winchester) disks have a much larger capacity (usually from 5 to 50 MByte). A Winchester disk unit contains several circular aluminium plates covered by a very thin magnetic layer. Read and write operations are carried out by the use of a special R/W head. The distance between the head and the aluminium plate is only about several μm. These disks are useful for storing system and application programs, data base files, etc. Finally, let us consider the following question: Is it possible to implement kinematics and dynamics of a 6-joint robot arm on a single-processor, low-cost 16-bit, general purpose up-to-date microcomputer?

In order to solve the inverse kinematic problem of a typical 6-joint manipulator, it is necessary to carry out about 100 multiplications and 100 additions. The computing time for the entire kinematic module is in the range of 20 to 40 ms depending on the speed of 16-bit microcomputer. The discrete servoregulators implemented within the dynamic module (we shall consider this in the next section) require an additional 10 to 20 ms. The sampling and scaling process of various transducer signals requires further time. We see that all of these modules require from 40 to 100 ms when implemented on 1 cpu. As the sampling time for a typical industrial manipulator must not exceed 20 ms, it follows that we have to use at least two parallel processing units. Also, taking into account the strategical control level (with a robot language interpreter/compiler, etc.), it becomes clear that it is more suitable to have three parallel processor boards.

The hardware structure of a 3-processor control system is presented in Fig. 5.6. Processor modules cooperate through the common memory unit and the multibus. The connection between the controller and the actuators (D.C. motors, hydraulic and pneumatic actuators and others) is realized through digital-to-analog (D/A) convertors. 12-bit convertors are commonly used because it is often enough to have 4096 discrete levels of analog signals.

Analog-to-digital (A/D) convertors are used for sampling analog signals (potentiometers, tachogenerators, force transducers, etc.). Special counters and registers are used for reading optical encoders which measure angular or linear displacements of robot joints. These transducers are digital high-resolution devices, often with 16-bit accuracy. Finally, the robot controller includes a I/O module with a large number of parallel channels. These channels are digital (level 0 or 1) and are used to read out digital sensors and to control various digital tools such as hydraulic presses.

5.3 Program Modules

We described in the foregoing section several basic hardware modules of an up-to-date robotic controller. In the following text we shall concentrate our attention on various software modules, kinematic algorithms and the servo-system implementation. Communication between these modules will be considered, too.

5.3.1 User-interface

The overall system control (manipulator, sensors, devices in robot environment) is performed at the strategical level of the controller. The most important part of

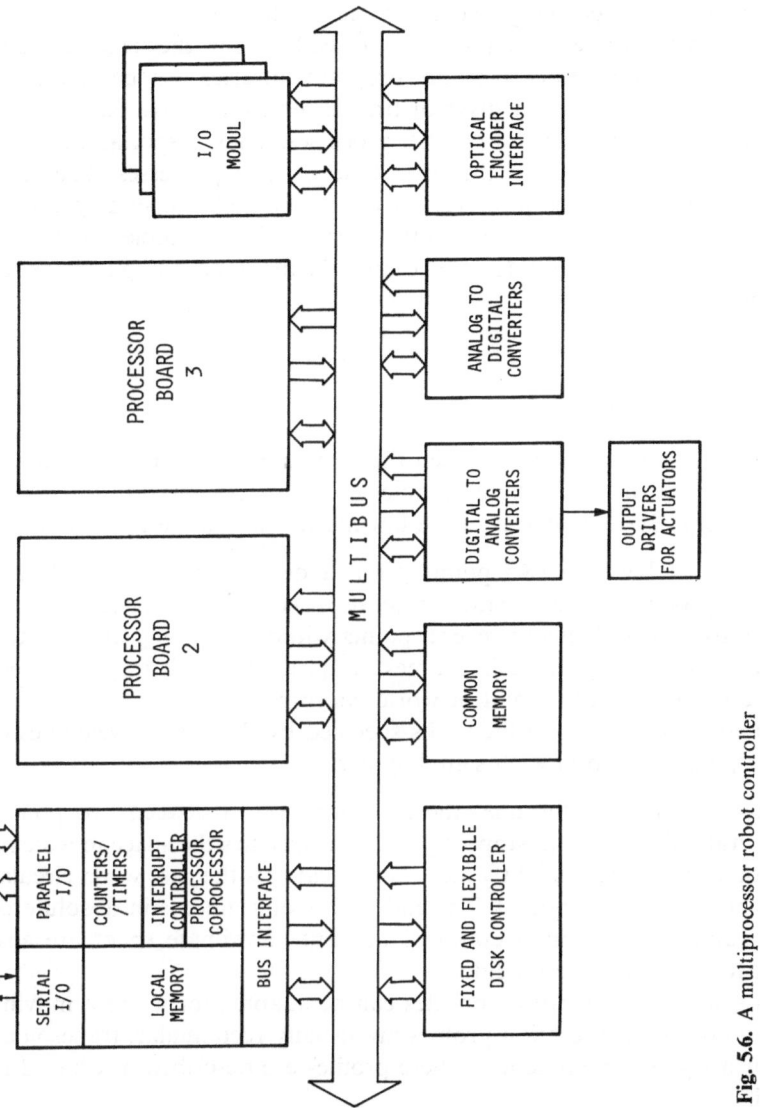

Fig. 5.6. A multiprocessor robot controller

the strategical level is the user-interface. This module is located on the first processor (Fig. 5.6). It has several functions. First, it controls robot motion by sending the commands of motion to the kinematic module. Then, it carries out signal processing by scanning various external sensors. It then makes communication possible between peripheral machines and devices located in robot environment or attached on the robot arm (hand). The user terminal, front panel, teaching unit, etc. are also connected to this module and finally, it makes the communication between other computers possible. The basic part of the user-interface is the robot-language interpreter. The robot commands are obtained from user-written programs, or directly from the teaching unit throughout the teaching process. Robot-program libraries can be located in EPROMs, Winchester or floppy disks or transferred from a host computer.

User interface controls several parallel processes. Thus, a special operating system must be available to synchronize these parallel processes. The complexity of this module varies from very low to very high, depending on the desired sophistication of the robot control system. For example, a painting robot needs a very simple user interface because the main job is to play back the recorded robot motion.

5.3.2 Kinematic Module

Let us consider communication between the user interface and the kinematic module. We pointed out that these modules exchange information through a common memory. Basic functions of the kinematic module are the following.

1. Linear motion between two points given a desired time or speed. The motion may be interpolated either in joint or world coordinates.
2. Continuous motion between several points without stopping at the intermediate points, given the desired times or speeds for each trajectory segment, performed either in joint or world coordinates.
3. Manipulator motion along the paths specified by the desired velocities of the manipulator hand at each sampling period.

Thus, the user interface determines the type of motion (between two points, between several points without stopping, etc.) and specifies the space in which a linear motion is performed. Slightly more complicated is the case when circular motion of robot hand is required. The end points of segments in absolute or relative coordinates either in operational or joint coordinate space are also transmitted to the kinematic module.

Various velocity (acceleration) profiles can be adapted for a robot moving between two points. Acceleration profiles are usually rectangular, trapesoidal, parabolic or a square sine function. These profiles are described in Chap. 2 in more detail.

As opposed to robot motion between two points, continuous motion between several points can be implemented without stopping at intermediate points. This type of motion is useful in arc-welding, obstacle avoiding tasks, etc. In the neighbourhood of intermediate points, the velocity changes in accordance with some polynomials (e.g. third-order parabolic functions) which protect the manipulator from abrupt changes ("jerk") in acceleration and velocity. Of course, some positional error between the actual position of the robot arm and the position of intermediate points is allowed. Thus, the robot hand moves near the intermediate points, as shown in Fig. 5.7.

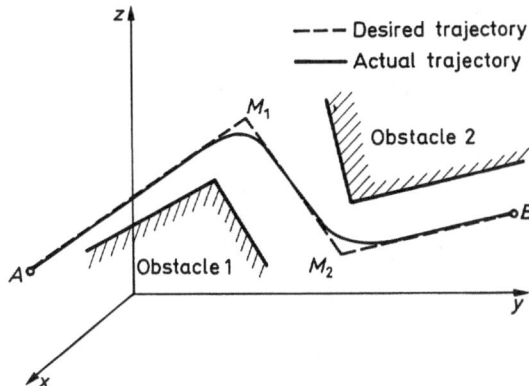

Fig. 5.7. Motion between several points without stopping at intermediate points

Manipulator motion along the paths specified by desired velocities of the manipulator hand is a very useful type of motion during the "learning" process. Then, the user interface transmits the desired velocities either in joint or operational space coordinates to the kinematic module at each sampling period. The operational space coordinates can be defined both by an inertial frame or a frame attached to the manipulator hand. In the second case the hand moves along the attached frame axes, or about them. These commands are relative to the instantaneous position and orientation of the hand. During the "learning" process the user forces the hand to move in such a way by activating a joystick or dedicated keys on a teaching unit. An example of a dedicated keyboard is given in Fig. 5.8.

During the robot motion the higher control level accepts information about the manipulator position both in joint and operational space coordinates. Thus, the kinematic module is solving the direct and inverse problem during the robot motion. This makes replanning the task possible during execution if necessary.

The number of numerical operations (additions, subtractions, multiplications, divisions, square roots, sines, cosines, etc.) represents the essential parameters of an algorithm (numerical complexity). Numerical efficiency of robot kinematics and dynamics depends on the number of joints and relative disposition of the joint axes, etc. We pointed out in Sect. 5.2 that the floating-point arithmetics, including some trigonometric calculations and square roots, are carried out within specialized mathematical coprocessors.

In order to have a clear insight into the numerical complexity of robot kinematics, let us consider two typical industrial robots: The first, shown in Fig. 2.1 (p. 19), has three degrees of freedom, while the second is a six degree-of-freedom robot (Fig. 5.9). Both of them have kinematic chains for which the inverse kinematic solution in analytical form can be found. The number of numerical operations for both robots are given in Table 5.1 (see p. 173). We see

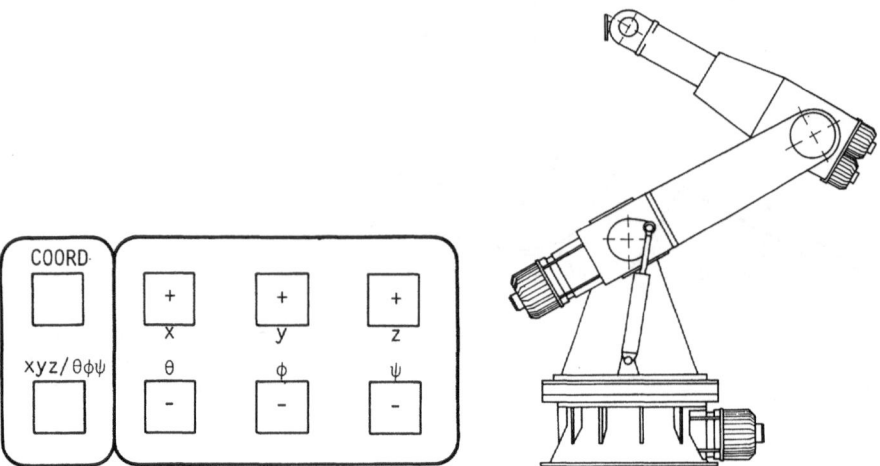

Fig. 5.8. A segment of the teaching unit keyboard (COORD-choice of coordinates, $+/-$ positive or negative direction of motion)

Fig. 5.9. Arthropoidal six-degree-of-freedom robot

that the complexity is practically the same for the direct and inverse kinematic problem. Either the direct or inverse kinematic problem for this six-degree-of-freedom robot can be solved within 5 to 10 ms by the use of an up-to-date 16-bit microcomputer. Although the computing time for the overall kinematic module is nearly double because a series of other operations must be performed as well. These are for example recognition of commands coming from the upper level, transmitting the commands to the lower control level (dynamic servoregulators), checking whether the commanded points are reachable by the manipulator hand, checking whether the commanded accelerations are too high, avoiding singular points, etc.

5.3.3 Dynamic Module

The purpose of dynamic servoregulators is to form appropriate control signals for the actuators which drive manipulator links. This signals should force the manipulator hand to move along prescribed desired trajectories. In the previous section we described the process of computation of the joint coordinates. Kinematic module sends in each sampling period the data about desired joint coordinates (q_i^0), velocities ($\dot{q}_i^0$) and eventually accelerations ($\ddot{q}_i^0$) to the dynamic module through the common memory (Fig. 5.6). Here, q_i^0 and $\dot{q}_i^0$ represent the nominal desired values of the coordinate and velocity of the joint i. Actual values of q_i and $\dot{q}_i$ are obtained from a position/angle transducer (potentiometer, shaft-encoder, etc.) and velocity transducer (tachgenerator) attached at each joint connection. Actual and desired joint coordinates and velocities differ from each other, because of different servoregulators characteristics (static and dynamic

		+/-	*	/	√	sin/ cos	atan
	DIRECT PROBLEM	14	16	9	0	8	3
	INVERSE PROBLEM	14	23	2	3	6	3
	DIRECT PROBLEM	47	62	9	0	14	3
	INVERSE PROBLEM	38	69	4	3	8	6

Table 5.1. Numerical complexity of the direct and inverse kinematic problem

errors, damping, etc.). The quality of tracking depends on the computational complexity of the dynamic module. If we use simple servosystems with constant positional and velocity feedback gains, the tracking quality might not satisfy our requirements because they do not take into account non-linear dynamic effects of the robot arm. An example of such regulators is described in Chap. 4 (see PID regulators, Eq. (4.28)). Compensation of the dynamic effects (variable inertias, centrifugal and gravity forces, etc.) may ensure a much better tracking quality, but it makes control laws more complicated. As it is often necessary to compensate the non-linear effects, microprocessors are used at the executive level.

A digital servosystem with a feedforward compensation is shown in Fig. 5.10. The feedforward block is aimed at forming the nominal control according to an approximate or even exact model of manipulator dynamics. The term "nominal control" is explained in Chap. 4 in detail. The other part of the control system shown in Fig. 5.10 represents discrete PID regulators. Each degree of freedom is accomplished with one PID regulator. The feedforward block can only com-

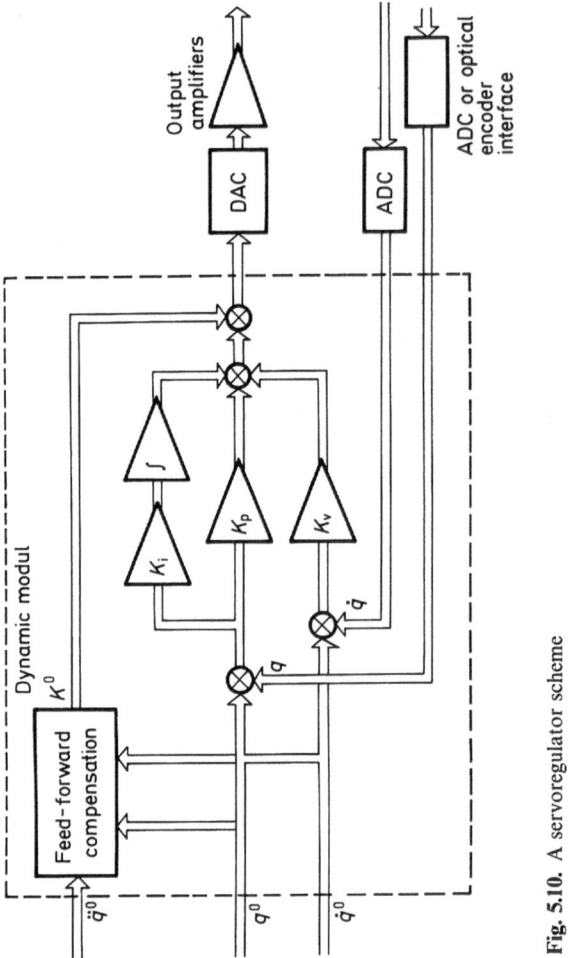

Fig. 5.10. A servoregulator scheme

pensate for the acceleration of each joint without taking into account the dynamic interactions between segments. But, if we want to compensate at least the gravitational effects, we have to take into account the entire system behaviour because the motion of one segment often affects the gravitational load of the other. The same holds for inertial, centrifugal and other effects. Although we want to compensate all dynamic effects, the computational burden can be very high. Thus, a compromise between numerical complexity and tracking quality must be found during design of a robot controller. This will be illustrated in the text to follow.

Consider a three degree-of-freedom manipulator shown in Fig. 2.11 (p. 30). We suppose that for some defined tasks it is enough to take into account the inertial and gravitational effects. By neglecting the Coriolis and centrifugal forces, the dynamic model of this robot reduces to

$$P_1^0 = H_{11}^0 \ddot{q}_1^0$$
$$P_2^0 = H_{22}^0 \ddot{q}_2^0 + H_{23}^0 \ddot{q}_3^0 + G_2^0$$
$$P_3^0 = H_{33}^0 \ddot{q}_3^0 + H_{23}^0 \ddot{q}_2^0 + G_3^0$$

with

$$H_{11}^0 = C_0 + [C_1 \sin q_2^0 + C_2 \sin(q_2^0 + q_3^0)] \sin q_2^0 + C_3 \sin^2(q_2^0 + q_3^0)$$
$$H_{22}^0 = C_4 + C_2 \cos q_3^0$$
$$H_{23}^0 = C_3 + C_5 \cos q_3^0$$
$$H_{33}^0 = C_3$$
$$G_2^0 = -C_6 \sin q_2^0 - C_7 \sin(q_2^0 + q_3^0)$$
$$G_3^0 = -C_7 \sin(q_2^0 + q_3^0)$$

Here, C_i represents the constants that depend on the geometry of the links, masses and moments of interia. From the model of the actuators (Eq. (4.3)), we get the nominal control in the form

$$u_i^0 = (J_R^i N_V^i N_m^i \ddot{q}_i^0 + (B_C^i + C_E^{i'}) \dot{q}_i^0 + P_i^0)/C_M^{i'}$$

where J_R^i, B_C^i, $C_E^{i'}$, $C_M^{i'}$ are characteristic parameters of the D.C. motor, while N_V^i and N_m^i represent the parameters of the reducer. By introducing the notation $d_a^i = J_R^i N_V^i N_m^i / C_M^{i'}$, $d_v^i = (B_C^i + C_E^{i'})/C_M^{i'}$ and $d_p^i = 1/C_M^{i'}$, we get

$$u_i^0 = d_a^i \ddot{q}_i^0 + d_v^i \dot{q}_i^0 + d_p^i P_i^0$$

This example shows that the computation for nominal control requires 23 multiplications and 16 additions/subtractions. If we took into account the Coriolis and centrifugal forces, we would get about 50 multiplications and 50 additions. For a six-degree-of-freedom robot these numbers are in the range of 100 to 1000. Thus, it is very useful to carry out an analysis of the influence of particular dynamic effects on the overall system behaviour, and to take into account only the effects that are necessary to achieve the desired tracking quality.

Let us now consider the part of the control scheme shown in Fig. 5.10 referring to PID regulators. As shown in Chap. 4, continuous-time PID regulators can be described by

$$\Delta u_i = -k_p^i(q_i - q_i^0) - k_v^i(\dot{q}_i - \dot{q}_i^0) + k_i^i \int_0^t (q_i - q_i^0) dt$$

Let us introduce the function

$$\varepsilon_i^l(t) = \int_0^t (q_i - q_i^0) dt$$

which describes the integral action of the regulator. The discrete time approximation of the integral functions is

$$\varepsilon_i^l(t_k) = \varepsilon_i^l(t_{k-1}) + [q_i(t_{k-1}) - q_i^0(t_{k-1})] T$$

with $t_k = kT$, $k = 1, 2$. Here, T is a sampling period much less than the time constants of the manipulator dynamics. Usually, T is between 1 and 20 ms. The discrete PID regulator can be presented in the form

$$\Delta u_i(t_k) = -k_p^i(q_i(t_k) - q_i^0(t_k)) - k_v^i(\dot{q}_i(t_k) - \dot{q}_i^0(t_k)) + k_i^i \varepsilon_i^l(t_k)$$

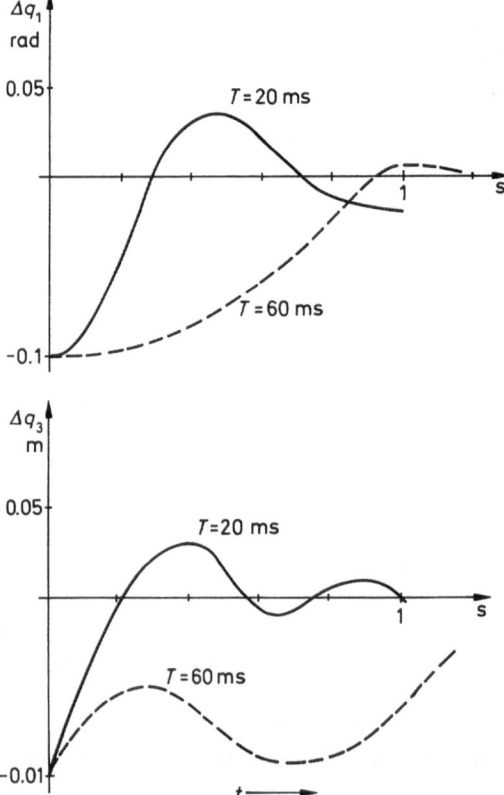

Fig. 5.11. Deviations from nominal trajectories

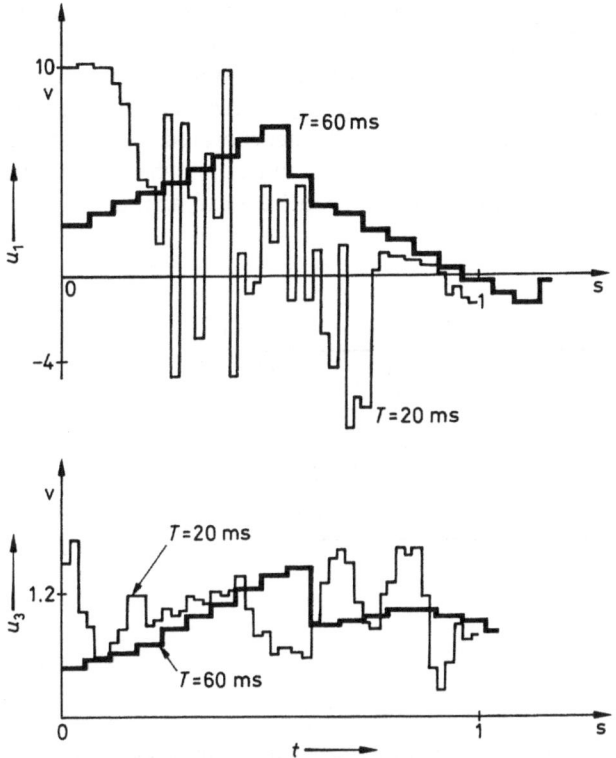

Fig. 5.12. Control signals during the robot movement

We see that in each sampling period 4 n multiplications and 5 n additions must be carried out. Here, n is the number of joints. It should be emphasized that in each period T actual joint coordinates and velocities must be sampled. The time interval T of course differs from the computational period of the kinematic module and is usually several times shorter. The value of the sampling period can influence the tracking quality of prescribed trajectories a great deal. In order to illustrate this, let us consider the experimental results obtained for the cylindrical robot shown in Fig. 4.5. The first experiment is carried out with the sampling time $T=20$ ms, while the second one with $T=60$ ms. In both cases, the PD regulator gains are computed for the same requirements on the tracking quality. The obtained feedback gains are given in Table 5.2. The deviations from the desired trajectories for the first and the third degree of freedom are shown in Fig. 5.11. The behaviour of the other degrees of freedom is practically the same as shown in Fig. 5.11. We see that the tracking quality is much better with $T=20$ ms than with $T=60$ ms. With $T>60$ ms the system is practically unstable, while for $T<20$ ms the behaviour of the system is nearly the same as with $T=20$ ms. Thus, we may conclude that $t=20$ ms represents an appropriate sampling period for this system. The control signals during the time evolution are shown in Fig. 5.12. We see that the control signals are constant during each sampling time, and change abruptly between them. Within the dynamic module except for the computation of control signals, a series of fault-detection functions that ensure high system reliability must be incorporated. For example, if the system goes out of the velocity or acceleration limits, the dynamic module must recognize that and stop the manipulator motion. This can happen if a joint sensor failure occurs.

Table 5.2. Parameters of discrete PD regulators

	1	2	3	4, 5	T [ms]
Positional gains	20.07	356.3	355.4	1.30	20
	10.0	4.1	100.0	0.37	60
Velocity gains	1.53	7.5	10.4	—	20
	4.03	0.3	10.0	—	60

5.4   Conclusion

This chapter has described the basic hardware and software of up-to-date robot control system. Special care has been paid to the organization of hardware, its speed of computation and basic elements at the strategical, tactical and the executive control levels. It has shown that for a six-joint manipulator an up-to-date controller must be composed of at least three parallel general-purpose low-cost 16-bit central processing units. The algorithms implemented at the kinematic and servosystem level of control are described.

The development of microprocessor technique, 32-bit microcomputers and vector-processors make the application of very complex algorithms at each control level possible. Algorithms at the strategical level are essential. These are trajectory planning algorithms based on artificial intelligence and processing of complex visual information, etc.

References

[1] Ortega, R. and Kelly, R., "PID Self-Tuners: Some Theoretical and Practical Aspects", IEEE Trans. on Ind. Electr., Vol. IE-31, No. 4, 1984.
[2] Components Quality/Reliability Handbook, INTEL, Santa Clara, 1986.
[3] iAPX 86/88, 186/188 User's Manual – Programmer's Reference, INTEL, Santa Clara, 1986.

[4] Vukobratović, M. and Kirćanski, N., Real Time Dynamics of Manipulation Robots. In series: Scientific Fundamentals of Robotics, Springer-Verlag, Berlin, Heidelberg, New York, 1985.

[5] Vukobratović, M., Stokić, D. and Kirćanski, N., Non-adaptive and Adaptive Control of Manipulation Robots, Scientific Fundamentals of Robotics, Springer-Verlag, Berlin, Heidelberg, New York, 1985.

Chapter 6

Industrial Robot Programming Systems

6.1 Introduction

The key requirement when introducing robots into manufacturing systems is the possibility of teaching them quickly to perform the desired task. Hence one of the main factors influencing the effectiveness of robots is their programmability. For this reason, practically all existing robot systems are equipped with tools that facilitate, more or less efficiently, the programming of some specific sets of robotic tasks.

A typical configuration of a robot system is outlined in Fig. 6.1. It consists of a manipulation mechanism, robot controller, commanding panel, portable unit for manual control and teaching and, optionally, a video terminal for textual programming.

As a rule, the manipulator is designed as an active spatial mechanism with six revolute or prismatic joints that are servoed by D.C., hydraulic or pneumatic motors. The joints are equipped with internal sensors which usually return feedback information about their positions and velocities to the servo-system. The last segment of the manipulator carries an end effector in a form of a gripper, a welding gun, etc.

The most important part of the system is the robot controller. Its heart is a microprocessor which controls the manipulator motion, processes information obtained from the sensory system, and controls the operation of attached peripherals, e.g. CNC machines, conveyors etc. The operation of the controller is performed in accordance with an earlier prepared robot program, i.e. a sequence of instructions which describe individual movements and other operations that should be executed during the robot work. Therefore, the controller can be viewed as an interpreter of the robot program.

The panel is a special device with commanding keys and visual displays intended for monitoring and communication between the user and the system during accomplishment of the robot task. The portable manual control unit (often called teaching box or pendant) is a miniaturized control device which enables execution of individual primitive operations immediately after the operator's request. The operations could be moving the gripper to desired positions and orientations, or recording current positions of joints etc. Requests are usually imposed via functional keys or with a joystick. The imposed operations can be stored in an internal memory of the controller, so that the whole sequence of operations can be automatically reinterpreted later. Such a programming procedure is often denoted as robot teaching and is effective in a range of simple robot applications such as painting or spot welding.

In the case of a more complex task which, for example, is characterized by a need for intensive processing of the irregularities, trajectory optimization, or automatic avoidance of some obstacles in a working space, it is absolutely necessary for the robot programmer to have at his disposal more powerful tools, enabling either interactive or off-line programming in specialized robot languages. Preparation of robot programs in a robot programming language

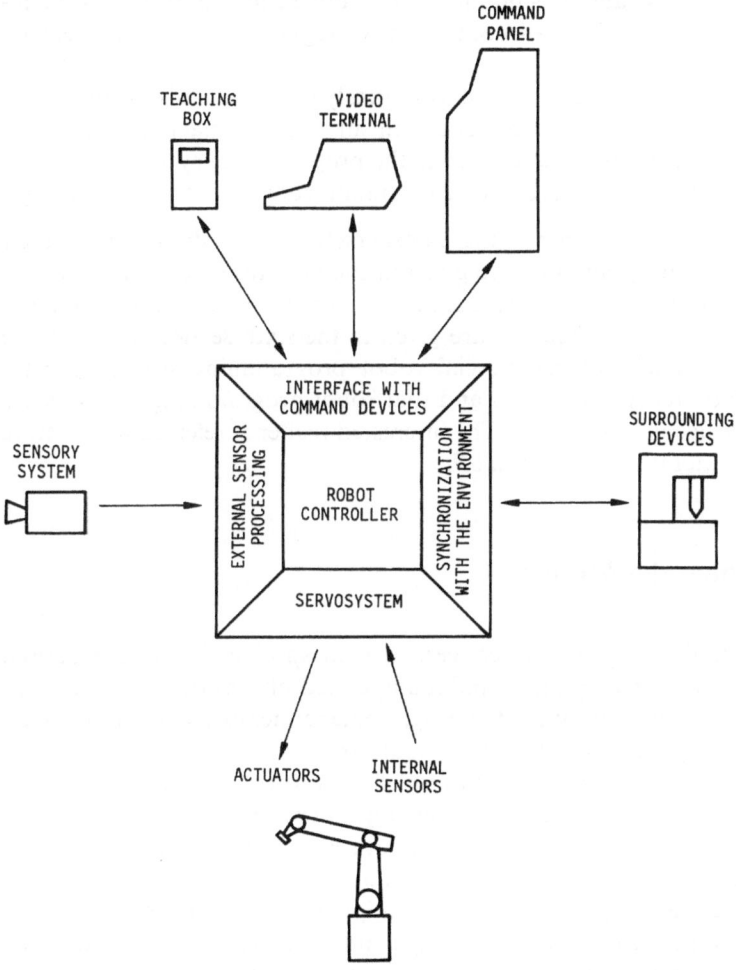

Fig. 6.1. A typical robot system configuration

can be achieved on a specially customized micro- or minicomputer, as well as a general purpose computer. It is also customary to have a robot programming system directly built into the robot controller.

The main reasons that make robot programming distinguishable from the programming of conventional numerical and similar tasks are the following:

1. Requirements for the effective description of manipulative work necessitates specific data structures representing the robot environment (basically, these are the structures which describe the positions of manipulated objects and the position of the manipulator itself), as well as the corresponding functions for handling such data;
2. The robot operation often takes place in an environment with numerous irregularities: inevitable positioning errors, dimension tolerances of manipulated objects, etc.; they all need intensive regular checking and sensory data processing;
3. The robot seldom performs its task completely autonomously; its operation must be almost always synchronized with surrounding machines, accessory devices and other robots, necessitating the programming system to support suitable synchronization tools which facilitate cooperative programming.

A more detailed description of these basic factors and a discussion of their influence on robot programming are given in the next three sections. The fifth section gives a short review of realizations of robot programming systems. Examples of some robot languages are given in the sixth section. It should be noted that performance of commercial robot programming systems are far below the already recognized needs and that robot programming methods are still under intensive investigations. The interested reader is referred to the more comprehensive literature in Section 6.7.

6.2 Describing the Motion

Robot programming is primarily concerned with specification of the desired space positions of the manipulator and manipulated objects during the robot's work. By writing a robot program, the programmer either explicitly or implicitly defines the space trajectories the robot should track in order to perform some useful job, such as changing the positions of some other objects in pick-and-place operations. A method that is used for trajectory specification depends to a great extent on the basic purpose of the robot system. In systems intended for simple manipulation, acceptable results are attained by specifying trajectories as sequences of the desired positions of the manipulator joints. In more complex situations, it is necessary that the programming system has an adequate mechanism for geometric data management, i.e. tools for space relationship data creation and manipulation.

Existing industrial robots are usually equipped with a positional control system that generates the actuating signals required to achieve the desired positions of the manipulator active joints. Therefore, from the point of the robot controller, it is natural to formulate the trajectory in the space of internal coordinates (joint angles), i.e. to define the trajectory as a sequence of manipulator configurations. However, this method can satisfy the programmer only in a limited number of situations for which there is a simple way to express all desired configurations and where the whole trajectory is fixed in advance.

Specification of the desired internal coordinates (or necessary changes in the coordinates) can be mainly accomplished by entering the corresponding numerical data into the system via the terminal keyboard. In an imagined robot language, the motion command could then have a form:

MOVE_JOINTS_TO $(q1, \ldots, qn)$

where $q1, \ldots, qn$ are selected positions of the joints.

Such direct entering of numerical data about the joint positions is often an unsatisfactory way of expressing the trajectory. It is sometimes very hard to imagine a position of the gripper which will correspond to given joint angles (unless the manipulator is of the cartesian or cylindrical type). For this reason, it is customary to select the internal coordinates in another way, using the robot itself: after the manipulator has been guided by the operator to a desired configuration, the configuration data can be read from robot internal sensors and stored in the controller memory. Configuration data are recorded either automatically in regular time intervals, either after the operator's explicit requests during the guiding. The method is usually called teaching by showing or teaching by guiding.

Teaching by showing methods are very popular because of their simplicity, but they also have significant disadvantages. First of all, the operator's capability of visual alignment is restricted, so that the guiding has to be combined with other methods whenever precise positioning and, especially, precise orienting of the gripper are requested. Second, the trajectory often includes a number of points whose space relationships are known in advance. Apparently, there are many more suitable ways of formulating such known relationships than to "show" all of them to the robot.

As an example, Fig. 6.2 shows a palletizing problem: parts conveyed by a moving belt should be placed by the manipulator in a pallet with 5×5 beds. For the sake of simplicity, suppose that the conveyor stops whenever a part comes in the fixed position PART where it will be picked by the manipulator. In order to ensure safe manipulation, the gripper first has to pass through the approaching position P1 and then to move to the gripping position H. After closing its fingers, the gripper carrying the part should lift to the position P1, move to position P2 behind the bed BED[i, j], put the part down, open the fingers again and finally retreat to the position P2.

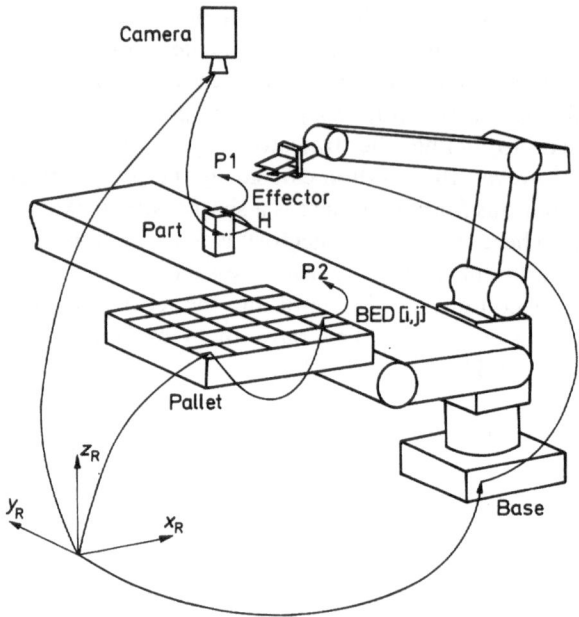

Fig. 6.2. Palletizing task

The whole procedure is repeated until the pallet is filled. Its programming by pure teaching by showing requires guiding the robot hand through a total of 150 positions. Obviously, it is desirable to reduce the number of configurations that have to be showed to the robot.

Another drawback of pure teaching by guiding is that the actual manipulator configurations must not deviate from the preplanned configurations. However, in a more realistic situation, some working positions are not known in advance and are partially or completely determined by some sensors during the robot operation.

In the example from Fig. 6.2, the programmer would not be able to know the actual trajectory if the picking position of the parts were determined by a camera. In such a case, it is necessary to have tools that are more suitable for representing the positions of the objects and their interrelations. This purpose is served by the external coordinates: position of an arbitrary rigid body is described by connecting an auxiliary coordinate frame to it, and by defining a position of that auxiliary frame to some reference frame. A relative position of one object in relation to another is described in a similar manner. For example, BASE in Fig. 6.2 can represent the position of a frame connected to the manipulator base in relation to the reference frame, while EFFECTOR can represent the position of a frame connected to the manipulator effector with respect to the frame connected to the manipulator base. The camera determines the relative position PART of the part visual centre with respect to its own reference frame CAMERA, etc. Relative positions between the frames are usually defined in terms of cartesian coordinates and Euler's angles, by the equivalent translation and rotation vectors, or via homogeneous transformations.

During the manipulator operation, some relative positions can change and others stay fixed. It is often useful to formulate the unknown position of a particular object by composing the position of some other object which is determined by sensors, and the fixed relative position between the objects. For example, it can be assumed that position H in Fig. 6.2 of the effector grasping the part, relative to the visual centre of the part, should always remain fixed. If we denote the composing operator by "+", the absolute gripping position may be expressed as CAMERA + PART + H. If, on the other hand, we express the absolute position of the effector as BASE + EFFECTOR, then the following kinematic equation must hold at the grasping moment:

BASE + EFFECTOR = CAMERA + PART + H

In a similar manner we may introduce other operations. If X describes a position of some object in relation to the reference frame, we may denote a position of the reference frame relative to that object by $-X$. Instead of writing $X + (-Y)$, we may introduce the binary operator "$-$", and write $X - Y$. The semantics of binary operations "+", "$-$" and unary "$-$" is apparent and it can be easily described in terms of whichever suitable external coordinates.

From the last kinematic equation, the robot controller is able to compute a position of its effector relative to its base as:

EFFECTOR = $-$ BASE + CAMERA + PART + H

and to determine necessary joint angles using a kinematic model of that particular manipulator. Thus, in a system that supports such operations, the motion command may have a form:

MOVE_TO (CAMERA + PART + H)

and a sequence of commands for placing a part to its bed may look like:

```
READ_FROM_SENSORS (VISUAL_SYSTEM, PART)
MOVE_TO (CAMERA + PART + H + P1)
MOVE_TO (CAMERA + PART + H)
CLOSE_FINGERS
MOVE_TO (CAMERA + PART + H + P.1)
MOVE_TO (PALLET + BED[i, j] + P2)
MOVE_TO (PALLET + BED[i, j])
OPEN_FINGERS
MOVE_TO (PALLET + BED[i, j] + P2)
```

On the other hand, a position of the pallet can be determined by the described teaching by guiding. Using a teaching pendant, the programmer may guide the manipulator carrying the part until the part comes into bed (1, 1), and he may then assign the current manipulator configuration to the variable PALLET.

Let us assume that our imagined robot language supports operation "$*$" having a scalar as its first operand and the position of a frame as a second operand. The application of operation "$*$" returns the position of a new frame with its translation and rotation vectors computed by multiplying the translation and rotation vectors corresponding to the old frame (the second operand) by the scalar (the first operand). Returning to the palletizing example, and recalling that PALLET represents the position of bed (1, 1), we may now express the relative positions between the successive beds along the axes i and j as:

$$Di = 0.2 * (BED51 - PALLET)$$
$$Dj = 0.2 * (BED15 - PALLET)$$

where the positions BED51 and BED15 correspond to beds (5, 1) and (1, 5), which can be determined by the same teaching procedure applied to the determining PALLET. Finally these equations make it possible to calculate the position of an arbitrary bed with respect to the pallet as:

$$BED[i, j] = (i - 1) * Di + (j - 1) * Dj$$

The number of similar operations enabling space data manipulation is one of the most important characteristics of modern robot languages.

Besides evident advantages, representation of the positions with external coordinates has some drawbacks. First, transformation of external coordinates into joint positions is not unique for many types of manipulators. In cases where it is necessary to achieve a particular manipulator configuration, the programmer must use special configuration commands (which are applicable only to non-redundant manipulators, i.e. manipulators with not more than six degrees of freedom) or even switch to programming in internal coordinates. Second, external coordinates may overspecify the configuration. For example, in a robotized manipulation with a cyclindrical object, the angle of rotation about its axis of symmetry is often immaterial. Nevertheless, the programmer usually has to specify all six external coordinates.

Whenever the density of trajectory points is large and there is no need for high positioning accuracy, movement of the manipulator from one configuration to another can be implemented by the uncoordinated motion of its joints (so called point-to-point motion). However, such situations are rare and a coordinated joint motion is frequently required. Modern systems usually support coordinated motion with a linear change of either external or internal coordinates, and commands for selecting a desired type of coordinated motion are included in their control languages. When selecting the motion type, the programmer must bear in mind that the motion with a linear change of internal coordinates is more profitable in terms of energy consumption. On the other hand, the resulting trajectory cannot be easily visualized and it can be very unpleasant when a motion in an environment with obstacles is programmed.

In order to avoid obstacles in the working environment, some intermediate points (often called via points) are often inserted between the starting and ending points of the trajectory. If the manipulator has to pass through the via points precisely, the resulting motion can be extremely slow. For this reason, most systems support a so-called continuous motion in which the robot realizes a smooth trajectory in the vicinity of via points. Programmers of such systems have at their disposal commands for selecting a type of motion (continuous or not), and, occasionally, commands for controlling deviations of the realized trajectory from the nominal.

6.3   Sensory Data Processing

Robot operation takes place in an environment with numerous irregularities. Positions of manipulated objects, and corresponding manipulator configurations, often cannot be determined accurately in advance. Dimensions of manipulated objects always deviate from their nominal. Manipulator positioning accuracy is limited. Due to operation errors and failures it is possible that positions of manipulated objects are significantly displaced from planned positions, or that the objects are even outside the domain of the robot reachability at the planned time.

Assessment of all possible irregularities is an extremely tedious job, but the appearance of an irregularity can result in damage and significant production delay. Inadequate handling of irregular conditions is an important reason why many robotized sites are unreliable and is probably the main obstacle in a wider use of robots in industry.

Combating these irregularities can be overcome in two different ways. An obvious approach is to eliminate their sources as much as possible. For this specially designed beds, feeders, limit and proximity switches, and other mechanical devices which disable deviations from preplanned positions are widely used. However, there are difficulties: a change in the robot task might require reconfiguration of supporting devices that can last for several days, thereby decreasing flexibility of the whole system. Readaptations of the robotized site should therefore be small whenever frequent changes in production are expected. It is preferable to have a robot which operates by taking into account changes in its environment, using its sensory system.

Sensors are usually classified as internal and external. An internal sensory system includes sensors which are mounted in manipulator joints and usually give information about their positions and velocities. These sensors are denoted as internal because they are mainly used by the robot controller software system in order to ensure accordance between the achieved and preplanned manipulator trajectories. By contrast with internal sensors information obtained from an external sensory system is not used directly by the robot control algorithm. Instead, it is available to a user-written robot program, where an appropriate control command can be selected depending on the state of the sensors.

A particular robot system can be equipped with a range of differently constructed sensors. Also, the number and types of sensors included in the system may vary from application to application. On the other hand, it is useful to unify a way of controlling their operation and to make robot programs dependent on their physical nature as little as possible. To this end, in most robot programming systems, a system procedure (or a set of procedures) is assigned to every particular sensor (or to a class of sensors) at the time of the system generation. These system procedures initiate sensory operations, read raw data from sensors, scale the data and perform other related preprocessing, and finally send the processed data to the robot program. Every such pair (sensor, system procedure) is assigned a number or a symbolic identifier, so that a sensory operation (e.g. reading information from a sensor) can be specified in the robot program as a simple system call which includes an identifier of the pair (sensor, system procedure), and a (symbolic) address of the memory block where the sensory information will be transferred.

Necessary sensory signal preprocessing can be relatively simple as an analog-to-digital conversion and scaling. Conversely it can be very sophisticated, as in the case of image processing where it may take several seconds to process a moderately complex scene. When the preprocessing lasts too long, breaks in the execution of the main robot program are avoided by overlapping, i.e. by parallel processing: a distinct processor (or a set of processors) is allocated to sensory processing, and the control of such peripheral processors is performed by the robot program running on the main processor of the controller. In such systems commands for reading the sensory data are split into more elementary subcommands like a request for the peripheral processor to initiate selected operation(s), a request to return a status of the previously started operation, etc. At the same time, the job of the robot programmer becomes more complex because the programmer is now responsible for synchronization of sensory processing with manipulator motion.

Robot programming is influenced by the use of sensors for two other reasons. First, the manipulator trajectory is determined from the sensory information. In most cases, sensors enable determining of only some parameters of the necessary manipulator configurations. Thus, robot programming systems must support some space data manipulation primitives enabling the programmer to complete the configuration data. Second, sensors enable checking some of the external conditions and then, depending on the result of the test, to stop some action that is in progress or to start some new action, that is, they are used to change sequential flow of execution of robot programs in order to make the robot behaviour adaptable to external changes.

In earlier systems, external condition checks and conditional branches from the sequential execution flow were almost always joined to commands relating to some steps of the robot task. For example, a gripping command had the characteristic form:

CLOSE_FINGERS (distance, number_of_tries, error_label)

in which a jump to an error recovery sequence starting from the error_label was requested if the specified number_of_tries ended in failure (the gripping success was tested by measuring a distance between the gripper fingers after every attempt and by comparing the measured distance with specified value). However, the concept of binding the branching possibilities exclusively to a pre-defined set of operations is mostly abandoned today. Modern robot pro-gramming systems offer much more flexibility thanks to control structures being accepted by conventional programming languages, i.e. structured conditional and repetitive statements, subroutines, etc.

External conditions may be checked synchronously, i.e. at particular points of the robot program, or may be checked in parallel with the execution of particular segments of the main program. Asynchronous monitoring is desirable whenever it is necessary to stop the whole sequence of operations as soon as an irregularity occurs. Also, execution of a robot command can take much longer than a time interval after which the system must respond to a change in some external signal. In such cases, it is not always satisfactory to check an external condition before or after execution of the command, and asynchronous moni-toring is needed.

Asynchronous monitoring is applied to an important type of motion denoted as guarded motion. Conceptually, a guarded motion command might have the form:

MOVE_TO (position, external_condition, interrupt_procedure)

in which a movement to a destination position with simultaneous monitoring of the external_condition is specified; if the condition becomes satisfied, the motion is stopped and the control is transferred to the interrupt servicing procedure. An opportunity to specify the guarded motion is important when it is required, for example, to bring the manipulator hand above some imprecisely located object: in such a case, it is necessary to monitor signals from the robot touch sensors simultaneously with the motion, and to stop the manipulator as soon as collision is detected by the sensors.

Besides motion without external restrictions (so called free motion) and the already described motion in which external signals are used as a motion terminating condition, it is often desirable to incorporate compliant motion, i.e. a motion with compliance to external (mechanical) restrictions. The need for compliant motion is characteristic for assembly operations. An insertion of a cylindrical part in a hole might require a large insertion force and a large torque around the axis of insertion. On the other hand, forces and torques acting along all other axes should be small, so that both the manipulator and the assembly parts are protected from deformation and breakdown that can occur due to imprecise positioning. In order to fulfill an active compliance using the robot's sensory system, a tight coordination of sensory data processing and servo control must be realized. Unfortunately, the algorithms proposed so far are either unsatisfactory or insufficiently fast. For this reason, a passive compliance is frequently used. Passive compliance assumes a reduction in forces and torques along particular axes with the help of a passive mechanical construction with six degrees of freedom that is inserted between the last active joint and the gripper of the manipulator. However, recent investigations in this field promise that the appearance of commercial systems supporting an active compliance is not far, what will bring new opportunities to assembly programming.

6.4   Synchronization

A need for overlapping between the robot controller and its peripheral sensory system, thereby necessitating their synchronization, was briefly touched upon in the previous section. In reality, the controller interacts with a much larger number of external processes. In order to perform its job efficiently, the robot must communicate with its operator as well as with the surrounding machines and other devices. For example, the robot should be able to send information describing its status to the commanding panel. The operator must be able to stop the manipulator or to change its operation in an arbitrary moment. A more sophisticated example is the robotized manufacturing system in which several numerically controlled machines process mechanical parts, which are further transferred through the system by conveyors. The machines are served by robots which also perform quality inspection and, eventually, assemble the finished parts. The accurate operation of such a system requires exchanging hundreds of control signals and responses between its active components.

Frequent and extensive interaction with external processes brings a significant increase in the complexity of controller software. A robot controller must be able not only to send and receive commands and other signals, but must also monitor input signals in parallel with motion control. Robot programming is further complicated in that besides describing the manipulation task, the programmer has to specify which external conditions and events should be monitored when and how, and he has to decide which operations should be performed as a response to an asynchronous external event.

Asynchronous monitoring can be implemented by assigning a specialized processor to this purpose. This processor then receives a list of external conditions that will be checked, scans corresponding sensory signals and after detecting a satisfied condition, generates a hardware interrupt for the main processor of the controller. Another possibility is to use a single processor and to share its time between monitoring and motion control. In that case, the system software temporarily suspends execution of the main robot program at regular time intervals (for example, every 20 ms) and switches the processor to scan a part of the monitored condition list. If any of the conditions become satisfied, a software interrupt is signalled.

As a response to the interrupt, a user-written interrupt servicing procedure is executed. However, during the interrupt servicing, a new interrupt may occur. On the other hand, all interrupt signals are not of equal importance. For example, the condition "hydraulic system fault" is much more significant than the condition "new part has been conveyed to its picking position". For this reason, the programmer should have the opportunity to protect its interrupt routine from "undesirable" interrupts. This protection can be achieved by, for example, assigning priority levels to interrupts (i.e. to monitored conditions). In that case, the interrupt servicing may be interrupted only if some event of higher priority takes place, and all events of lower priority are forced to wait for completion of the interrupt servicing routine.

The rate of change of monitored signals, as well as the required response time, may vary significantly. Conversely, the nature of the robot task may require monitoring many external conditions simultaneously. The large number of monitored signals can lead to the controller overload resulting in an insufficient scanning speed or even a disjointed motion. In such cases, help from the programmer is required: by carefully selected signal scanning frequencies, he may assure reliable operation of the whole system.

One of main questions arising in programming production systems with several robots and numerically controlled devices is how to work out the synchronization of parallel processes that are executed during the system operation, and how to express the solution in a unified and well-documented manner. This problem was noticed earlier in computer operating system programming, where a range of software tools for interprocessing synchronization and communication, such as guarded regions, semaphores, mailboxes etc., was proposed. Application of such mechanisms in robot programming is demonstrated in several experimental robot programming systems. For example, system AL, developed at Stanford Artificial Intelligence Laboratory, supports parallel programming of two robots. However, use of concurrent programming techniques is still infrequent in robotics due to the diversity of existing controllers and their programming languages. Every particular robot type is equipped with its own characteristic programming system. Basically similar operations or data structures take various forms for different robots, and programmers then have to describe the already complex algorithms using extremely primitive tools. Lack of standards for exchange of information and commands between active components of robotized manufacturing systems is still an open problem and several working groups and committees in the world have put great effort into the development of standardized communication protocols. Acceptance of a common standard will facilitate a better portability of robotic software, as well as easier programming.

6.5   Implementations of Robot Programming Systems

A robot program preparation procedure is strongly dependent upon the complexity of the robot task and is significantly influenced by the capacities of the given robot and its programming system. The first industrial robots from the early 1970s have frequently been programmed manually. Manual programming is not really programming in the conventional sense. It comprises setting different stops, switches and relays in the robot controller and can therefore be effectively used only for programming short and simple robotized operations.

In applications like painting and especially seam welding, the most important issue is to achieve the correct trajectory profile. The most suitable programming method in such applications is a variant of the earlier mentioned teaching by showing what is usually known as the walkthrough method: a programmer, or more precisely, an experienced technician, manually moves the manipulator tip through a sequence of desired positions. Due to the significant weight of the manipulator, teaching is usually performed using a pilot, i.e. a model of the manipulator in natural size. The pilot is equipped with joint position sensors, enabling it to read configuration data and to record them to the magnetic or the paper tape. Later, during automatic operation, memorized configurations are read from the tape and sent to the robot servo. Speed of movement in automatic operation can be controlled independently, so that the technician is not obliged to take special care about speed of motion in the course of teaching. Such systems behave like recorders and are often denoted as playback systems due to their capability to reproduce once recorded (or "learned") trajectories.

The majority of commercial industrial robots are programmed using so-called leadthrough method that assumes the use of a portable teaching pendant. The pendant enables the programmer to guide the manipulator to the desired positions and to store the corresponding motion commands in the controller memory. The programmer also has an opportunity to edit stored sequences and to insert some special-purpose commands between the motion commands. These special commands can be trajectory control commands or they can be requests for executing characteristic operations like opening/closing the gripper fingers, waiting for an external signal, generating an output signal etc. Such systems also usually contain built-in parametrized procedures that facilitate programming more complex manipulation tasks, for example, palletizing. The user is then required to impose parameters specific to the given problem, e.g. to specify the number of beds along the axes of the pallet, to guide the manipulator to characteristic positions, etc.

The need for robotization of complex assembly tasks demanded much more efficient programming tools. It caused a significant growth in robotic software and the appearance of specialized textual programming languages during the 1970s. Some important advantages of textual programming regarding teaching by showing are as follows.

1. Geometric data manipulation capabilities, supported by the majority of textual languages, enable calculation of the necessary manipulator configurations rather than showing them to the robot.
2. Textual languages facilitate off-line programming and the robot is therefore released to perform some other productive job during the preparation of new robot programs.
3. Textual languages make it easier to use CAD systems in robot programming, and to simulate and optimize robot operation off-line.
4. Textual robot languages offer the possibility of easier transporting of robot programs from one manipulator type to another.
5. Robot programs written in textual languages are usually much better documented.
6. Specific data structures and structured commands, supported by a majority of textual robot languages, facilitate readability and understanding of robot programs and shorten the programming cycle time.
7. Most robot languages provide subroutines, functions and other constructs that can serve as extensions to the basic language commands; the programmer with such application-related extensions can then concentrate on the robot task rather than minor details.

The first widely popularized experimental robot language was WAVE, which was developed in 1973 at Stanford University Artificial Intelligence Laboratory. However, it took almost six years before the appearance of the first commercial industrial robot language: it was the VAL system, developed by Unimation Inc. The VAL system was intended for programming the PUMA family and the Series 2000 and 4000 robots, and has remained the most popular robot language up to now. From 1979, a range of big robot manufacturers have offered robot control systems supporting textual programming. At the same time, there has been a further growth of robot teaching systems. For example, the ASEA robot programming system, which can be classified as an advanced teaching system, supports simple flow control commands, facilitates limited use of sensors, makes it possible to get listings of stored robot programs, etc.

Programming languages, according to their implementations, are commonly divided into interpretable and compilable. In the case of interpretable languages, language implementation comprises development of an interpreter, i.e. a program which accepts as input a source program written in the given language, and performs the computations and other actions implied by the program. In the second case, the source program is first translated by a compiler into the machine language of the implementation computer, so that the translated program can be later executed on that computer. Thus, in the case of compilable languages, the role of the interpreter takes on computer hardware.

The main advantage of the first language implementation method is that interpretation of the program can start immediately after it has entered into the computer. This is important whenever an extensive testing is required and frequent changes in the program are expected, as it is in robot programming. One source command may correspond to several hundred, or even thousands of machine commands, enabling high compactness of the program to be achieved. On the other hand, compilation of the source program into the machine program assures its significantly fast interpretation, an essential requirement in the range of robotic applications where it is necessary to execute thousands of floating-point operations per second. Also, many compilers support separate compilation which shortens the compilation time when only small portions of the program are expected to be frequently changed. The programmer breaks the source program into less or more independent modules, i.e. into program parts which are then separately compiled in object modules. Using a special program, denoted as the linker, the object modules are linked in a ready-to-load module which can be later loaded into the computer core memory and then executed.

As both implementation methods have advantages and disadvantages, it is rare to find pure interpreters or pure compilers. Instead, in the case of interpretable languages, it is common to have the interpreter operation divided into two phases. During the first phase, the source program is translated into some suitable internal form which is then interpreted in the course of the second phase (the part of the interpreter that performs the first phase is called the compiler, and the part that performs the second phase is called the pure interpreter). In the case of compilers, source commands that imply more complex processing (for example, coordinate transformation or motion control) are not directly translated in long sequences of machine instructions. Instead the compiler generates system calls to the corresponding built-in system procedures. Thus, complex commands are interpreted by the system software rather than translated into machine code and executed by the hardware.

The system software of almost every robot controller incorporates some tools for robot programming. Such robots are autonomous in the sense that their programming and operation can be performed (at least in principle) without external equipment. In a multirobot system consisting of autonomous robots, every controller might interpret its own program stored in its own memory (and probably written in its own language), and their synchronization might be achieved via sensory signal lines. However, if their tight coordination is needed, it is then desirable to have a common programming language enabling control of the whole system from one place. If such a system is available, the operation of individual robot controllers can be non-autonomous in the sense that they can be supervised by a central computer which also serves as a program development system. In the extreme case, controllers may only be allowed to interpret individual robot commands one by one, and to return status reports to the central computer.

Autonomous robot systems are usually programmed using teaching-by-guiding methods which are sometimes combined with low-level textual programming (see Fig. 6.3a). The main feature of these systems is their interactive programming capability: due to the fact that the programming system is a part of the robot controller, motion commands can be executed as soon as they are entered, and the programmer may immediately see the effect of every command entered. In simpler cases, where the main task of the programmer is to guide the manipulator through some fixed trajectory, the advantage of the interactive programming is obvious.

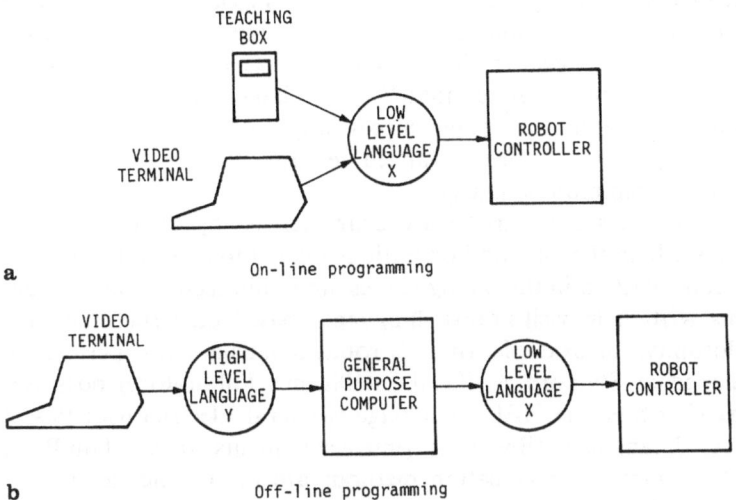

Fig. 6.3 a, b. Robot program preparation methods

Main shortcomings of these systems are due to unproductive engagement of the robot during programming. Also, limited computer resources of robot controllers are often insufficient for implementing more sophisticated programming languages. These disadvantages can be eliminated by introducing software systems for off-line robot programming that can be implemented on almost any general-purpose computer. The off-line programming systems are essentially compilers which translate textual source programs to object programs that are interpretable by robot controllers (Fig. 6.3b).

A natural base for the classification of textual robot languages is their method of expressing typical robotized operations. Some languages force the programmer to describe a desired action in much more detail than others. Thus, three programming levels can be distinguished: motion programming, structured robot programming, and task-oriented programming.

Languages from the first level are primarily concerned with manipulator motion. Their typical representative is the already mentioned VAL system. Motion commands incorporated in VAL describe manipulator movement from one configuration to another, i.e. every command specifies one trajectory segment. Segment end points are mostly determined by guiding methods. Stored configurations are assigned symbolic names, they can be displayed for user convenience as cartesian coordinates or joint angles, and be modified by a restricted set of operations (typically a translation by some small amount). The flow of execution of VAL programs can be controlled by using integer control variables and integer arithmetics combined with simple branch commands, by using repetition commands, etc. A primitive form of user-written subroutines is also supported. VAL's simple input and output processing capabilities provide a relatively primitive means for communication with the external sensors and the surrounding peripherals. However, despite all that has been said about its primitive capabilities, it must be noted that VAL is a language from the middle of the 1970s, and was the first language successfully used in the industry. Its successor, VAL-II, possesses more advanced features that make it a true structured robot programming language.

Languages from the second group are characterized by their structured statements for describing the data and operations related to a given robot task. The robot programs written in these languagues are reminiscent of the conventional programs written in well-known languages based on structured programming philosophy: for example, the AL robot language (an experimental language developed at Stanford Artificial Intelligence Laboratory) possesses ALGOL syntax elements, the HELP language (General Electric) uses Pascal constructs, the AML language (IBM) flow control statements are based on PL/I, etc. Although their motion specification methods are mostly the same as in languages from the first group, their powerful space relationship expressing capabilities and their constructs encouraging structured programming mean a qualitative jump in robot programming. A majority of these languages enable to determine the positions of manipulated objects using a visual system. Also, some of them support concurrent programming, facilitating in this manner the programming of multirobot systems.

The main disadvantage of these languages from the first two groups lies in their requirement for detailed specifications of all elementary operations, data structures, methods for monitoring events in the outside world, responses to irregularities and other external events, i.e. an explicit procedural programming. A novice user can be lost in a mass of details, and efficient programming in these languages demands a professional programmer. This deficiency is being worked on in high-level task-oriented robot languages which are still a subject of research, and whose representative is AUTOPASS, from the Thomas J. Watson Research Centre. New advanced robot programming systems are aiming for descriptive programming i.e. task description in a manner which is common in non-robotized production. For example, the most general command in AUTO-PASS is the PLACE statement by which the programmer may specify placement of a manipulated object on or in another object ("on" implies a relatively open goal position, whereas "in" implies that the goal position is enclosed). Due to its world modelling capabilities, the system is able to automatically determine feasible pickup points and suitable gripping forces, to calculate a safe trajectory, etc. High-level user instructions in such systems are automatically translated into sequences of more elementary operations that can be directly executed by the robot controller. However, transformations of the task-oriented commands into low-level manipulation commands can result in a disagreement between the intended action and a generated code. In such cases, interaction between the system and the user is necessary, and the importance of additional simulation and debugging tools is therefore significantly increased.

6.6 Robot Programming Examples

For illustration, let us examine a simplified problem of inserting pins into holes in some assembly process, as shown in Fig. 6.4 – a pin insertion task. The pins are acquired from a feeder which ensures that all pins are picked up from the fixed position. Manipulator configurations during this course of insertion can vary due to tolerances of the assembly parts. As the insertion trajectory is not precisely known at the programming time, it should be determined using the robot sensors. One possible strategy for using the sensory information is by a spiral search around the hole: if the sensors do not confirm that the pin is seated in the hole, succeeding insertion tries are made at break points of a quasi-spiral path in a plane normal to the direction of insertion (see Fig. 6.4). Once the pin has fallen into the hole, the insertion should be performed as a motion with compliance to external forces.

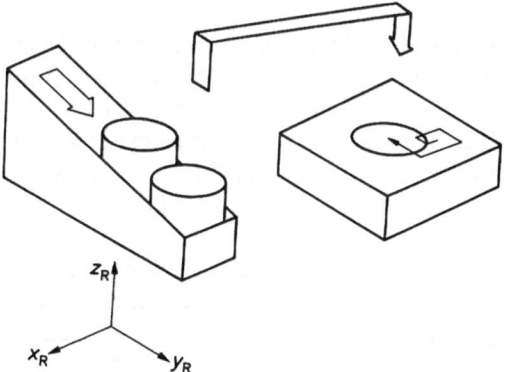

Fig. 6.4. Pin insertion task

In the following pages, we shall see how this insertion task can be programmed in two different robot languages, WAVE and AML.

The basic philosophy implemented in WAVE is that only minor deviations from the planned trajectory can be expected during the robot operation. With this assumption, it is possible to transform a source WAVE program into an executable object code that contains precalculated joint angles and dynamic effects as a function of time. During the execution, deviations from the plan are monitored by internal sensors, which are used by the WAVE executive to detect errors and modify subsequent sections of the plan.

Positions and orientations of manipulated objects are specified in WAVE as cartesian coordinates and Euler's angles, and are transformed internally into homogeneous transformation matrices. Besides the transformation matrices, the WAVE system also has two other data types: vectors, and integers serving as loop counters. Variables are defined and assigned initial values with commands TRANS (for transforms), VECT (for vectors) and ASSIGN (for integers). Predefined identifiers X, Y and Z denote unit vectors along axes of the reference frame, and NIL denotes the nil vector.

The basic motion command is MOVE, which takes as an argument a symbolic identifier of the destination transform. Its execution is performed in three steps: it starts by moving the hand vertically away from its initial position, continues by moving it to the position above the destination, and terminates by approaching from the vertical direction (a simple, but often efficient, collision-avoiding mechanism). Relative changes in the hand position may be specified by the CHANGE command, which is executed in the same way as the first step of the MOVE command, and which has five arguments: unit vector of translation, scalar distance of translation, unit vector of rotation, angle of rotation, and time (if the last argument is omitted, the system automatically calculates the necessary time). Guarded motion is specified by inserting the STOP command before MOVEs and CHANGEs. Arguments of STOP command are force and torque values; if the force or the torque acting at the robot tip exceeds the specified values, the motion is stopped. Opening and closing the gripper are specified by using OPEN and CLOSE commands in which a desired (or expected) finger opening is specified.

In the WAVE system, a relatively simple flow control mechanism is implemented. Errors that can be detected during the robot operation are numbered, and SKIPE and SKIPN commands are provided to check whether a particular error is detected or not. SKIPE causes a skip over the next command in sequence if its argument is equal to the number of detected error, and SKIPN causes the skip if its argument does not match the error number. Commands SKIPE and SKIPN may be combined with the JUMP command performing a jump to the error recovery block or a jump to the next step of the robot task. A command SOJG is provided for repetitive execution of a sequence of commands. It has two arguments: a loop control variable and a label. The execution of SOJG includes decrementing and testing the control variable; if the value of the variable remains positive, a jump to the label is performed.

Contrary to its poor program control capabilities, WAVE provides some very high-level functions that are rarely found in commercial robot programming languages. For example, a spiral search may be specified with the help of two commands: SEARCH, which allows selection of the search parameters, and AOJ, which performs a jump to the next search step. Compliant motion is also supported: commands FREE and SPIN are provided for specifying translatory and rotational degrees of freedom of the manipulator in relation to the reference frame.

Now that we have some brief knowledge of the WAVE commands, let us return to the problem from Fig. 6.4. The whole insertion task can be divided into the following steps:

(a)   moving the hand to the pin picking position (free motion with the obvious requirement to have determined a safe trajectory preventing collision between the manipulator and surrounding objects);

(b)   closing the gripper's fingers around the pin and checking whether the pin is successfully acquired;

(c)   placing the pin above the expected position of the hole (free motion as in step (a));

(d)   searching for the actual hole position (guarded motion is performed in every search step);

(e)   inserting (compliant motion).

A possible form of the WAVE program for this sequence of operations is outlined in Fig. 6.5. The two transforms defined in the program, PIN and HOLE, represent the corresponding planned positions. They can be assigned their actual values by filling the optional arguments of TRANS commands, or they can be determined by guiding the robot and recording its coordinates. The force vector FV is used as a motion terminating condition in spiral search and insertion operations. The command MOVE after the label NEWPIN specifies movement of the gripper to the pin position, whereas the six commands after the label PICKUP describe grasping the pin. The spiral search parameters are specified in the commands SEARCH and ASSIGN after the label FIND, and the search procedure itself is described by the next eight commands. If the search fails after ten unsuccessful tries, the last WAIT is executed, informing the user about the failure. Otherwise, the insertion is performed.

The AML system represents an essentially different approach to robot programming compared to WAVE. The WAVE system is based on off-line trajectory planning, so that it consists of a compiler (serving as a trajectory generator) and a relatively simple execution system. On the other hand, AML is implemented as an interpreter capable of immediately executing every command (AML is an expression-oriented rather than a statement-oriented language; every legal construct in AML produces a value, and AML commands are simply functions in which we are more interested in their side effects than in their return values). In contrast to WAVE, the AML language provides unusually flexible facilities for data manipulation. Its basic data types include

```
        TRANS PIN . . .              ; Pin position
        TRANS HOLE . . .             ; Initial guess of the hole position
        VECT FV 0, 0, −50            ; Force vector, 50 oz in the negative direction of z axis
NEWPIN:
        MOVE PIN                     ; Move to the pin position
PICKUP:
        CLOSE 1                      ; Try to pick up the pin (expected opening equal or greater
                                       than one inch)
        SKIPE 2                      ; Error 2, opening less than expected
        JUMP FIND                    ; Success, find the hole
        OPEN 5                       ; Failed, open the fingers
        WAIT MISSING PIN             ; Write error message and wait for operator
        JUMP PICKUP                  ; Try to pick up again
FIND:
        SEARCH X, Y, 0.1             ; A search in the x-y plane of 0.1 inch steps with the first step
                                       in the x direction
        ASSIGN TRIES 10              ; Allowed number of search tries
LOOP:
        MOVE HOLE                    ; Move to modified hole position
        STOP FV, NIL                 ; Stop on 50 oz.
        CHANGE Z, −1, NIL, 0, 0      ; Try to go down one inch
        SKIPN 23                     ; Error 23, failed to stop
        JUMP INSERT                  ; The hole is found, insert the pin
        SOJG NTRIES, NEXT            ; The hole is not found, if the allowed number of tries is not
                                       exhausted jump to the next search step
        JUMP ERR                     ; else signal the error
NEXT:
        AOJ LOOP                     ; Add search increment and try again
INSERT:
        FREE 2, X, Y                 ; Two degrees of freedom along x and y axes
        SPIN 2, X, Y                 ; Two degrees of freedom about x and y axes
        STOP FV, NIL                 ; Stop on 50 oz.
        CHANGE Z, −2, NIL, 0, 0      ; Make insertion
        OPEN 5                       ; Open the fingers
        WAIT END OF TASK             ; Wait for new assembly parts
        JUMP NEWPIN                  ; Repeat the whole operation
ERR:
        WAIT NO HOLE                 ; Write error message and wait for operator
```

Fig. 6.5. Pin insertion program in WAVE

scalars and aggregates. Scalar data may be integer, real, character strings, references (pointers to other data), identifiers and labels. Aggregates are very similar to S-expressions in LISP and can be used to describe complex data structures. The aggregates are formed as ordered sets of scalars and/or other aggregates that are separated by commas and enclosed in parentheses, "$\langle$" and "$\rangle$". The system provides a rich repertoire of operations on both scalar and aggregate data objects. For example, besides the usual arithmetic ($+, -, *, \ldots$), relational (EQ, NE, LT, . . .), and logical operators (AND, OR, NOT, . . .), the system provides an unary operator "$" which, applied to an identifier, creates a new object of identifier type, a binary concatenation operator " # " facilitates concatenation of the aggregates, etc. Additional flexibility is provided thanks to a simple semantic for defining the extension of operations from scalar to aggregate objects. For example, $\langle 1, 2 \rangle * \langle 3, \langle 4, 5 \rangle \rangle$ gives as a result $\langle 3, \langle 8, 10 \rangle \rangle$.

The control elements of AML include the usual constructs of structured programming (IF ... THEN ... ELSE, WHILE ... DO ... , etc.) and subroutines. Inside a subroutine, variables may be defined as static (STATIC declaration mode) or dynamic (NEW declaration mode). Optionally, variables may be assigned initial values inside their declarations. Parameters may be passed to subroutines via lists of arguments or via global variables. Every subroutine may return a value: the return value is specified with the help of a built-in RETURN command.

AML commands are simply predefined subroutines provided for data manipulation, motion control, sensory data processing, etc., and there is no difference in form of these commands and calls to user-written subroutines. In this manner, the system can grow. Initially, new commands may be implemented as AML subroutines and, later, when greater computational efficiency is required, these subroutines may be recoded and incorporated into an extended AML system without changes in their functional behaviour. The availability of fundamental subroutines, enabling to control the operation of the AML interpreter, is of essential importance. Such subroutines are CLEANUP, which permits setting a subroutine exit trap, ERRTRAP, which is used to set a subroutine error handler, etc. The fundamental routines are of valuable help whenever there is a requirement to enlarge the AML repertoire with additional commands. For example, although the AML system does not provide the general compliant motion command, this function could be implemented as an AML subroutine (naturally, with a slower speed of execution).

The basic motion command is MOVE, which has four arguments. The first argument is an aggregate of indices of joints that are involved in the motion, and the second argument is an aggregate of their target positions (the AML system is intended for programming the cartesian robots, so that there is no difference between external and internal coordinates). The joints are numbered from 1 to 7 (the seventh joint denotes fingers of the gripper) and, for user convenience, they are assigned symbolic identifiers JX, JY, JZ, JR, JP, JW and JG. The third argument of the MOVE command is an optional aggregate of indices ("sensor monitor numbers" in the AML terminology) assigned to external conditions that are monitored during the motion, and whose fulfilment causes termination of the motion. The fourth optional argument is an aggregate of trajectory parameters: speed of movement, acceleration, deceleration and the settling time. The command DMOVE is similar to MOVE, with the difference that its second argument denotes incremental changes of positions rather than the destination positions of joints.

It is interesting that even the MOVE command may be implemented in AML using the more primitive functions AMOVE, which returns control immediately after the beginning of the motion, and WAITMOVE, which waits for completion of the motion. The implementation could be:

```
MY_MOVE:  SUBR;
          APPLY ($AMOVE, PARMS);
          WAITMOVE;
          END;
```

making use of the primitive system APPLY, which enables a call AMOVE with a program-generated set of arguments. Actual arguments for AMOVE are taken from the system variable PARMS, in which the AML interpreter stores actual parameters of a call to MY_MOVE.

Besides simple sensor input and output operations under direct control of user programs, AML provides commands for asynchronous monitoring. Specifying external monitored conditions is achieved with the command MONITOR. In a simplified MONITOR call, four arguments are supplied: an aggregate of sensor indices, a type of test (value inside the limits, value outside the limits, etc.), a lower and an upper test limit. After MONITOR is called, the system returns a set of indices identifying the monitored conditions, and begins reading the specified sensors and performs specified tests in parallel with the execution of succeeding AML commands. Results of the tests may be examined with QMONITOR command, and a termination of monitoring some selected conditions may be requested with ENDMONITOR command.

Use of static variables may sometimes improve the readability of AML programs. Declarations of several such variables are given in Fig. 6.6.

```
SUCCESS: STATIC −1;                 −− completion results
FAILURE: STATIC 0;

HAND: STATIC ⟨JX, JY, JZ⟩;          −− translatory motion
VERTICALLY: STATIC ⟨JZ⟩;            −− vertical motion
FINGERS: STATIC ⟨JG⟩;               −− opening/closing fingers

SAFE_DISTANCE; STATIC 0.5;

TOLERANCE: STATIC 0.05;

SlOWLY: STATIC ⟨0.3⟩;               −− 30 percent of max. speed
```

Fig. 6.6.   Declarations of static AML variables

The examples of using these variables in programming some elementary operations are displayed in Fig. 6.7. Subroutine APPROACH from Fig. 6.7 can be used to lower the manipulator gripper close to an object whose z-coordinate is imprecisely known. The argument of the subroutine is the expected position of the part; after completion of the motion, the subroutine returns information whether the attained position matched the expected. Subroutine GRASP from Fig. 6.7 closes the robot fingers around a part with a specified diameter, and returns its completion status.

Subroutines APPROACH and GRASP make use of the guarded motion with a stopping force of 5 oz. This value is used as an actual argument in subroutines PINCH_FORCE and TIP_FORCE which complete the par-

```
PINCH_FORCE: SUBR (F);                    -- subroutine for completing
                                          -- parameters for pinch force
                                          -- sensor monitors

    RETURN (⟨⟨SLP, SRP⟩, 1, 0, F⟩);
    END;

TIP_FORCE: SUBR(F);                       -- subroutine for completing
                                          -- parameters for tip force
                                          -- sensor monitors

    RETURN (⟨⟨SLT, SRT⟩, 1, 0, F⟩);
    END;

RETREAT: SUBR;                            -- subroutine for lifting up
                                          -- the gripper

    DMOVE (VERTICALLY, SAFE_DISTANCE);
    END;

APPROACH: SUBR(P);                        -- subroutine for falling down
                                          -- the gripper
    FMONS: NEW APPLY ($MONITOR, TIP_FORCE (5 * OZ));
    CLN: SUBR;
        ENDMONITOR (FMONS);
        END;
    CLEANUP ($CLN);
    MOVE (HAND, P + ⟨0, 0⟩ # ⟨SAFE_DISTANCE⟩);
    DMOVE (VERTICALLY, − SAFE_DISTANCE, FMONS, SLOWLY);
    RETURN (NOT QMONITOR (FMONS));
    END;

RELEASE: SUBR;                            -- subroutine for opening the
                                          -- gripper

    DMOVE (FINGERS, SAFE_DISTANCE);
    END;

GRASP: SUBR (D);                          -- subroutine for grasping an
                                          -- object
    FMONS: NEW APPLY ($MONITOR, PINCH_FORCE (5 * OZ));
    RESULT: NEW INT;

    MOVE (FINGERS, D/2 − TOLERANCE, FMONS, SLOWLY);
    RESULT = QMONITOR (FMONS);
    ENDMONITOR (FMONS);
    IF RESULT EQ FAILURE THEN RELEASE;
    RETURN (RESULT);
    END;
```

Fig. 6.7. AML subroutines performing some elementary robot operations

ameters of the corresponding force sensor monitors. It is assumed here that the robot fingers are equipped with touch sensors; AML global identifiers for pinch force sensors of the left and the right fingers are SLP and SRP respectively, and the identifiers for the corresponding tip force sensors are SLT and SRT. Indices "1" in the return values of PINCH_FORCE and TIP_FORCE specify that the stopping conditions become satisfied if either the (left or right) sensor value exceeds the range 0 to F.

Subroutines APPROACH, RETREAT, GRASP and RELEASE may be further used as building elements in programming more complex functions that implement particular steps of the insertion task in Fig. 6.4. Let us assume that the objects relevant to this task (e.g. pins and holes) are described by aggregates, and that every such aggregate has three components: expected position, diameter and height (or depth). If we decide to use distinct subroutines for obtaining the selected features of the objects, as shown in Fig. 6.8, then the functions that implement realize picking a pin, finding a hole, and inserting, may be programmed as in Fig. 6.9.

```
POSITION: SUBR(PART);
    RETURN (SELECT (⟨1⟩, PART));
    END;

DIAMETER: SUBR(PART);
    RETURN (SELECT (⟨2⟩, PART));
    END;

HEIGHT: SUBR(PART);
    RETURN (SELECT (⟨3⟩, PART));
    END;
```

Fig. 6.8. AML subroutines for selecting a feature of an object

AML subroutines like PICKUP and INSERT_PIN in Fig. 6.9 are actually high-level robot commands (clearly, Fig. 6.9 shows their very simplified form). Now that we are armed with such commands, programming the insertion task becomes almost negligible (Fig. 6.10).

Figures 6.6 to 6.10 illustrate at the same time the significant opportunities provided by AML and other modern robot programming systems for developing appropriate application-dependent library programs. Modern programming systems are usually designed to meet the requirements of their most demanding users which may further expand the basic language and customize the system to a particular application. In this way, the same robot programming and control system is able to meet the demands of a wide class of users, from programmers of application packages to operators.

```
PICKUP: SUBR(PART);

    RESULT: NEW INT;
    APPROACH (POSITION (PART));
    RESULT=GRASP (DIAMETER (PART));
    RETREAT;
    RETURN (RESULT);
    END;

FIND_HOLE: SUBR(F, STEP, DELTA_H, MAXTRIES);

    DELTA: NEW ⟨0, 0, −DELTA_H⟩;
    NTRIES: NEW 0;
    SEARCH_DIR: NEW 1;
    FMONS: NEW APPLY ($MONITOR, TIP_FORCE (F));
    RESULT: NEW INT;

    DMOVE (VERTICALLY, −DELTA_H, FMONS, SLOWLY);

    WHILE (RESULT=QMONITOR (FMONS)) AND NTRIES LT MAXTRIES
        BEGIN
            DMOVE (VERTICALLY, DELTA_H/2);
            DELTA (SEARCH_DIR)=STEP*(NTRIES=NTRIES+1);
            DELTA (SEARCH_DIR=3−SEARCH_DIR)=0;
            DMOVE (HAND, DELTA, FMONS, SLOWLY);
        END;
    RETURN (RESULT);
    END;

INSERT_PIN: SUBR (PART, HOLE);

    FMONS: NEW APPLY ($MONITOR, TIP_FORCE (5*OZ));
    RESULT: NEW INT;

    APPROACH (POSITION (HOLE)+⟨0, 0⟩ # ⟨HEIGHT (PART)⟩));
    RESULT=FIND_HOLE (5*OZ, 0.1, 0.1, 10);

    IF RESULT EQ SUCCESS THEN
        BEGIN
            DMOVE (VERTICALLY, −HEIGHT (PART), FMONS, SLOWLY);
            RELEASE;
        END;

    RETREAT;
    RETURN (RESULT);
    END;
```

Fig. 6.9. AML subroutines performing characteristic steps of pin insertion

```
RESULT: NEW INT;
PIN: NEW ...
HOLE: NEW ...

RESULT = PICKUP (PIN);
WHILE RESULT EQ FAILURE DO
    BEGIN
        BREAK ('*** missing pin ***', EOL);
        RESULT = PICKUP (PIN);
    END;

IF INSERT_PIN (PIN, HOLE) NE SUCCESS THEN
    DISPLAY ('*** no hole ***', EOL);
```

Fig. 6.10. Pin insertion program in AML

6.7   Additional Reading

Topics related to robot programming are too numerous to mention in this text. More details about open questions in this area as well as a more comprehensive bibliography can be found in the excellent Lozano–Perez review [1].

Descriptions of languages WAVE and AML, used in the programming examples in the previous section, were taken from Paul [2] and Taylor *et al.* [3]. Similar to these two systems, other robot programming languages may also differ drastically from one another, both in their syntactic forms and implemented algorithmic solutions. Competitive reviews of experimental and commercially available robot systems give Bonner and Shin [4] and Gruver *et al.* [5]. However, one may only get a complete insight into the characteristics of any particular robot language by carefully studying the languague manual and practising on a real robot.

In this chapter, the main attention was placed on the methods used for robot programming, as opposed to questions relating to methods for implementing the robot programming systems which were only sporadically touched upon. As a good introduction to programming languages on design methodology, books by Pratt [6], and Tennent [7], may be used. Lewis II *et al.* [8], Aho and Ullman [9], and Gries [10], cover compiler development techniques in detail. Readers interested in design and implementation of interactive programming languages can benefit from the tutorial by Brown [11]. A detailed account of language constructs relevant to robot programming is given in Blume and Jacob [12]. Paul [13] describes in some detail specifics relating to the development of software for robot controllers. A description of a simple robot programming system implemented in Pascal is given by Blume and Jakob [14].

References

[1] Lozano-Perez, T., "Robot Programming", Proc. IEEE, 1983, Vol. 71, No. 7, pp. 821–841.

[2] Paul, R., "WAVE: A model based language for manipulator control", Industrial Robot, March 1977, pp. 10–17.

[3] Taylor, R.H., Summers, P.D., Meyer, J.M., "AML: A Manufacturing Language", Robotics Research, 1982, Vol. 1, No. 3, pp. 19–41.

[4] Bonner, S., Shin, K.G., "A Comparative Study of Robot Languages", Computer, December 1982, pp. 82–96.

[5] Gruver, W.A., Soroka, B.I., Craig, J.J., Turner, T.L., "Industrial Robot Programming Languages: A Comparative Evaluation", IEEE Trans. Syst. Man Cybernet., 1984, Vol. SMC-14, No. 4, pp. 565–570.

[6] Pratt, T.W., Programming Languages: Design and Implementation, Prentice-Hall, Englewood Cliffs, 1975.

[7] Tennent, R.D., Principles of Programming Languages, Prentice-Hall, Englewood Cliffs, 1981.

[8] Lewis II, P.M., Rosenkrantz, D.J., Stearns, R.E., Compiler Design Theory, Addison-Wesley, Reading, 1976.

[9] Aho, A.V., Ullman, J.D., Principles of Compiler Design, Vol. 1: Parsing, Vol. 2: Compiling, Addison-Wesley, Reading, 1977.

[10] Gries, D., Compiler Construction for Digital Computers, Wiley, New York, 1971.

[11] Brown, P.J., Writing Interactive Compilers and Interpreters, Wiley, Chichester, 1979.

[12] Blume, C., Jakob, W., Programming Languages for Industrial Robots, Springer-Verlag, Berlin, 1987.

[13] Paul, R., Robot Manipulators: Mathematics, Programming and Control, MIT Press, Cambridge, 1981.

[14] Blume, C., Jakob, W., PasRo-Pascal for Robots, Springer-Verlag, Berlin, 1985.

Chapter 7

Sensors in Robotics

Robots, as flexible machines capable of performing different tasks, are being used more and more in industry, but for some tasks (assembling mechanical parts, for example), it appears that in order to operate successfully robots need feedback information about the state of environment. The necessity for increasing robot adaptability demands the introduction of sensors' information in control algorithms together with elements of artificial intelligence to gain a higher degree of independence of the robot. In practice, only positional sensors are used as an obligatory part of a robot, with the purpose of feeding the robot control system with information about the state of the robot itself – internal coordinates and their time derivatives.

Although industry has accumulated experience in technology of production and the application of sensors, attempts have shown that some specific features of the robotic systems disabled straightforward (or slightly modified) application of these sensors on robots. The main difficulties are an insufficient robustness of sensors to operate in the industrial environment (temperature range, humidity, vibrations, etc), and control algorithms that are to use the sensor information in real-time. These problems are the subject of intensive research, but in many cases there are still no satisfactory solutions which can be applied in practice.

The chapter is organized in the following way:

In Sect. 7.1 a review of the positional sensors is given, and performances of the most widely used positional sensors, potentiometers, encoders and resolvers, are compared.

In Sect. 7.2 force sensors, tactile and ultrasonic sensors are presented. These sensors were chosen because they are, along with vision, explored for the application on robots more than other sensors.

In Sect. 7.3 robot vision is discussed. The reason is that robot vision in robotic systems is the subject of intensive research, and there are a great number of practical results already in use in industry.

7.1 Positional Sensors

Positional sensors play an important role in the control of manipulation robots. They provide information about the robot position, and are used in forming the control signals for servoes. In this part resolvers, encoders and potentiometers are presented in more detail as widely used in practice. After a brief description of each sensor the comparison of the most important features of sensors is given.

7.1.1 Resolvers

Resolvers (and synchros) belong to electromechanical positional transducers. The resolver is a rotary transformer, where the primary coils are on the rotor and the secondary coils are on the stator. The secondary coils consist of two coils displaced 90 degrees on the shaft, while synchro has three secondary coils displaced 120 degrees on the shaft.

The principle of the resolver (Fig. 7.1) operation is similar to the transformer: an alternate voltage is applied on primary coils and the secondary voltage is measured. The ratio of the primary and the secondary voltage is the function of the shaft position, and does not depend on the amplitude and frequency variation of the primary voltage, temperature, electrical load of the stator, humidity etc.

The typical accuracy of the resolver is on average three arc minutes, but it can be significantly increased by the use of "reduction". The "reduction" is achieved electrically, using multiple pairs of poles on the stator. When using n pairs of poles, resolution and accuracy is increased n times, thus, for $n = 16$ the resolution would be $3'/16 \sim 11''$. Since one shaft revolution corresponds to n periods of amplitude change of the electrical output, one independent pair of poles is needed to determine in which period the angle is measured.

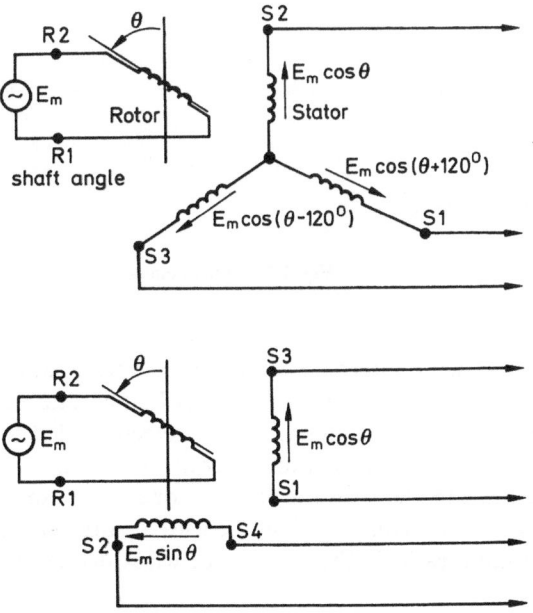

Fig. 7.1. The principal scheme of a resolver

Because electromechanical transducers are converting the angle into the voltage amplitude, they cannot be applied in the digital control systems without appropriate convertors. There are now sufficiently reliable and cheap components on the market that perform the conversion of the voltage amplitude into digital information about the position and even provide a digital tacho information.

7.1.2   Encoders

Shaft encoders are angle positional sensors which, because of the specific construction, convert the angle position directly into digital information. There are several types of encoders (optical, contact, magnetic), but in practice optical encoders are mostly used, having some advantages over others.

Optical encoders may be absolute and incremental. An incremental encoder detects the quantum of the change of the angular position, and an absolute one shows the absolute angular position. The encoder consists of the disk (see Figure 7.2) which is connected to the rotating shaft.

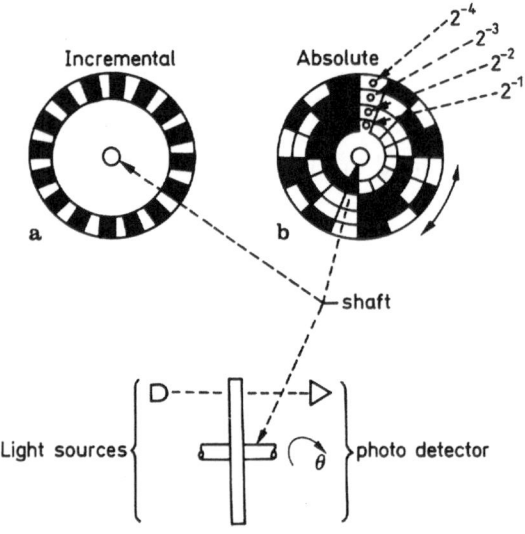

Fig. 7.2. Optical encoder

The disk has concentric ring(s) with alternate transparent and opaque segments. On one side of the disk there are light sources (usually LEDs), and on the other side light detectors are situated. The signal from the detector is led to the electronics for the signal conditioning and then to the control system.

The advantage of the incremental encoder over the absolute one is the use of only one pair of light source and detector, while the absolute uses n pairs, where n is its resolution in bits. In the same ratio is the number of wires needed to transmit the signals. Because of the more complex construction, absolute encoders are more expensive and less reliable than incremental ones. The advantage of the absolute encoder over the incremental one is that it does not accumulate the error induced by disturbances, and after reconnecting to power supplies it shows the absolute position at once, while the incremental encoder has to be brought into the referent position first.

7.1.3   Potentiometers

Potentiometers (Fig. 7.3) are the simplest positional transducers. They consist of a circular resistor (made of wire, carbon or some other resistive material) with the sliding contact on it connected to the rotating shaft. The resistance between the connections of the potentiometer is the function of the relative position of the shaft with the slider to the circular resistor. This signal is converted into digital information by means of the appropriate measuring electronics.

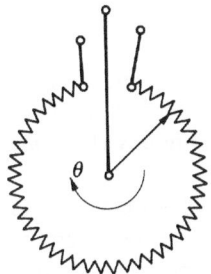

Fig. 7.3. Potentiometer

7.1.4   Performance Comparison

7.1.4.1   Mechanical Characteristics

The only moving parts of the resolver which might be the source of errors are the contact rings on the rotor. Since the information is given by the ratio of the primary and secondary voltages, the changes of contact resistance do not influence the accuracy. The transduction is in fact contactless, which makes the resolver very suitable in the environment where the vibrations are present. Resolvers are also resistant to stress and various forms of mechanical strain.

Mechanical coupling between the light source and the detector is somewhat problematic with encoders because, for example, a 12-bit encoder has 4096 segments, and if the source and the detector are not in the appropriate mutual position, the light source is obscured, which implies loss of information. Encoders are very sensitive to stress and vibration, since there is a possibility of breaking the disk. Vibration of the disk is transmitted to the output signal (jitter).

The contact resistance between the slider and the resistive ring has a random value, and since it is in serial connection with the resistance carrying the information, it directly influences the accuracy of the potentiometer. Vibrations can make this contact unreliable.

7.1.4.2   Environmental Conditions

It has already been mentioned that environmental conditions like temperature, humidity and dust practically never influence accuracy of the resolver.

Optical encoders operate in a narrow temperature range because of limitations introduced by LEDs and optical detectors. Humidity may cause deformation of the disk and dust influences the optical part of the sensor and may significantly degrade its performance.

Temperature changes the resistance of the resistive ring and deforms the slider, which does not return to its initial position after the temperature cycle. Higher temperatures may cause oxidation of the slider and further degrade the performance of the potentiometer.

7.1.4.3   Installation

The size of the sensor is important when considering installation. The average size of the resolver with 12-bit accuracy is 2 inches in diameter, the encoder with similar features is twice that and the potentiometer has approximately the same dimensions as a resolver.

The output resistance of the resolver is very small which yields immunity to electromagnetical disturbances and it is possible to use longer leads between the sensor and the rest of the system. The output resistance of the encoders is much higher, and since the signals are weak, the signal conditioning electronics must be in the vicinity of the sensor. It has already been mentioned that there is a problem with the number of wires needed to transmit the signal from the encoder. There is a similar problem with potentiometers, and, in addition, the output resistance is by definition variable, so that the resistance of the leads lessens the accuracy of the measurement.

All the sensors require signal conditioning electronics.

7.1.4.4   Accuracy

Resolvers and encoders have approximately the same static accuracy (the measurement when the shaft is still); potentiometers which achieve appropriate accuracy are so expensive that they are not used. When considering the application of sensors on the fast rotating shafts, commercial resolvers can work up to 6000 rpm, and encoders 750 rpm (this limitation is due to the response time of the optical detectors). In these applications a phase error is induced with resolvers because of the time delay between the information signal and the actual position, but is compensated in the signal conditioning electronics. The use of potentiometers in this case is severely limited because of unreliable contact and the lifetime of the sensor, defined as the maximal number of rotations, diminishes.

7.1.4.5 Reliability

As signal conditioning electronics have approximately the same complexity and reliability, any differences that may appear are on the level of the sensors themselves. When considering the meantime between failures and the total lifetime, resolvers are the most suitable, then encoders with particular light sources (and the particular price) and lastly potentiometers.

All this leads to the conclusion that for industrial purposes the use of resolvers (Fig. 7.4) as positional sensors are recommended because of their better resistivity to the environment, robustness, accuracy, and the best performance/price ratio.

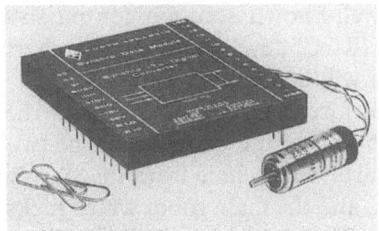

Fig. 7.4. Resolver with the signal conditioning component

7.2   Environment Sensors

The robot control usually drives the robot from an initial into the final position, through a certain path in the space. This kind of control usually does not take into account contact between the robot and the environment. For some tasks, e.g. assembling mechanical parts, introducing information about reaction forces that appear in contact into the control algorithm may be sufficient for the task to be completed successfully and reliably (see Section 4.6). Tactile sensors, which usually represent an array of minute force sensors, provide additional information about properties of the object in the robot's gripper. Application of ultrasonic systems will also be discussed in this Section.

7.2.1   Force Sensors

When force sensors were taken into consideration for use on robots, various force and torque sensors which usually measured one component of the vector had been already in existence. There were many solutions and methods of transducing the force/torque into an electrical signal.

A common force sensor consists of an elastic mechanical part (spring, cane) which deforms when the force is applied and thus transduces the force into deformation. The second part of the transducer transforms the deformation into the electrical signal (strain gauges, inductive transducers, photoelectric detectors, etc).

There are other ways of transducing the force, e.g. transducers based on the piezoelectric effect, magnetostriction (the change of magnetical permeability as the consequence of deformation of the magnetic circuit), etc.

Information about reaction force and torque consists of six scalar components. This can be obtained by measuring the forces and torques in the joints of a robot, or, more often, by measuring in the manipulator's gripper directly. There are many solutions described in the literature about special constructions of the robot gripper that have incorporated force sensors. The information obtained is introduced in feedback and then the actuators correct the values of internal coordinates. Better performances can be obtained if, by means of a special construction, a passive compliance of the robot gripper (without the engagement of actuators) is achieved. The well-known solution for passive compliance is the Remote Centre Compliance (RCC) developed by researchers in the Charles Stark Draper laboratory which will be described later.

An example of the successful application of force sensors on robots is the HI-T-HAND Expert-1 system [3]. Figure 7.5 shows the cooperation of two robots, an auxiliary one which picks up the parts with a hole from the supply device and brings them to the working position, and the main robot which picks up the shafts and performs the insert operation. The arm and the gripper of the main robot are connected via flexible construction with four strain gauges

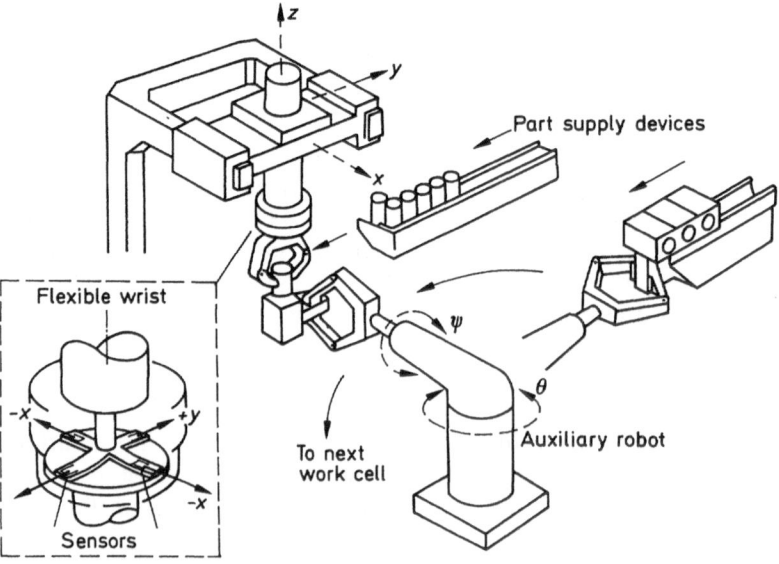

Fig. 7.5. HI-T-HAND Expert-1

capable of detecting the displacement in XYZ directions. Both robots are connected to the single controller, which uses the sensor information for correcting the centre position and direction.

The control algorithm (Fig. 7.6) consists of three phases:

1. *Contact.* The shaft is brought into the vicinity of the anticipated hole position and pushed with a specified force.
2. *Search.* Because of the elasticity of the construction, the shaft is slanted and when the shaft tip reaches the hole the spring forces push it into the hole. This is detected, and the search is over.
3. *Insertion.* By means of the sensor signal from X–Y plane, the correction of the axial shaft position is made and the pressing force is controlled to be in specified limits to prevent locking.

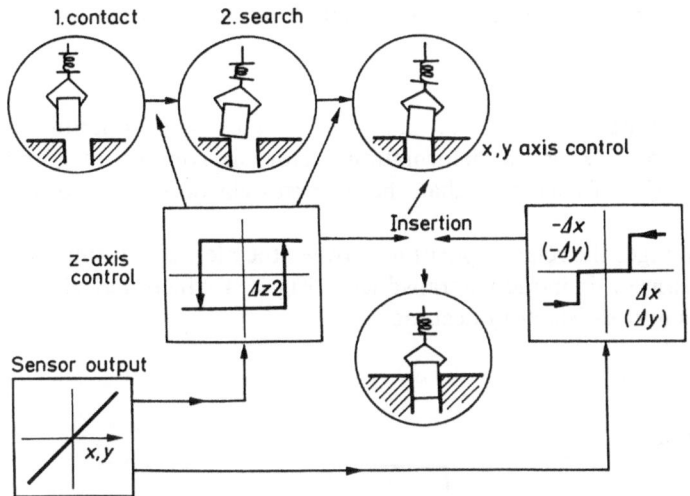

Fig. 7.6. Control algorithm

The main idea of the algorithm is that the motion of the robot which reduces the reaction forces to zero enables successful assembly. The following experimental results were obtained with this system:

(a) the successful insertion is possible if the initial axial distance between the shaft and hole is less than 3 mm.
(b) the time for task completion does not depend on the initial error mentioned.
(c) the comparison between the robot and the human worker: the robot completed its task on average in 1 second with very small dispersion of time; the worker needed 0.5 to 4.5 sec with approximately uniform distribution of time.

In the case of the Expert-1 system the algorithm used does not require precise force measurement. The force sensor [4] enables precise force and torque measurement. The basic principles for the sensor construction were:

(a) applied forces are converted into deformations which are measured;
(b) the number of measured deformations is six, and they are mutually independent, so that the components of the force and the torque can be uniquely defined;
(c) elastic deformations of the sensor should be as small as possible in order not to lessen the accuracy of the robot.

The sensor (Fig. 7.7) consists of two rigid plates connected with the deformable elements.

One plate is attached to the last segment of the robot and the other carries the gripper. The assumption is that the deformations are small so that there is a linear relationship between the force being measured and the deformations:

$$[FT]=[R] \ [D]$$

where vector $[FT]$ contains three force and three torque scalar components acting upon the sensor, $[D]$ is the deformation vector (six components), and $[R]$ is the 6×6 matrix (of elasticity) that characterizes elastic features of the sensor.

Applied force changes the relative position between the plates. Deformations are measured by means of six inductive transducers (Fig. 7.8), which were chosen for their robustness and satisfactory accuracy.

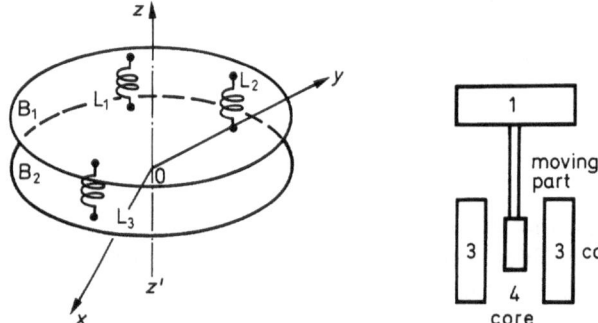

Fig. 7.7. Six degree-of-freedom force/torque sensor   **Fig. 7.8.** Inductive positional sensor

The sensor must be calibrated first so that the coefficients of the pseudo-elasticity matrix $[PF]$ are determined, using the relation:

$$[FT]=[PF] \ [SD]$$

where [SD] is the vector of signals measured on the inductive transducers. The shape of the prototype sensor is shown in Fig. 7.9. The IBM force sensor has a somewhat simpler construction. It consists of several standard modules (Fig. 7.10) that measure the corresponding torques. The orthogonality of the modules' axes makes the coupling between modules less than 1%, and simplifies the calculation of the reaction force and torque.

If S_i denotes the bending torque acting on the i-th component of the sensor (these torques are measured by strain gauges) then referring to the Figure 7.10,

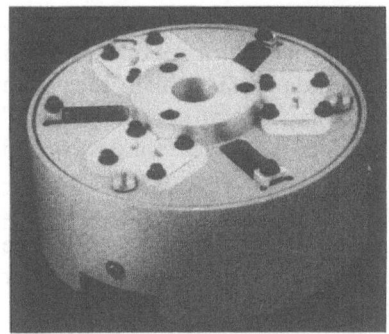

Fig. 7.9. Prototype of the six degree-of-freedom force sensor

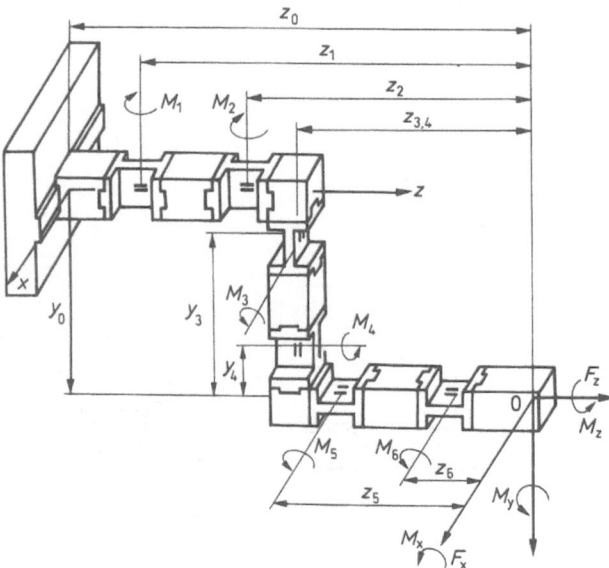

Fig. 7.10. Modular force sensor

the following equations are obtained:

$$F_x=(S_1-S_2)/(z_2-z_1) \ , \quad F_y=(S_6-S_5)/(z_6-z_5)$$
$$F_z=[S_j+F_y(z_0-z_3)-M_x]/(y_0-y_3)$$
$$M_x=[-S_5(z_0-z_6)+S_6(z_0-z_5)]/(z_6-z_5)$$
$$M_y=[S_1(z_0-z_2)-S_2(z_0-z_1)]/(z_1-z_2)$$
$$M_z=S_4+F_x(y_0-y_4)$$

The previously mentioned RCC belongs to a class of passive components developed for the task of assembling mechanical parts. These components are usually placed between the gripper and the last segment of a robot. They add degrees of freedom to the gripper that are not actuated. The main idea with compliance components is that reaction forces appearing at contact 'control' these additional degrees of freedom in such a manner that assembling is carried out. Good features of such components are simplicity of operation and simplification of the control system so that the operating speed can be achieved. The disadvantages are that the mechanical parts being assembled must usually be specially prepared, resulting in the impossibility of reprogramming some other task which thereby limits the use of the robot.

RCC (Figure 7.11) absorbs the errors of the mutual position of the assembling parts independently compensating both lateral and angular errors (Figure 7.12).

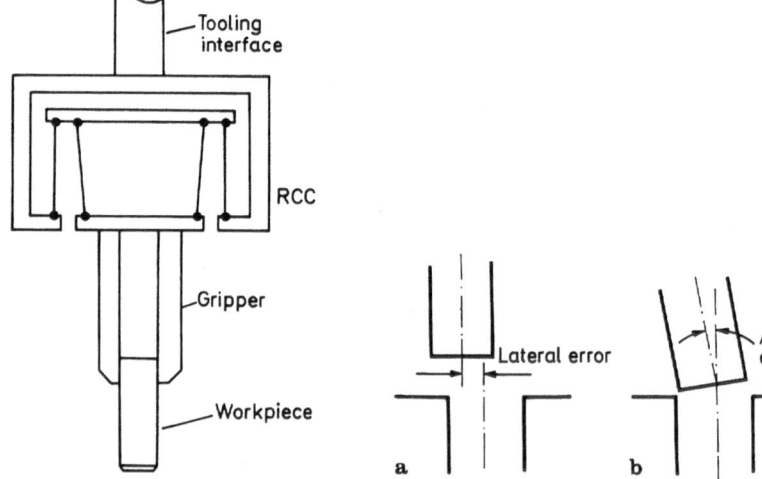

Fig. 7.11. Remote centre compliance

Fig. 7.12 a, b. Definition of lateral **a** and angular **b** error

Lateral reaction forces act on RCC so that it only translates the gripper, and when the tip of the shaft comes to the hole contact forces produce the torque that acts only upon the angular part of RCC thus assisting the insertion.

Further development of RCC led to IRCC (Instrumented RCC, Fig. 7.13), with the main purpose of incorporating passive compliance, but equipped with both force and torque sensors that provide useful information for the control system, enhancing the operating capabilities of the robot [6]. The principle of measurement is similar to that of the sensor from the Figure 7.9, differing only in the method of measuring deformations: instead of inductive, optical sensors are used. Sources of infrared light are fixed to one plate in the vicinity of the elastic elements, and two-dimensional sensors detect the beam. Before use, the pseudo-elasticity matrix is determined by calibration, and then the linear relation between the contact force/torque and the signal from optical detectors is used.

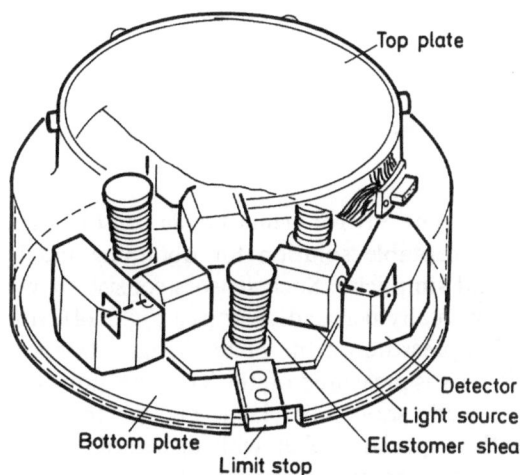

Top plate

Detector
Light source
Bottom plate   Elastomer shear
Limit stop

Fig. 7.13. The main components of IRCC

7.2.2 Tactile Sensors

Tactile sensors imitate the sensitivity of human fingers. They are situated in the gripper. Their purpose is to give information to the control system about the position and the orientation of the object in the gripper, its shape, the distribution of the forces acting on it and slipping of the object (speed and the direction). With this information the object from the specified class can be recognized and the mechanical properties be determined by the degree of the deformation of the object, etc.

In sensor systems on robots the role of tactile sensors is important when the object is in the gripper because it is then out of sight of the present robot vision. Their application specifies that they must be in contact with the object, often in extreme industrial conditions (high temperature, humidity). An additional limiting factor is the value of the contact force between the sensor and the object, so that tactile sensors are still used only for laboratory experiments. Research has sought to find suitable technology that ensures reliable operation of sensors in the industrial environment, and to achieve a certain level of performance (response time, resolution, sensitivity).

Tactile sensors consist of the matrix of tactile elements – taxels, each of them representing a miniature switch or a force sensor. Information is collected from each taxel and then a tactile image of the object is made, which can include distribution of the forces along the surface of the sensor, pressure, deformations etc.

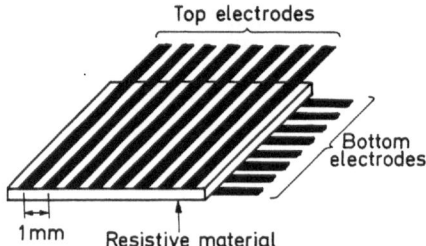

Top electrodes

Bottom electrodes

1 mm   Resistive material

Fig. 7.14. A typical tactile sensor

At present, when operating in different environments tactile sensors with resistive array represent the most acceptable solution for obtaining tactile information, both for performance and simplicity. Such sensors consist of two plates with groups of parallel leads, mutually perpendicular (see Fig. 7.14) with the resistive material in between. The pressure on the plate changes the local resistance properties. The local resistivity is measured on the 'crossing' points of the leads. Surfaces which carry the leads should be elastic enough to allow deformation. Early experiments with a special rubber with dispersed carbon particles did not give satisfactory results. The discrimination of pressure levels acting on a taxel was so bad that practically only two levels of pressure were distinguished and the sensor behaved as if it consisted of a matrix set of switches. Further research found more convenient materials (elastomers) which enabled recognizing eight pressure levels. A suitable resistive material should have a large change of resistivity in the pressure function (to increase resolution and decrease 'cross-talk' among taxels), resistivity to mechanical overload, small hysteresis, good repeatability, stability and a low fatigue. A typical dependence in material resistance for applied pressure is shown in Fig. 7.15.

A sensor must be accompanied by an analog signal conditioning device and a digital circuit that controls the scanning of tactile array and transmits the information to the control system. Typical scanning rate for the 16×16 taxel matrix is 4 kHz (taxels are scanned at 1 MHz).

Tactile sensor based on magnetostriction (the effect of changing the magnetic permeability of the material being deformed) was developed at North Carolina

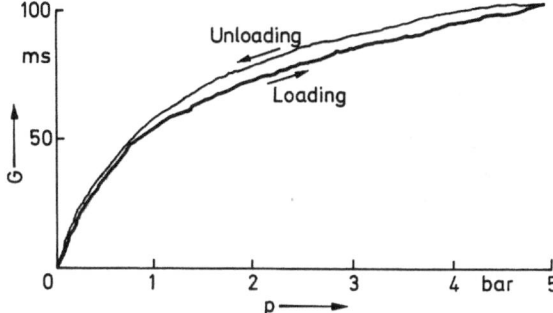

Fig. 7.15. Dependence of material resistivity for applied pressure

State University [7]. Since there are no moving parts (Fig. 7.16), the sensor is reliable and robust. It consists of 16×16 taxels, each of which represents a miniature transformer. An A.C. current is applied to the primary coils, and the output is measured after the multiplexer. In this experiment only two levels of the output voltage were used, so the binary picture was produced. The tactile picture was scanned at the rate of 1 kHz.

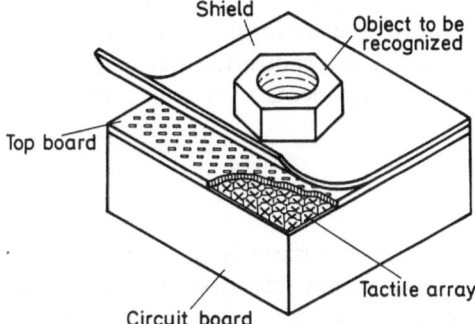

Fig. 7.16. Mechanical structure of the tactile sensor

The magnetoelastic material that was used showed promising results when the temperature sensitivity (a common problem with resistive elastomers), dynamic range, sensitivity, linearity and small hysteresis were considered.

An interesting tactile sensor (Figure 7.17) with an electro-optical force transduction was developed in the Lord Company, USA [8]. The tactile matrix has 8×8 taxels. The forming of the tactile picture has two phases: first, the force acting on a taxel is converted into the deformation (the displacement of the elastic element), and then the displacement is measured by the optical transducer and converted into the electrical signal. The other phase produces some difficulties because the mechanical and electrical properties of the emitter-detector pairs is not uniform over the sensing array, so it has to be compensated through the use of appropriate hardware and software. A pairing process is employed for matching these pairs and in software the output is changed for the correction factor (obtained during the calibration) of the taxel.

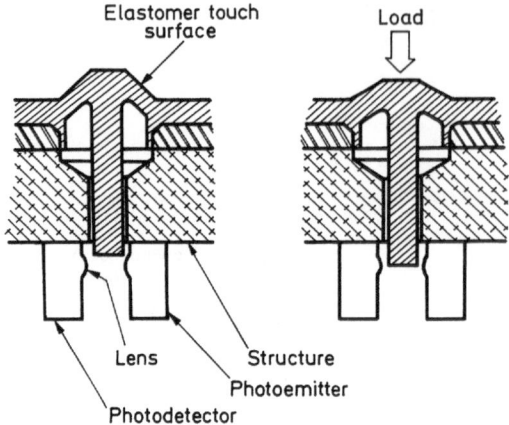

Elastomer touch surface

Load

Lens

Structure

Photoemitter

Photodetector

Fig. 7.17. Element of a tactile sensor

An interesting experiment is integrating a tactile sensor in the robot's gripper. Construction of a sensor using VLSI technology is described by Raibert and Tanner [9]. The sensor integrates the tactile transducers based on resistive elastomers and the electronic part for parallel signal processing resulting in an increased operating speed. In the experiment the sensor consisted of the 6×3 tactile matrix, and its characteristics, reliability and robustness were explored. It is clear that in the foreseeable future the components with bigger and more dense tactile matrices can be expected.

7.2.3 Ultrasonic Sensors

The application of ultrasonic sensors on robots has recently become the subject of much intensive research. It encompasses two main activities: determination of the distance from the robot to some object in the neighbourhood and forming the picture of the environment with information collected. Ultrasonic sensors have the advantage over robot vision in that they provide 3D information about the environment and do not require special illumination in the working space, but their disadvantage is in more complex signal processing.

They were initially applied on mobile robots for detecting obstacles. One moveable sensor or, more appropriately, fixed distributed sensors on the robot scanned the environment and the control system formed a 2D picture of the space where the obstacles were represented as polygons.

An ultrasonic sensor consists of two main parts: the source (emitter) of the ultrasonic waves and the corresponding detector. The range of operating frequencies is from 20 kHz to 300 kHz. The principle of operation is based on the measurement of the time delay between emission of the ultrasonic impulse and detection of the reflected impulse. When the speed of wave propagation is constant, that time delay defines the length of the wave path. Impulses are usually modulated (phase, time modulation, etc.) in order to be discriminated in the detector. The range of distances in which ultrasonic sensors can scan the environment is lower bounded by the geometry of the sensor and from the other side by the attenuation of the waves in the propagation medium. In the air the range is usually between 0.3 m and 10 m. However, difficulties occur with robots which operate through the medium of air. Attenuation is larger if the wavelength is small, but on the other hand better resolution requires shorter wavelengths. Local fluctuations of the air parameters (density, temperature, turbulence) impair picture quality. These problems did not occur to such extent when ultrasound was used for depth measurement, in medical tomography or materials analysis. There are no commercial components on the market intended for a straightforward application on robots and research is consequently aimed at devising such components that can be easily integrated in the gripper with appropriate operability.

An example of using an ultrasonic sensor for the detection of objects in a 3D environment is presented in [10]. The working space being scanned has the form of parallelepiped, and the sensor is situated in vertex (Fig. 7.18).

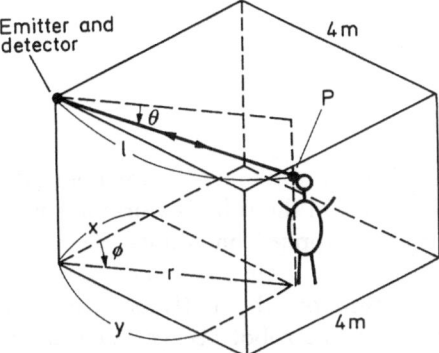

Fig. 7.18. Scanning of the space with an ultrasonic sensor

The sensor used in the experiment consisted of 21 emitter–detector pairs distributed as 2D array with an ultrasound frequency of 40 kHz. The emitter is placed in the focus of a small parabolic mirror in order to obtain the planar wave and diminish the dispersion of ultrasonic energy. The experiment consisted of scanning the working space with an appropriate angular resolution and detecting the persons in it. Two methods of scanning were applied: mechanical and electronic. The first method employed two step motors for changing the direction of the sensor. It gave satisfactory results, but showed some disadvantages: 85% of the time consumed was in operation of the motor and the motors produced parasitic noise and enlarged the size of the sensor itself.

Electronic scanning (Fig. 7.19) is based on producing the planar wave in the desired direction by the superposition of the waves from individual emitters having the appropriate (programmable) time delay.

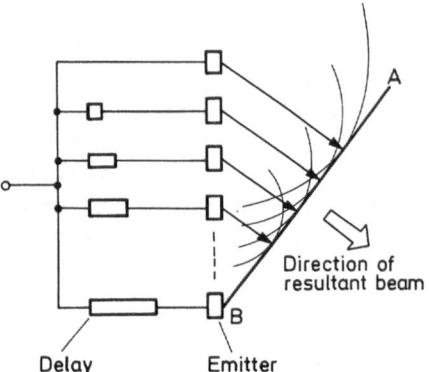

Direction of
resultant beam

Delay   Emitter   Fig. 7.19. Electronic scanning

This method of scanning is much faster than the mechanical one and the sensor has no moving parts.

The use of ultrasonic systems as robot sensors is not limited to the distance measurement. They can be used in combination with some other sensors (e.g. tactile), but the primary purpose is 3D scanning of the environment.

7.3   Robot Vision

The role of robot vision is to feed the control system with information about the shape and the spatial distribution of the objects in the robot's working space. The existing vision systems cannot be directly applied on robots. Electronic hardware that should support the use of vision on robots is still under development. The vision sensors provide 2D information about 3D enrivonment. Therefore, they have to be properly placed on the robot because such information has to be preprocessed (extraction of elements required from the picture). Direct analogy with human eyes suggests that the camera should be placed above the working area of a robot, but from several aspects it is more convenient that the sensor be connected to the gripper or the last segment of the robot. It simplifies the transformation of coordinates, and because the sensor is in the working space near the observed object the necessary resolution is less. The small resolution shortens the picture processing time, enabling the robot to operate with a sensor in real-time. Vision sensors are also used for distance measurement with a laser and the appropriate optical detector.

7.3.1   Illuminating the Scene

To operate successfully, vision sensors require a proper working space illumination. The factors that determine the scene are the illumination, the observed object, the environment, the sensor used and the robot itself.

The object influences the picture by its shape (holes, shades), optical properties (coefficients of reflection, absorption, transparency) and the degree of surface finish. All these require the appropriate light sources (punctual, fluorescent) with convenient space distribution.

The environment of the object may pose difficulties concerning contrast, and the reliable detection of edges determining the object. The change of light direction can solve the problem, but some important elements may be obscured.

The sensors influence the picture shape by its spectral characteristic (the distribution of sensor sensitivity by light wavelengths) and the spectral characteristics of the optics used for picture focusing.

The light sources connected to the moving parts of the robot may impair the dynamic performances of the robot. Sudden movements, vibrations and inconvenient positions can cause the destruction of the light sources or the robot can obscure by its movements some elements of the scene important for the proper operation which implies the need for multiple light sources or an *a priori* knowledge of the working space.

7.3.2   Special Vision Sensors

First applications of the vision on robots employed commercial cameras, slightly modified in order to be more robust and to allow reliable connection to the robot. Later, the whole line of specialized components (CCD arrays, DRAM components with a transparent cover over the silicon chip, fibre optics, semiconductor detectors, etc.) were incorporated into vision sensors because they were more suitable for robot vision application (easier integration with gripper, robustness, reliability).

Dynamic Random Access Memory (DRAM) components, specially made for use as optical detectors, appeared on the market. The light penetrates through the transparent cover to the silicon chip. Memory cells are set to such a value that they are electrically charged. Cells that were illuminated become discharged (because of the photoelectric effect) and the picture projected on the chip directly converted into the binary information available to the control computer for further processing. Different levels of gray are obtained by changing the exposition time. In an experiment (see Fig. 7.20) with these components [11] a system with resolution 256×128 was made, which, including the optical system (lenses, blenda), had a size of $2.5 \times 2 \times 3$ cm.

This resolution is satisfactory when pattern recognition is considered, the whole sensor being much cheaper than the appropriate one with a vidicon tube or CCD arrays.

Fibre optics represent a new technology that is being intensively applied on vision sensors. Such sensors usually consist of a bundle of optical fibres (Fig. 7.21), with one side in the gripper directed to the observed object and the other connected to an optical detector. The object must be illuminated by some external light source or a part of optical fibres is used instead, transferring the

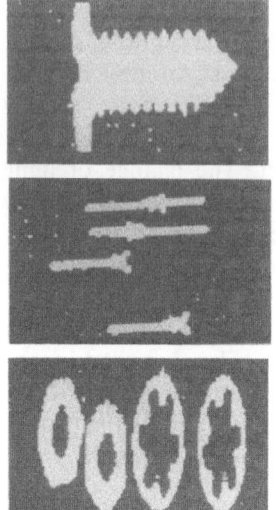

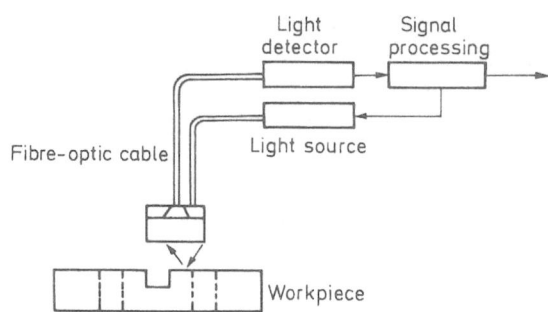

Fig. 7.20. The picture of objects obtained with DRAM sensor

Fig. 7.21. A vision sensor with fibre optics

light of some distant light source. The picture (Fig. 7.22) of the object is obtained on the matrix consisting of the fibre endings and is then transferred to the other end of the CCD detector, the linear detector array, or a photodiode matrix, which convert the picture into electrical signals. Advantages of this method for obtaining the picture are that the present level of the fibre optics technology allows the production of small sensors with high resolution, easily integrable with the robot gripper, and since they have a negligible weight, the presence of the sensor does not influence the dynamics of the robot. The coupling of the other fibre end to the photodetector is mechanical or focusing optics are used. CCD components have some advantages over vidicon tubes: they are integrated circuits that beside the detector contain signal conversion electronics. They have no retention, the excellent positional accuracy of each detector ensures precise

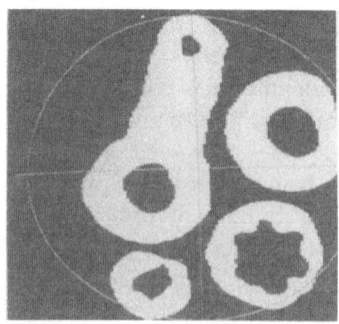

Fig. 7.22. Pictures of objects obtained with a fibre-optic sensor

measurements and the spectral sensitivity has the maximum in visible and infrared spectra range.

7.3.3   Lasers

The application of lasers with appropriate optical detectors on robots is mainly directed to distance measurement which can be easily extended to 3D scanning of the working space. Lasers that are present on the market are not suitable for use on robots because of the sensitive and complex mechanics needed for directing the laser beam and the high price of the device. Nevertheless, the possibilities offered by lasers indicate a wide range of applications in future robotics.

A simple technique for distance measurement using a laser is presented in Fig. 7.23.

Laser source S emits a narrow beam perpendicularly to the axis of rotation of the mirror M on its surface. The mirror rotates with angular speed ω, and the reflected beam, according to the laws of reflection, rotates with 2ω. The lens L,

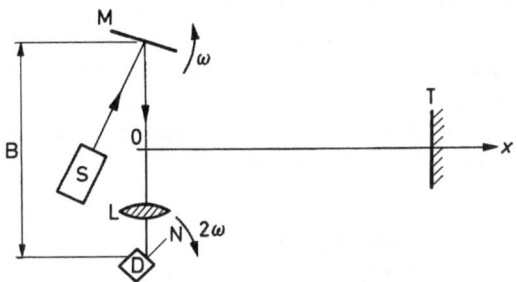

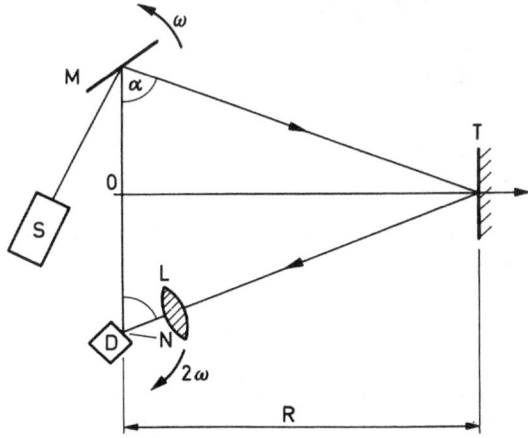

Fig. 7.23. Distance measurement with a laser

focusing the beam on the detector D, rotates with 2ω, and is synchronized with the mirror so that the angle α between the line B and the reflected beam is equal to the angle between the optical axis of the lens L and line B in every time instant. The supposition is that the scanned object has the appropriate coefficient of reflection such that the place of the beam incidence can be detected. According to Fig. 7.23, the distance R is given by relation:

$$R = B/2 \tan \alpha$$

The angle α is calculated by measuring the elapsed time t between the path source–mirror–detector and the path source–mirror–object–detector. The speed of rotation of the reflected beam 2ω implies:

$$\alpha = 2\omega t$$

A detailed analysis shows what relations among ω, R and B result in some predefined accuracy, if the minimal measurable time delay is known. It follows from the second relation that the stability of the speed of rotation has negligible influence if it is stable at small time intervals, comparable to t.

The described method of the distance measurement should only present the basic idea, while in practice more complex systems with complex optics (usually only mirrors) are used with more reliable synchronization of the laser source and detector which enables more precise time measurement and rotation of the scanning plane so that a 3D picture of the environment can be obtained (Fig. 7.24). Linear CCD arrays are usually used as detectors.

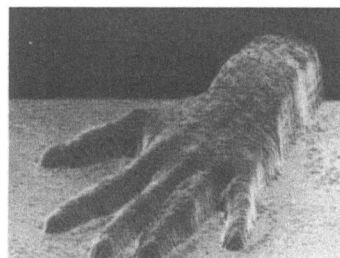

Fig. 7.24. Reconstruction of the shape of objects scanned with a laser

In applications where it is necessary to detect the edge of an object the laser systems that generate laser plane are used. The discontinuity of the line indicates the edge (Fig. 7.25).

It is possible to determine the slope of the plane and the shape of an object using the laser planes.

The use of lasers in robotics offers other possibilities, too, but the appropriate technology for a wider application is still being developed.

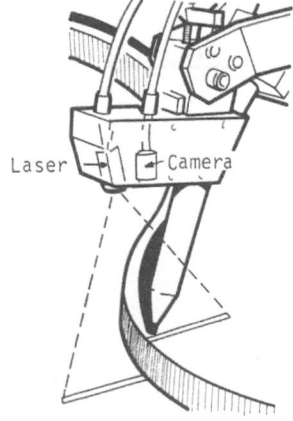

Fig. 7.25. Edge following with a laser sensor

References

[1] Cullum, M.W., "Angle-Sensing Transducers for Shaft-to-Digital Conversion", North Atlantic Industries Technical Bulletin 121, 1980.

[2] Davis, S., "Digital Measurement of Shaft Position: Synchros and Resolvers or Encoders", Computer Design, February 1976.

[3] Goto, T., Inoyama, T., Takeyasu, K., "Precise Insert Operation by Tactile Controlled Robot", Proc. 2nd Conf. on Industrial Robot Technology, Bedford, 1975.

[4] Piller, G., "A Compact Six-Degree-of-Freedom Force/Torques Sensor for Assembly Robots", 12th ISIR, Paris, 1982.

[5] Whitney, E.D., Nevins L.J., "What is the Remote Centre Compliance (RCC) and What Can It Do?" 9th ISIR, Washington, 1979.

[6] DeFazio, L.T., Seltzer, S.D., Whitney, E.D., "The IRCC Instrumented Remote Centre Compliance", Industrial Robot, 1984.

[7] Luo, C.R., Wang, F., Liu, Y., "An Imaging Tactile Sensor with Magnetoresistive Transduction", Int. Conf. on Intelligent Sensors and Computer Vision, Cambridge, USA, 1984.

[8] Rebman, J., Morris, A.K., "A Tactile Sensor with Electrooptical Transduction", 3rd Int. Conf. on Robot Vision and Sensory Controls, Cambridge, USA, 1983.

[9] Raibert, H.M., Tanner, E.J., "Design and Implementation of a VLSI Tactile Sensing Computer", Int. J. of Robotic Research, Vol. 1, No. 3, 1982.

[10] Kuroda, S., Jitsumor, A., Imari, T., "Ultrasonic Imagining System for Robots Using an Electronic Scanning Method", Robotica, Vol. 2, 1984.

[11] Whitehead, G.D., Mitchell, I., Mellor, V.P., "A Low Resolution Vision Sensor", Journal of Physics, 1984.

[12] Pugh, A., Robot Sensors, IFS Publications Ltd, UK, 1986.

[13] Sveider, A., Gorineviskii, D., Lenskii, A., Mozevelov, S., Force–Torque Sensors for Robotic Systems (in Russian), Preprint of the Institute for Problems of Information Transmission, USSR Academy of Science, 1984.

Chapter 8

Elements, Structures and Application of Industrial Robots

8.1 Basic Postulates of Industrial Robot Design

A great majority of contemporary industrial robots, i.e. their mechanical part, frequently referred to as the manipulator, are in the form of bigger or smaller articulatory ("anthropomorphic", "arthropoid") or portal ("gantry") hoists. These working machines, mostly used for manipulating workpieces, are frequently known as "manipulation robots", although this is not applicable in a great number of cases (welding, grinding and drilling with a carried electrical tool, etc.). The normal end or manipulator tip is in the form of a terminal organ for gripping in the form of the so-called gripper which can have a diversified design, which will be discussed below.

Figure 8.1 illustrates a typical contemporary industrial robot in vertical articulated configuration, with electromechanical drives and a carrying capacity

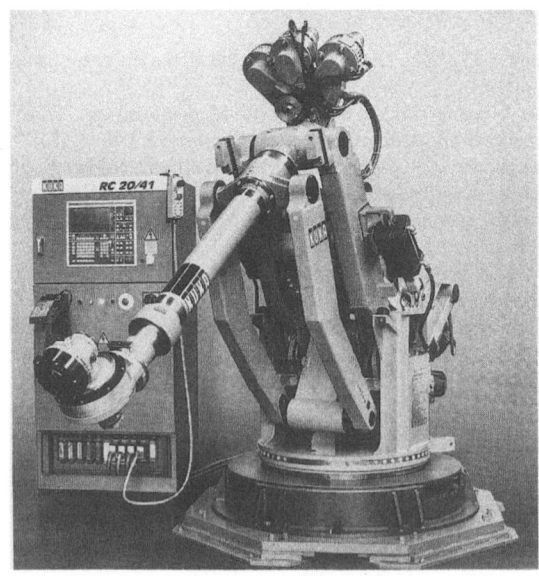

Fig. 8.1. Contemporary industrial robot ("KUKA"–FR Germany)

of 60 kg of the IR 601/60 type from the West German firm of Kuka. It can be seen that massive segments of the robot are in the form of cast, large cross-section boxes for rigidity, and large bearings, a massive pedestal and the far positioning of the three gripper drives can be seen.

8.1.1 Specifications of Industrial Robots as Mechanical Structures

Lay-out and design of the mechanical part of industrial robots (manipulators) demands the analysis of the state of motion and the acting forces, i.e. of the necessary holding and working forces, then forces in the joints and the driving forces aimed at dimensioning the carrying elements, joints and drives (actuators). Manipulators with active joints generally possess a structure of unbranched open kinematic chains with one degree of freedom per joint (V-class kinematic pairs). Each active joint is equipped with one drive whereby the joint can rotate or move translatorily. In that case the robot structure becomes branched into a closed kinematic chain (see Fig. 8.1) and the calculations become rather complex but it is possible to carry them out by means of automatic algorithms using the corresponding software packages.

The motion sequences for the individual drives which in the first place are characterized by the acceleration and deceleration phases are obtained during

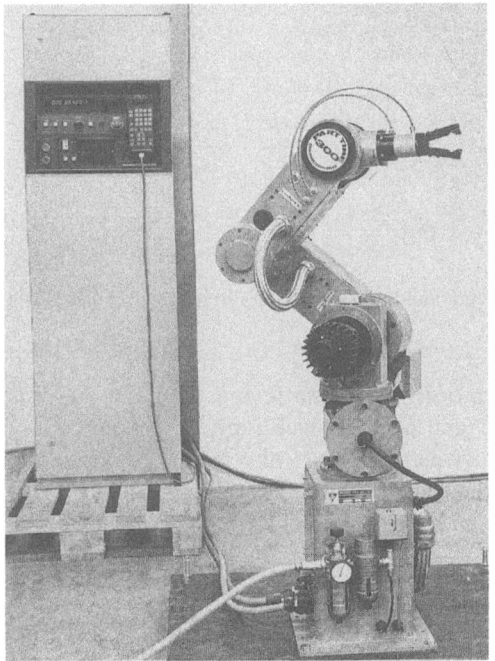

Fig. 8.2. Robot with actuators in joints ("Dainichi-Kiko" – Japan)

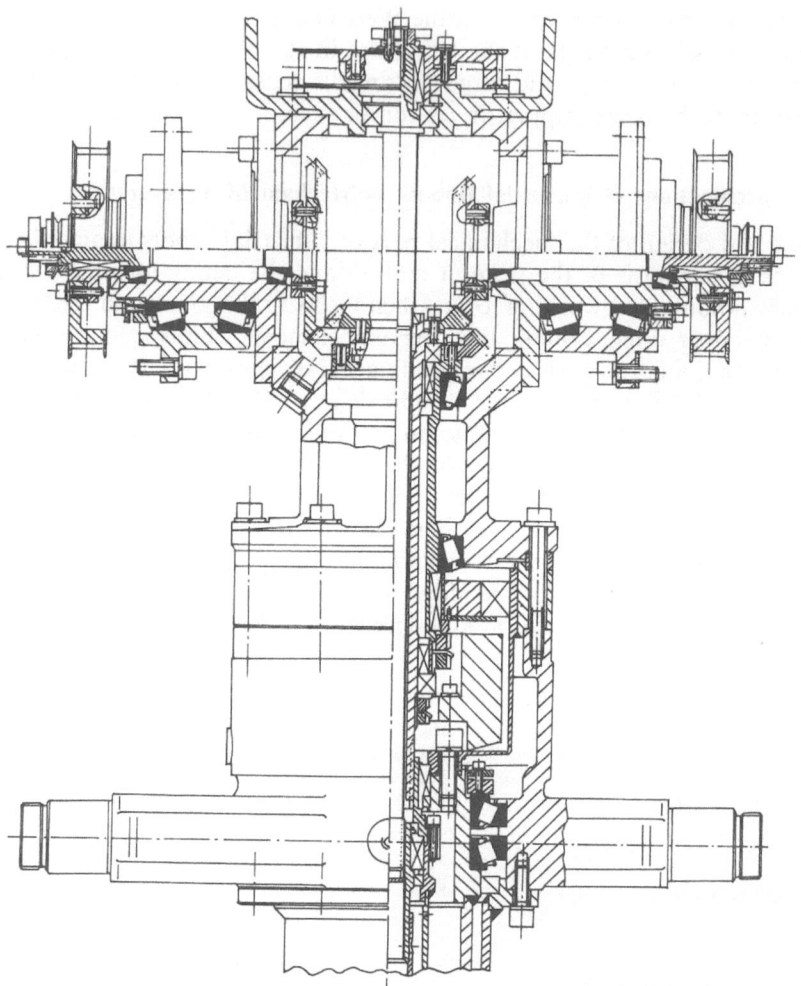

Fig. 8.3. Contemporary solution of central column and "shoulder" joint bearings of industrial robot

working. Hence care should be taken during kinematic and dynamic analysis about the characteristics of the drive (actuator) of each joint.

In the case of controlled (servo) drives, care must also be taken about the action of the individual control loops in the drive functions, as for example with the P.T.P. control, where the prescribed drive functions are realized by means of control loops.

The drives (actuators) of the industrial robot mechanical part are diverse complex mechanisms. Because the real robot segments possess inertia and elasticity and the joints are acting as driving forces (torques) and friction forces (torques), it can be concluded that the manipulator kinematic chain represents a complex dynamic system with inertiality, elasticity and damping.

With modern industrial robots, various types or drives find their use: electric, hydraulic and pneumatic. These drives are distributed on the mechanical robot part in different ways, but more and more drives are being placed directly into the corresponding joints (Fig. 8.2, Dainichi-Kiko 300) thereby excluding any drives which are weak. On the other hand, drives outside the joints transfer their motion in the same way by means of mechanical reducers or levers. In fact, with industrial robots, a demand has arisen for compact drives and driving units containing a driving electric servo-motor-reducer, feedback elements regarding position and speed (encoders and tachogenerators) and, most often, the brake. The reducers of such driving units have a high reduction ratio (a few hundred, even thousand) are of minimal dimension and have the smallest possible play. Play augmentation in the driving chain causes instability and oscillations, which is reflected unfavourably in the settling time and positioning accuracy.

From the standpoint of mechanical engineering, the structures of industrial robots are used on sub-assemblies of tool machines or hoists (cranes), but differ by a much smaller mass and more precise execution which necessitates smaller gravitational and inertial forces and moments with substantially less of play (dead zones). An example is given in Fig. 8.3, representing the design solution of the main central robot column and the "shoulder" joint, as well as their bearings.

8.2   Design Solutions of Main Robot Sub-assemblies

8.2.1   Pedestals of Robots

Pedestals (bases) of robots exist in two basic versions:

1. standing or hanging,
2. portals (gantry types).

Pedestals of the first version, beside the basic function of carrying the whole "robot arm", very often in the case of electrohydraulic robots contain the hydraulic power source as in the known robots of that type the Unimate 2000 and 4000. At the present time industrial robots are electrically driven, with the example of a modern robot pedestal (Cincinnati-Milacron T3-756) being given in Fig. 8.4, executed in the form of a massive cast box with a cover, fastened by means of screws and containing the bearings of the main column with the drive unit, consisting of a servomotor-reducer with a tachogenerator and encoder (Fig. 8.5). In the production of this type of pedestal, which primarily serves as a stable base for erection and fastening to the floor the weight of the pedestal is not considered, because it is only important when the robot is being carried and mounted in place. The case is quite different with elements of the robot "arm" – these are now mostly the "aftarm" and "forearm" of the so-called

Fig. 8.4. Pedestal of standing robot ("Cincinnati Milacron" – USA)

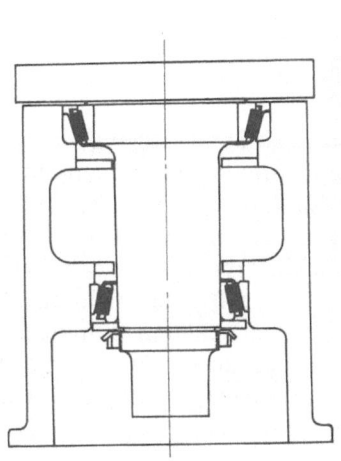

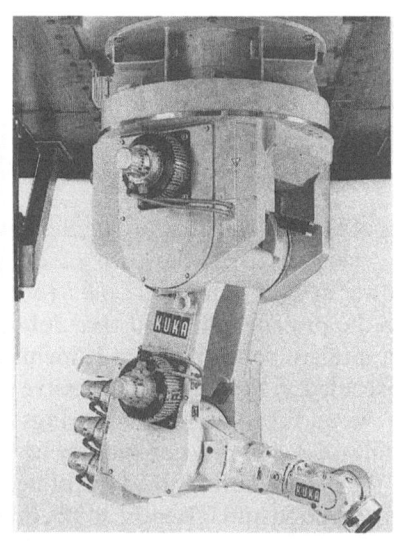

Fig. 8.5. Solution of robot pedestal

Fig. 8.6. Pedestal of hanging robot ("KUKA" – FR Germany)

articulated or anthropomorphic, or arthropoid configuration, which will be discussed in more detail.

Again the case is different with robots which because of their task and lack of space or better access to the workpiece, (e.g. welding "from above") are suspended to hang in an inverted position, as illustrated by the Kuka robot in Fig. 8.6. The pedestal has been produced by welding together thick sheet elements, thus achieving greater rigidity together with an acceptable moderate mass.

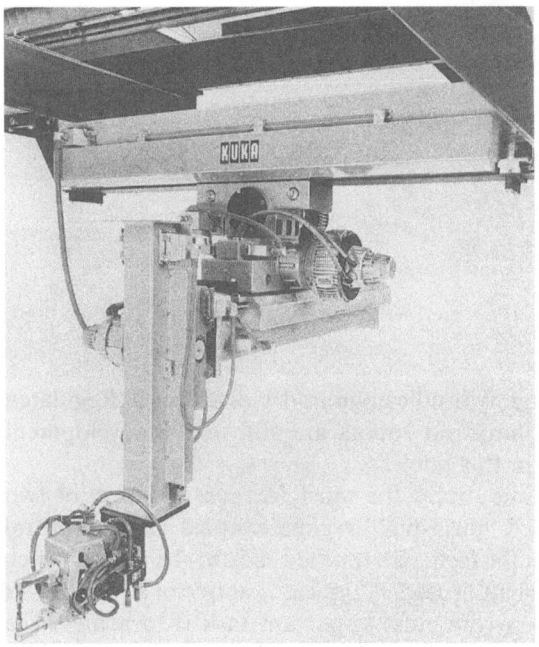

Fig. 8.7. Gantry (portal) type robot ("KUKA" – FR Germany)

Other versions of the portal or "gantry" type exist and much experience can be gained from these designs. A typical solution is illustrated in Fig. 8.7, showing the manipulator in the so-called T-T-T configuration working in the Cartesian coordinate system. Along the portal, a "cat" is moving, ensuring the transversal motion of the robot "arm". In Fig. 8.8 an example has been given of one of new robot types with electromechanical drives from the Swedish firm ASEA, representing a portal robot (the toothed rack for transversal motion drive can be clearly seen) with an universal joint of the "arm" so that this robot works in a spherical coordinate system moving as a whole along a linear transversal path (along the portal). Such a solution ensures unobstructed access from above to the workplace where mounting of precise sub-assemblies can take place.

"C"-pedestals should also be mentioned which present a stiffened version of the column for linear vertical motion but as previously mentioned, this solution is being used less in modern robot design.

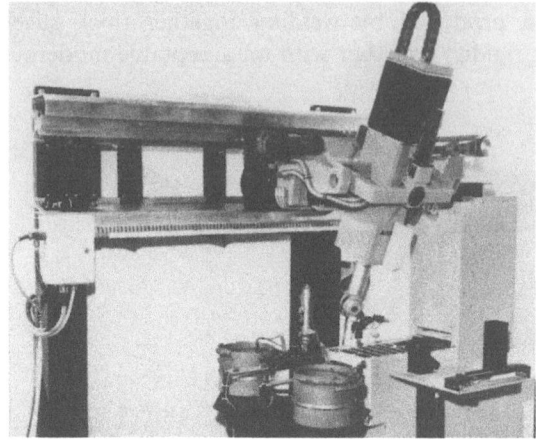

Fig. 8.8. Gantry robot with universal joint (ASEA – Sweden)

8.2.2   Angular Drives

With industrial robots electric and hydraulic angular drives are used. Regulated (servoed) pneumatic drives for industrial robots are still in the development phase and will not be reviewed in this book.

In the case of electro-hydraulic drives the most frequent solution of two cylinders working in the so-called "push-pull" regime coupled via the central part of the common piston rod in the form of a toothed rack to the pinion which is an integral part of the output shaft is used. A typical example of such angular drive is shown in Fig. 8.9, whereby the most important task is to achieve the coupling of the toothed rack and pinion with the smallest possible play. This task is solved by very close machining and applying a spring pressing the toothed rack constantly against the pinion or v.v.; tooths are of first-class quality, properly generated, with the highest surface treatment obtained by thermal processes, grinding, lapping, etc.

Electromechanic angular drives play a significant role in modern contemporary industrial robots. Today, fast and lightweight electric servomotors have demanded improvements in new design of the existing development of mechanical reducers. Classical geartrain reducers using several reduction steps with the same number of gear pairs, along with the needed reduction ratio of approximately fifty to several hundred (even thousand in some cases), have not been used for some time. The basic reason was that they needed very small play at the output shaft which in the case of several reducing steps was in the order of only a few angular minutes (normally up to 3) which could not be achieved and

Fig. 8.9. Hydraulic angular actuator

which had a positional accuracy of under 1 mm; compared to a typical value with robots with a medium-size "arm" of about 1.5 m is ±0.2 mm.

Worm gears allow very high reducing ratios in one step; however, their mechanical efficiency is so low in the case of high reducing ratios that this is often not acceptable as regards energetic robot efficiency, which itself is still very low.

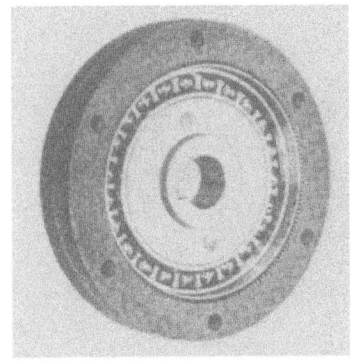

Fig. 8.10. HDUC unit
Fig. 8.11. HDUF unit

Robots with mechanical reducers marketed by Harmonic Drive (USA) have found widespread use working on the principle of deforming a pot- or ring-form element as the outer ring of a ball race. They attain a reducing ratio in one step of up to 1:320 (with some models even more). Figure 8.10 and 8.11 illustrate two basic models of the reducer type: pot-form HDUC and the flat HDUF. The drawback of this reducer type with robot designers lies in their limited angular stiffness due to the elastic element in the geartrain, which in some work regimes (sudden acceleration or braking) develops a state of high-frequency oscillations of the robot "arm" leading to prolonged settling time at the terminal points of the trajectory resulting in time loss while the robot settles down, notwithstanding the high medium robot velocity. However, angular drives produced in series in combination with reducers of this type (see the example given in Fig. 8.12) are widely used in the design of fast assembly robots of a smaller carrying capacity and also with some robots of a larger size and capacity as one member of the geartrain (e.g. drive of ballscrew with the robots ASEA, MOTOMAN, etc.).

Much has been gained by the CYCLO reducers of the European firm Infranor which, along with a more complex design, demonstrates significantly higher angular stiffness values compared to the previous solution and have acceptably small angular play at the output shaft. The working principle of this unusual reducer design is represented in Fig. 8.13: if the central excentre rotates in the sense of the arrow, the engagement point with the outer ring (internally toothed) travels from point M to N; whereby the disc centred in K rotates inside the ring in an inverse sense of rotation. In practice two systems moving angularly by 180°, are used. By means of small angular "interpretensioning" the necessary small angular play of the output shaft is achieved.

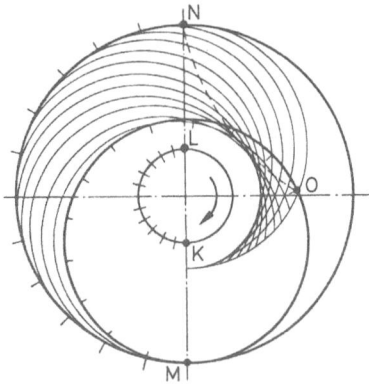

Fig. 8.12. Integrated electromechanical actuator (Harmonic Drive – USA)   **Fig. 8.13.** Work principles of CYCLO reducer

These reducers are also combined with driving servomotors, tachogenerators, encoders and frequently with brakes into complete angular drives, one execution of which is presented in Fig. 8.14.

Such angular drives are chosen from production and designed directly into the robot joints. However, due to the mass of these actuators such direct implementation into the joints is basically applied only for the first three robot degrees of freedom (basic robot configuration), while the gripper degrees of freedom, usually also three, are normally powered by means of push-pull or torsional rods with universal joints, chains, toothed belts, etc.

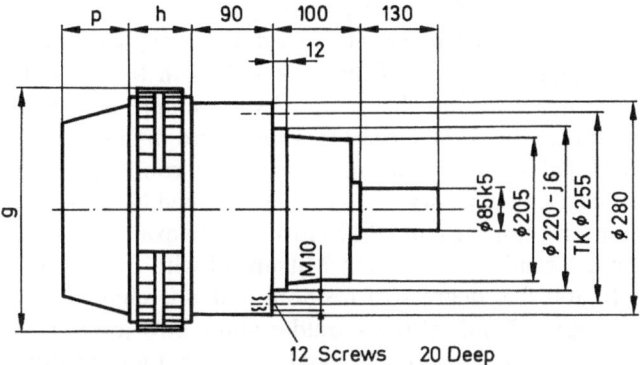

Fig. 8.14. INFRANOR-CYCLO actuator (INFRANOR – Switzerland)

The best solution according to work published is the angular electro-magnet (the Direct-Drive-System), in which any reducer is superficial giving a simpler and safer solution. Such angular electromagnets, or motors, are applied (in 1986) only in the case of a smaller assembly robot as in the case of the Adept Company (USA). This demonstrates the only basic deficiency in the form of a very complex control system, only able to control such a system in stable manner with high-torque and low inertia characteristics. They also still possess a rather high mass, so generally only the degrees of freedom near the robot pedestal can be driven directly by that type of actuator. This is also the case with the Adept-one and Adept-two robot of the above-mentioned firm, illustrated in Fig. 8.15.

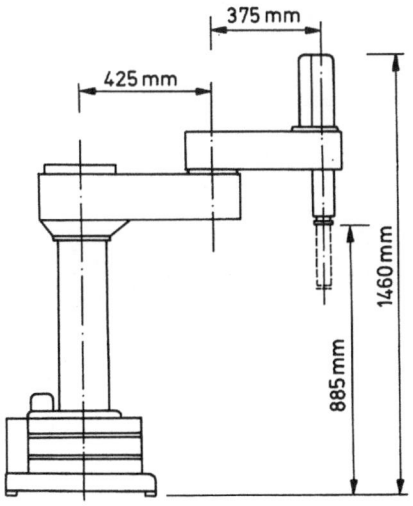

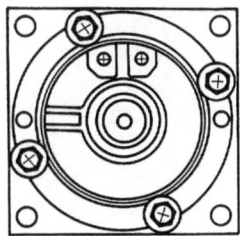

Fig. 8.15. Adept-One robot with direct drives (ADEPT – USA)   **Fig. 8.16.** Vane actuator

In the case of the hydraulic robot drives for powering the gripper degrees of freedom the so-called Vane-type actuators are often used (Fig. 8.16) which in the one-vane version work through an angle of up to 300° and in the two-vane version up to 150°. These actuators are basically of very simple design but demand high technology for the internal surface treatment and manufacture of the linear seals along the vane's edges. With smaller assembly robots (reaching 0.5 m) and with a smaller carrying capacity of up to 1 kg, toothed belts are used to a great extent for angular motions. Figure 8.17 illustrates an example of such a solution by the Japanese assembly robot from Mitsubishi with four toothed belts driving four out of total five degrees of freedom and one chain for the realization of the "parallelogram" link of the shoulder and elbow joint. These toothed belts enable motion drive without play and although they introduce some elasticity in the transmission disturbances are low due to the small mass of the robot segments and the workpiece. These transmissions do have considerable damping effects due to friction of the belt teeth along the belt wheels acting favourably on its control characteristics.

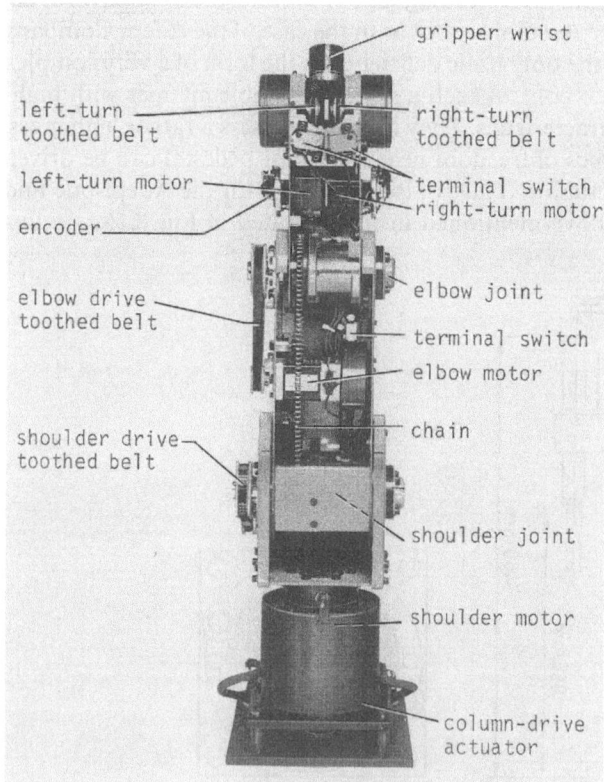

gripper wrist
left-turn toothed belt
right-turn toothed belt
left-turn motor
terminal switch
right-turn motor
encoder
elbow drive toothed belt
elbow joint
terminal switch
elbow motor
shoulder drive toothed belt
chain
shoulder joint
shoulder motor
column-drive actuator

Fig. 8.17. Angular actuators of Mitsubishi robot (MITSUBISHI – Japan)

8.2.3 Linear Drives

For linear (translatory) movements numerous solutions are at hand in robot design, featuring special characteristics in electrohydraulic and electromechanical drives. Thus Fig. 8.18 (a–e) the first, and Figure 8.18 (f–h) the second, possibilities are given for solutions to linear drives with robots.

Solutions in Fig. 8.18 (a–e) – hydraulic drives – the danger of play between the driving unit (hydraulic cylinder) and the executive robot does not exist, while in Fig. 8.18 (d) it does, so this must be taken into consideration using the available means: pinion laterally pressed against the rack, spring-loaded twofold pinion, etc. In the solutions for linear drives with electromechanical actuators (Fig. 8.18, f–h), in all three cases the problem must be taken into account; as in the case of the "ball-screw"–"ball-nut" coupling minimum play can be achieved, while for all linear guides the so-called "linear ball races" are used, in which, as with the "ball-screws", steel balls are permanently in circular motion, which significantly reduces friction along with minimum play in the order of 1–2 μm.

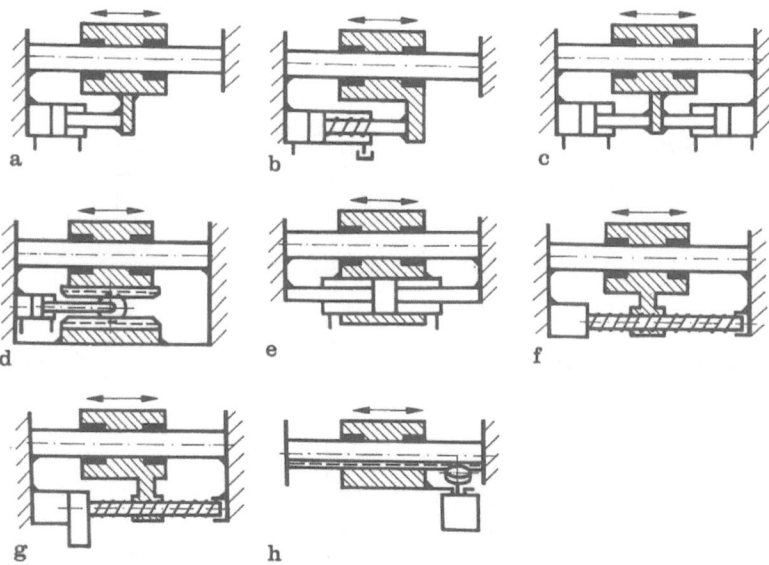

Fig. 8.18 a–h. Various solution of robot linear drives

Figure 8.19 presents a modern solution of a servocylinder for use with robots. However, contemporary solutions of electrohydraulic driving units use hydrocylinders with an integrated position sensor usually of the inductive type and with an integrally mounted electrohydraulic servovalve as illustrated in Fig. 8.20. Also very often modular linear units are used, consisting of one servocylinder and two parallel guides (Fig. 8.21). The guides, terminal micro-

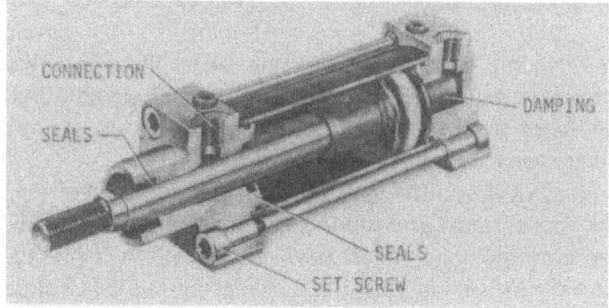

Fig. 8.19. Contemporary solution of servocylinder

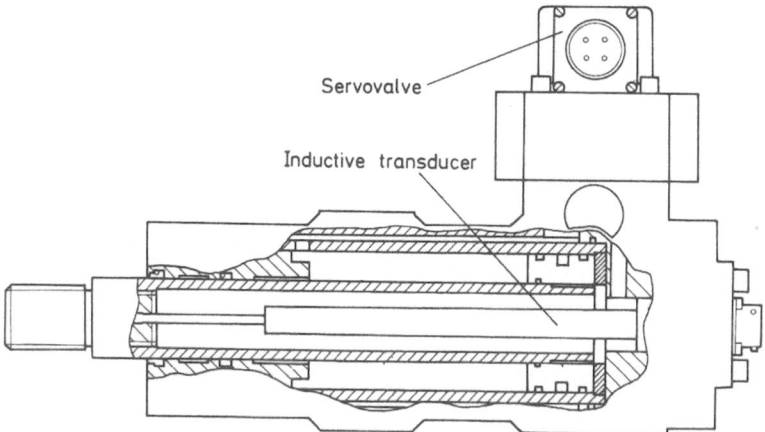

Fig. 8.20. Integrated linear actuator with servovalve and position transducer

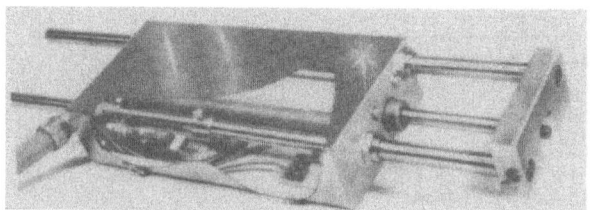

Fig. 8.21. Modular linear unit actuator

switches and electric terminals of the inductive position sensor can be depicted in Fig. 8.21.

A separate problem is encountered when combining the linear driving unit with angular drive which is often necessary in the case of robots working in the cylindrical coordinate system. Here it is necessary to complement linear play-free guidance of the linear unit besides rotating the whole entity at a certain angle. In the case of tubular elements moving inside one another this is achieved by grinding flat guiding surfaces along the same, and exactly produced and assembled, steel races rolling along these flat paths (as shown in Fig. 8.22) or by means of grooved wheels, usually eight, in the case of linear motion of a square element.

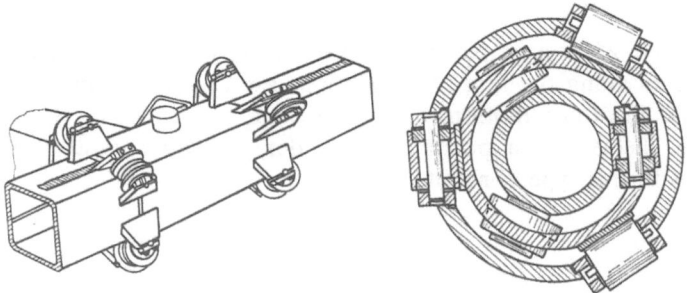

Fig. 8.22. Solutions of linear guides

Electromechanical linear units, when used for the combination, cause much greater design and technological problems. All the earlier problems with danger of play have linear angular and linear geared transmissions; here they appear jointly and require solving as described. A possible solution of the linear/angular unit is illustrated in Fig. 8.23, where the lower electric motor (EM 1),

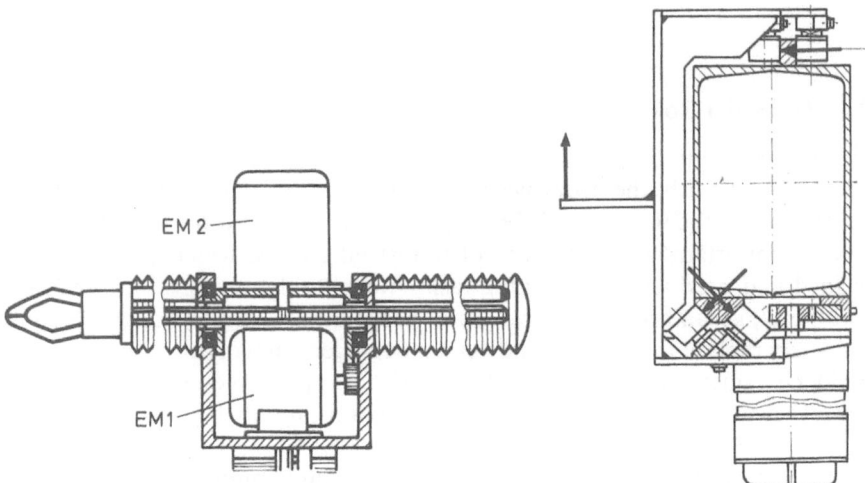

Fig. 8.23. Unit for linear and angular motion

Fig. 8.24. Linear unit of gantry robots

attached to the upper part of the central robot column, drives, via a toothed segment, the complete upper linear unit with its driving motor (EM 2), giving drive to the linear motion of the gripper.

Special solutions of linear drive appear with gantry (portal) robot types and here already tested solutions are applied, one of which is shown in Fig. 8.24. In this case too the driving unit is in the form of a servomotor-reducer connected via a pinion to a rack, while the linear bearings for linear motion are implemented by means of six needle-bearing rollers (three are illustrated in this section).

Finally, in Figure 8.25 an example of a precise assembly robot is presented, by which movements along the first two directions are realized by means of electric units while the third motion, also linear and perpendicular to the previous plane, is effected by a pneumatic cylinder, an almost universally adopted practice with this type of robot.

Fig. 8.25. SEIKO XY 2000
linear motion robot
(SEIKO – Japan)

8.2.4   Industrial Robot Grippers

Grippers, or terminal organs of industrial robots, are the parts which come into direct contact with the workpiece, or, which is often the case with some worktool when the workpiece is being processed or worked on. Hence grippers are very important working parts of industrial robots, which are still being designed and produced for working tasks. Regrettably, universal grippers still do not exist, i.e. an artificial human hand has still to be developed, which could completely replace this natural and very capable terminal organ. Nevertheless it should be remembered that industrial robot grippers frequently have to lift loads of several hundred, even thousand kilograms, totally beyond the capacity of a human hand.

The largest group of grippers is formed by simple models with pneumatic drives for gripping standard-form objects (e.g. those of similar dimensions or form, or both). These grippers are most frequently in the form of "scissors"

(Fig. 8.26) or with parallel motion of the gripper jaws (Fig. 8.27). They are in fact carriers of the jaw elements themselves which are attached by screws as can be seen in the figures. These added jaw parts are adapted to the form and material of the workpieces.

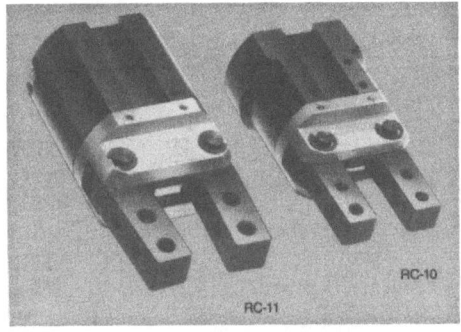

Fig. 8.26. Scissor-type grippers

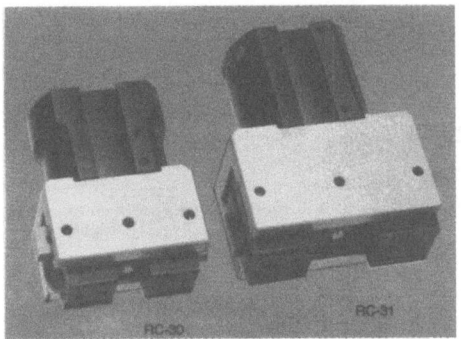

Fig. 8.27. Parallel jaw-motion grippers

Frequently gripping over two surfaces, or even along two lines, is not sufficient to ensure transferring and handling of the workpiece, as in the case of hard material which tends to slip out of the jaws. For these special solutions like the one from the Japanese firm Fanuc are applied for secure gripping of round workpieces of different diameters (see Fig. 8.28). By parallelogram linking of each of the three jaw elements, their movement which runs parallel to the gripped surface under a constant angle is ensured, so that twisting of the workpiece during gripping is avoided which can be important in some cases.

There are often unusual solutions for grippers handling fragile, breakable, weak or workpieces with a specially treated (e.g. polished) surface. Thus the gripper illustrated in Fig. 8.29 is executed in the form of two or more (usually three) "fingers" made of a special rubber compound which by using solution of an internal plane and external ribbed surface have the property of bending when subdued to internal air pressure and so gripping the workpiece "softly".

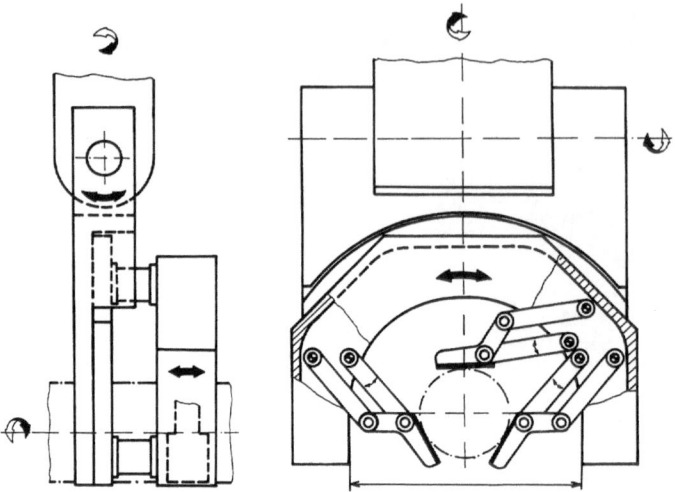

Fig. 8.28. Gripper solution for various diameter round objects

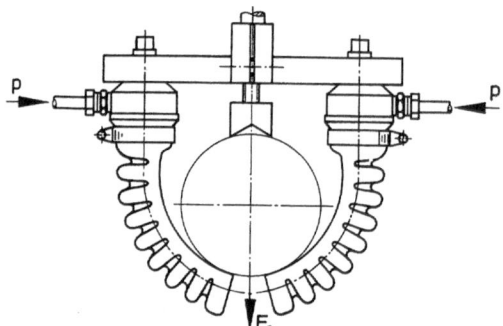

Fig. 8.29. Gripper with "fingers"
(SIMRIT – FR Germany)

For gripping large plane or slightly curved relatively lightweight workpieces, like glass, plastic or sheet metal plates, as well as big box-like products with large flat surfaces (plastic inlays for refrigerators, big boxes, etc.) vacuum grippers are used, the typical form of which is shown by Fig. 8.30. These so-called "vacuum-cups" are produced of various sorts of natural or synthetic rubber in order to be able to adapt to the form of workpieces. Thus for work in high temperatures silicon rubber is used. As vacuum source vacuum pumps are used, or injector systems, ensuring absolute vacuum of 0.2 bars, which is absolutely sufficient for this application. However, it should be stressed that these injector systems are uneconomical where air consumption is concerned. For gripping especially sensitive workpieces, for instance, eggs and lamps in the food and light industry, special sorts of vacuum cups in the form of soft rubber bellows have been developed, while for gripping heavier objects, e.g. car windshields, large and powerful grippers were developed.

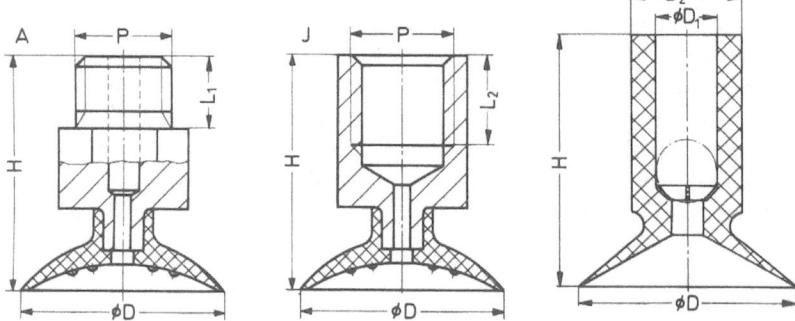

Fig. 8.30. Vacuum cups for grippers (SIMRIT – FR Germany)

To enable gripping two or more different work objects gripper carriers with several attachments and positions were produced. Figure 8.31 illustrates a good example of a gripper carrier (wrist) with two flanges (produced by Hirata, Japan), which are turned into the work position by means of a vane-type angular drive with two end positions. Via a toothed belt, this drive is transmitted to the carrier of the two flanges, along with an adjustment of the reduction ratio to obtain the angle of 90°.

A good example of the gripper carrier solution of the mini-robot D-Tran (Seiko, Japan) is shown in Fig. 8.32. A vane-type actuator has been applied enabling positioning of the gripper in the desired position. End positions are ensured by means of end stops.

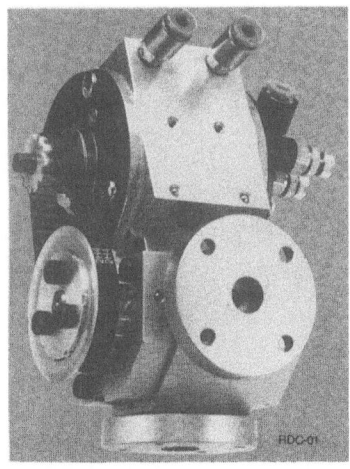

Fig. 8.31. Double-flange gripper drive (HIRATA – Japan)

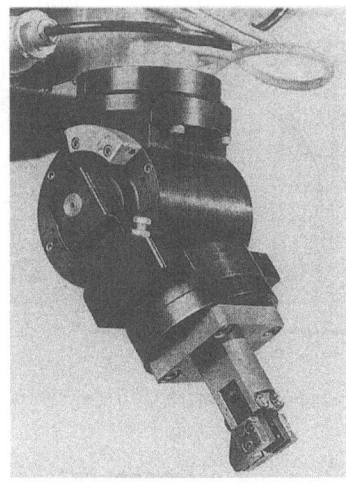

Fig. 8.32. Active gripper mount (SEIKO – Japan)

An example of the solution of the three degrees of freedom, typical for powering the carrier of the spot-welding gun with robots having electromechanical or electrohydraulic drive, is given in Fig. 8.33. It can be seen that three active degrees of freedom of angular type were realized, which in fact are coupled because when one is active, the two other must be compensated by means of supplementary control signals in order to achieve their autonomy. This is achieved relatively easily at the servocontrol level by means of algebraic addition of the position feedback signals of the individual degrees of freedom.

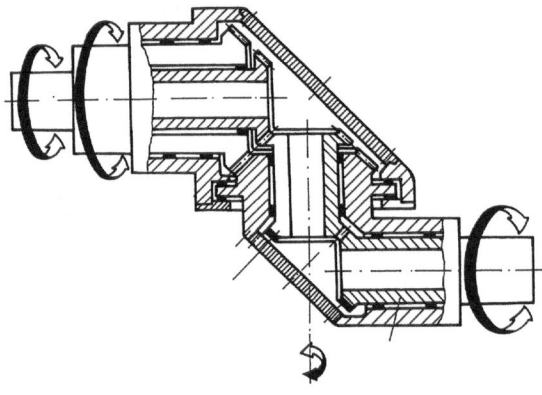

Fig. 8.33. Three active degrees of freedom of gripper mount

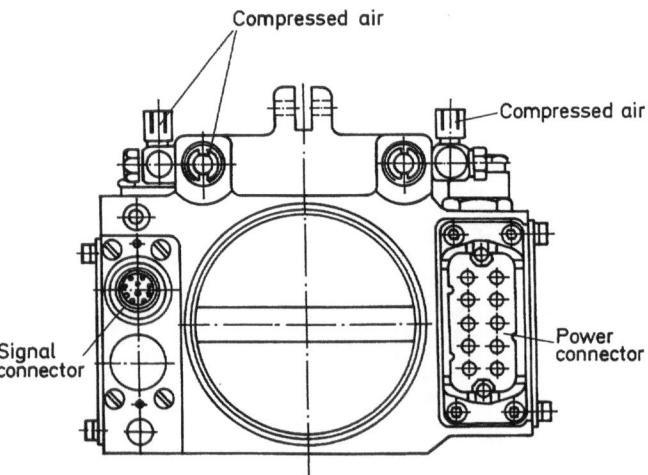

Fig. 8.34. Fast coupling of tool carrier (FEIN – FR Germany)

Modern solutions for quick tool change should also be mentioned. Figure 8.34 illustrates the solution produced by Fein-FRG. The tool itself from the same company – in this case a fast electric hobber for finishing the edges of castings and similar workpieces – is shown in Fig. 8.35. Carried tools (grinders, fast hobbers, drills etc.) are more frequently being applied in industrial robotics, due to less preparation being needed in the workplace and greater working efficiency compared to instances, where the workpieces are carried to the work machine and then worked on. Workpieces are often heavier than the robot's carrying capacity, in which case the described solution can be successfully applied.

Fig. 8.35. Carried tool on tip of robot arm (FEIN – FR Germany)

As an attempt to realize a gripper independently adjusting itself to the workpiece form, the Japanese proposition of an adaptive gripper with electric drive is presented in Fig. 8.36. By a special solution of steel cables and cable-wheels "finger phalanges" always arrive at the workpiece surface, but evidently only the convex one. The example of this design demonstrates the broad field of direction in design of industrial robot grippers.

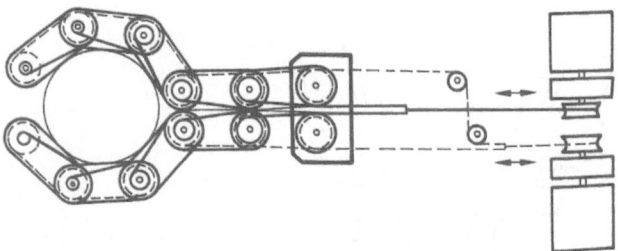

Fig. 8.36. Special adaptive gripper

8.3 Examples of Industrial Robot Applications

There is now seldom some activity in industrial production where industrial robots are not used. They have been successfully developed for welding electrical connections of microchips and large robots have been used for welding ship subassemblies. However, these examples can be regarded as exceptional in the sense, that their use is not universal. In the following Section examples will be given which have already acquired full civil status in modern industry.

8.3.1 Electrical Spot-Welding

This procedure for joining bigger assemblies of relatively thin, usually steel sheet has been robotized for almost 20 years for electrically spot-welding car bodies and today the car industry is still the largest user. Although for joining thicker, but more even thin sheets the process of electrical seam-welding is gradually being introduced having some advantages over spot-welding (stronger welds, continual joining of sheets, leakproof joints, etc.).

Robots for electrically spot-welding car bodies are now with electro-hydraulic drives, but more and more electromechanical ones are being introduced. These robots possess a significant carrying capacity of up to 100 kg, due to the mass of the spot-welding gun with cables of 50–80 kg. They demonstrate high motion speeds of up to 1.6 m/s due to fast cycle changes, along with high repeatability and positioning accuracy within the limits of 1 mm. The reach of the "arm" is usually 2.5 to 3 m, and they possess practically all six active degrees of freedom: three in the basic configuration and three in the spot-welding gun carrier, when fully orientated in space. That this is really necessary can be seen from Fig. 8.37 which shows a small but complex change of the work zone during the welding process of two adjacent window seams on a car body; two angular gripper positions can be seen from the figure, but bearing in mind that with practically all modern cars the rear side windows are oblique, a third angle change, i.e. degree of freedom, also becomes necessary.

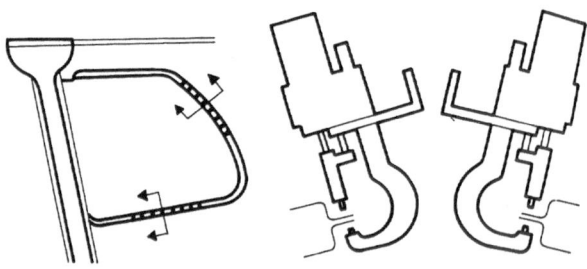

Fig. 8.37. Complex change of working zone

With this workprocess, which is usually on a transfer line, an extremely important role is played by synchronization of the work of the whole line, which can move in two modes:

1. Discretely, stopping at exactly predetermined points, at which robots usually in pairs weld the points of the subject group in this welding phase.

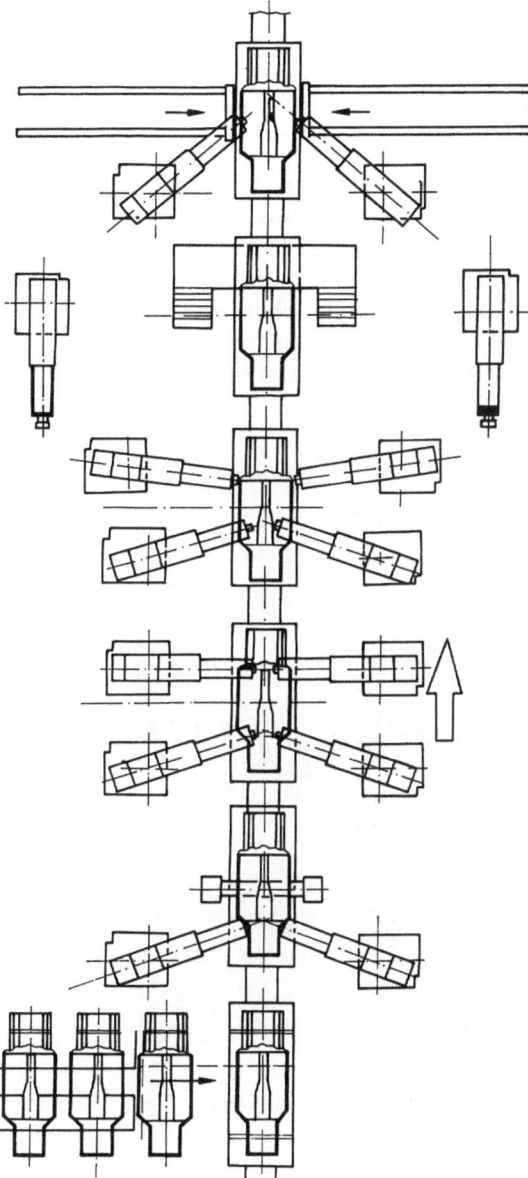

Fig. 8.38. Car body spot-welding line

2. Continually, without stopping, when the robots follow the workpiece (e.g. car body) and during motion weld the points from that phase which are within their reach; this procedure is much more complex from the control standpoint, which itself is clear, but further explanation is beyond the scope of this book. It is sufficient to mention that in the spot-welding gun

coordinate system of motion, the longitudinal coordinate (along the transfer line) is superimposed in each time instant by another (transfer) coordinate, which increases gradually with time. In this way the robot "arm" with the welding gun follows exactly the workpiece motion and exercises the spot welds at exactly the prescribed positions.

Figure 8.38 shows a typical transfer line for welding car bodies flowing from below, with 16 robots working on the line and two spare ones standing ready to replace some faulty robot in a time interval of only half an hour.

Such a transfer line for spot-welding car bodies in one of Ford's car factories with T^3 type Cincinnati-Milacron electrohydraulic robots is shown in Fig. 8.39. It can be clearly seen that in the workzone of robots the "arms" laden with heavy welding guns are moving constantly; access to this zone therefore is strictly forbidden.

Fig. 8.39. View of car body spot-welding line (Cincinnati – Milacron – USA)

We have already mentioned that the workprocess of electrical spot-welding is more and more performed by electromechanical robots. Figure 8.40 illustrates such an application, where the same two robots "attack" the subassembly of the car body floor (rear part) while one "hanging" model performs his part of the welding task from above. In this example the car body parts are inserted in the welding tools by hand and the welding itself is performed by robots, quickly and accurately.

Fig. 8.40. Subassembly spot-welding by three robots

Figure 8.41 shows the scheme of the first Yugoslav spot-welding robot installation in the Magnohrom factory in Kraljevo where a robot of the UMS-3 type welds about 40 spots on the basic box of a thermoaccumulative oven.

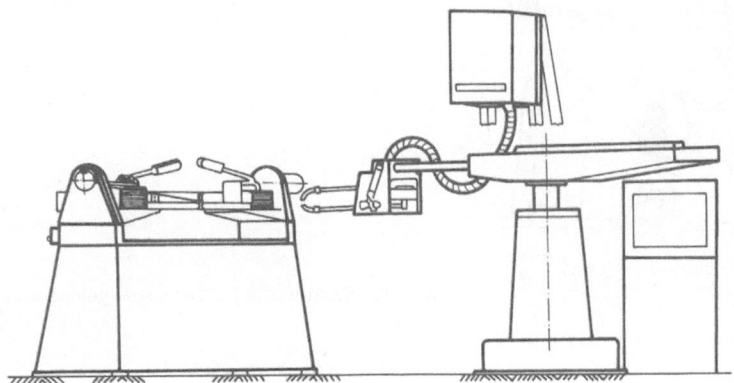

Fig. 8.41. View of the "MAGNOHROM" robot installation

8.3.2   Electrical Seam Welding

With this procedure of joining metal parts, mostly of steel or aluminum by welding, a continuous electrical arc between the central electrode is maintained.

This electrode is made of tungsten in the so-called tig-procedure, or it may be a wire of the same welded material in the so-called wig-procedure; the welding gun containing the said electrode is kept at a precisely defined distance from the sheets to be welded and on the exact path of the future seam. The gun serves also to conduct the protective gas (nitrogen or carbon-dioxide) to prevent oxidation of the hot seam immediately after welding. The gun itself must move along the seam (prescribed trajectory) continuously and smoothly, because in different circumstances the seam can be of irregular thickness, form and quality. This technological process requires the robot performance to be an absolutely smooth motion with a relatively low speed of a few 10 cm per minute along exactly prescribed trajectories which can be determined by programming, teaching or direct guiding (for instance by laser). With seam welding robots five degrees of freedom are sufficient. For orientation of the welding gun two are sufficient, because the gun is symmetrical to its longitudinal axis so this degree of freedom is unnecessary.

Figure 8.42 demonstrates a typical installation for electric seam welding equipped with a visual system for guiding the gun along the seam (ASEA, Sweden).

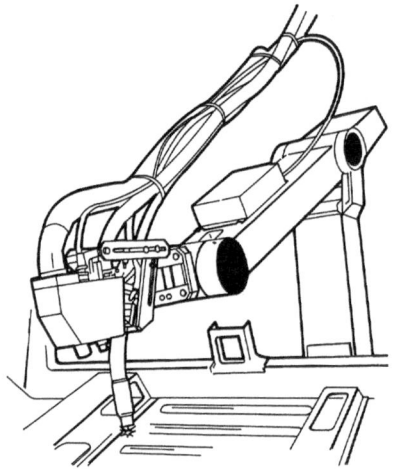

Fig. 8.42. Seam welding using visual guidance system

Figure 8.43 illustrates the case of a double workhead in the welding process, while Fig. 8.44 shows the same process but laser-guided. This laser device enables the robot to "see" the end of the seam if the welding process is to be continued at this point and to switch on the welding device automatically as well as the flow of the protective gas.

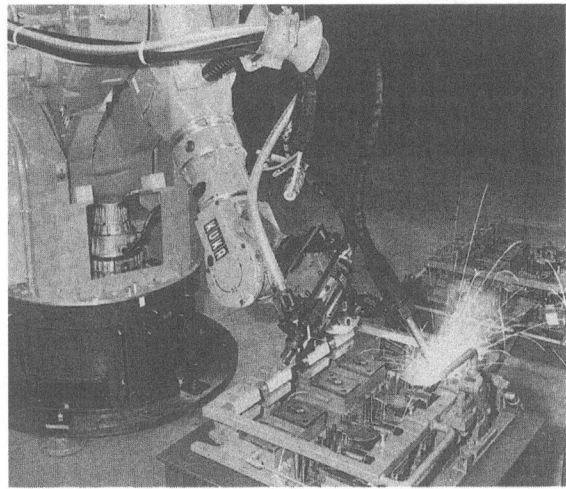

Fig. 8.43. Seam welding by robot (KUKA – FR Germany)

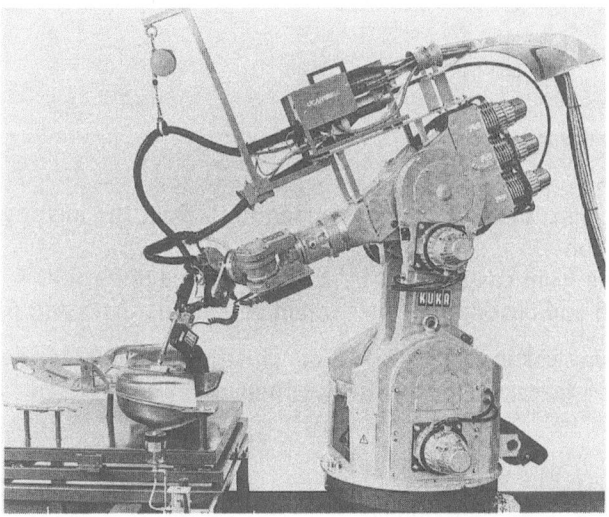

Fig. 8.44. Laser guiding device for seam welding (KUKA – FR Germany)

8.3.3 Pressure Die-Casting

Serving machines for pressure die-casting have been developed mainly to move the worker from a zone of possible heavy injuries, due to the danger that molten metal under very high pressure and temperature can seep between the halves of the closed casting tool and flow outside the machine. During the procedure normally light (aluminum) and zinc alloys are processed, so the melt temperature is a few hundred degrees Kelvin and any injuries sustained can be very serious, often lethal.

The role of the robot in this workprocess is threefold (Figure 8.45):

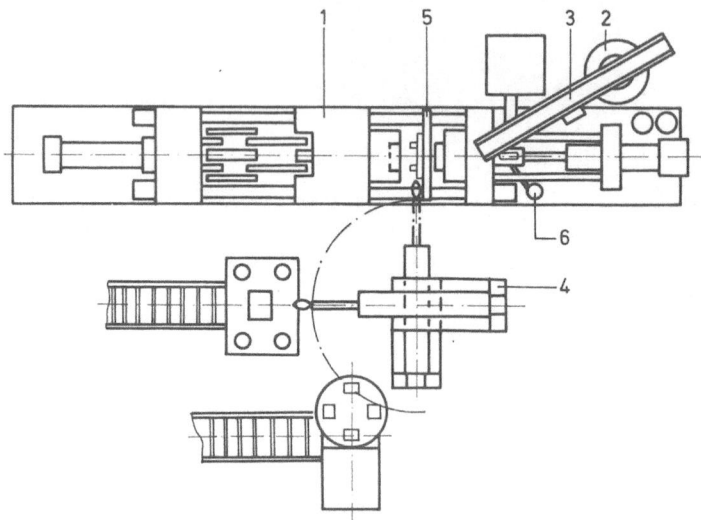

Fig. 8.45. Robots serving die-casting machine

1. Pouring molten metal into machine 1, transferring itself from the melting oven 2 by means of robot 3.
2. Taking finished casting from casting tool by means of extracting robot 4.
3. Oiling injection piston and casting tool by spraying manipulators 5 and 6.

In Fig. 8.46 the Yugoslav casting robot UMS-6L illustrated for taking the melt from the melting oven transferring it to the machine and pouring it in.

Fig. 8.46. Yugoslav UMS-6L robot working on die-casting machine

Figure 8.47 illustrates the process of taking finished castings by robot transferring them to the cooling vessel and then to the calibrating machine.

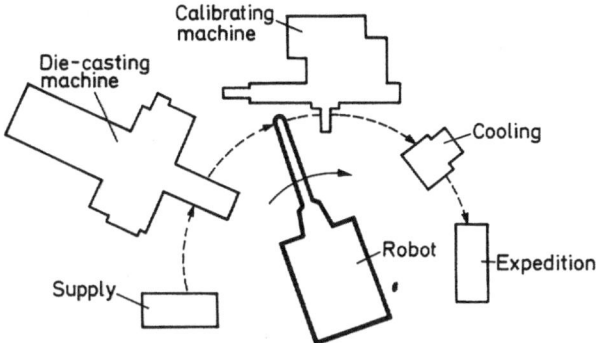

Fig. 8.47. Die-casting, cooling and calibrating process robotized

8.3.4 Spray Painting

An important part played by industrial robots is spray painting of workpieces with a paint or plastic layer which can be effected in various manners: by a flow of air under pressure, by high pulsating pressure of the paint itself in the spray nozzle, or by melting the paint from solid state at a high temperature and spraying it on with some corresponding device. The paint can be cellulose based (nitro), enamel, acrylic, polyurethane or aqueous. The robotization of spray-painting is mostly used in mass production of car bodies and parts of tractors, bicycles, motorbikes, and domestic appliances (refrigerators, ovens, etc.). Here too the basic aim was to remove the worker from this unhealthy and fatiguing workplace and make up for the task of workers willing to perform this job. Finally, by robot spray-painting a uniform quality of the product surface is attained which is sometimes more important than quality. Latterly, producers of quality cars have switched to robots for spray-painting car bodies. Undoubtedly, responsibility for development of this robot technology is due to the small Norwegian firm Nils-Underhaug, whose Trallfa robots even to-day more than 15 years after their first appearance, lead this field and are the most advanced of this type in the world.

Figure 8.48 shows the electrohydraulic spray-painting Trallfa robot sold under licence to firms in the USA and Japan. With this and other similar robots for this job, a very high positioning accuracy and repeatability is not required (± 3 mm suffices), but "smooth" progression of the programmed trajectories with a high speed of up to 2 m/s is required.

In most cases programming of these robots is performed by leading the robot tip-painting gun through a real work cycle by a qualified painter whereby these movements, containing all principal features (speed, pauses, returns, gun-on/gun-out periods, etc.) are being recorded in the robot memory.

The procedure is shown in Fig. 8.49, which illustrates guiding the real robot tip through the cycle, but with heavier robots and for easier operation a pilot-robot is applied with the same dimensions as the original robot and of (Fig. 8.50) a much lighter structure with potentiometers in the joints, the signals of which are recorded and memorized as described.

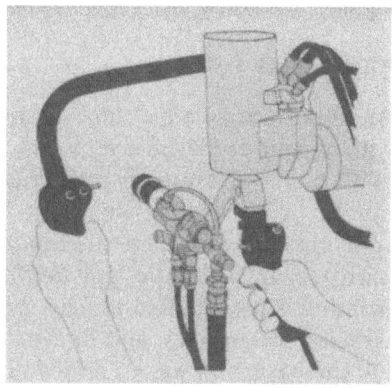

Fig. 8.48. "TRALLFA" spray-painting robot (NILS-UNDERHAUG – Norway)

Fig. 8.49. Direct "teaching" of robot

Fig. 8.50. Pilot-teaching robot

How complex the installation of a robotized painting booth is, is presented in the simplified sketch in Fig. 8.51. The main components of the installation are: 1, workpiece–car body; 2 and 3, spray-painting robots; 4, supply duct of conditioned air to the cabin; 5, lights; 6, moisture separators; 7, water collector; 8, floor; 9, waste paint collector; 10, servicing opening; 11, service gangway–gallery; 12, exhaust tubes; 13, fan; 14, inlet air heater; 15, flow damper; 16, inlet port; 17, conditioning cabin. Workpieces 1 (car bodies) move through the booth with a constant speed and robots 2, 3 perform painting during motion. Often more than two robots are applied in this workplace, where separately one or two robots are used for painting the roof and one for applying the plastic protective layer on the car body.

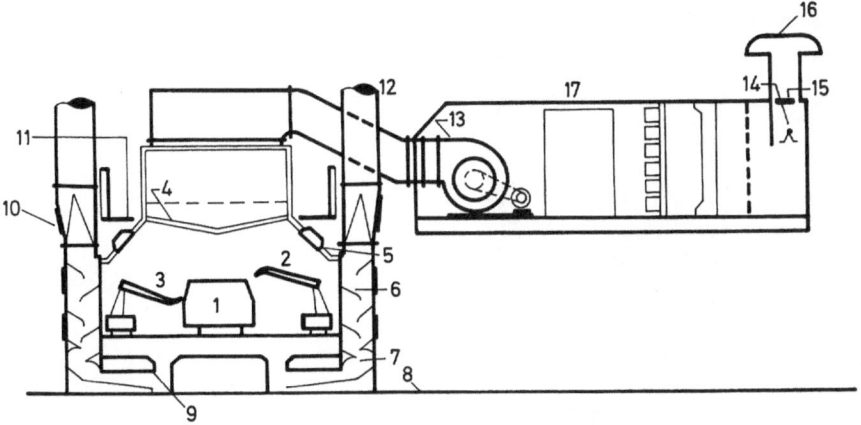

Fig. 8.51. Robotized painting booth in car factory

8.3.5 Machine Tool Serving

An important field of robot application in the metal-processing industry is the serving of machine tools. Serving one machine by one robot is still the most frequent method used, although it is being surpassed. The increased working speeds of modern robots, their reach and dexterity in new articulated, vertical or horizontal ("SCARA") configurations, their enhanced positioning precision and high repeatability in all gripper orientations, along with a short settling time in the trajectory terminal points, enable the serving of production cells consisting of several machine tools by only one robot. The common control system in this case serves for the control and supervising of the whole installation.

An example of serving one big excentre press of 6300 kN by the Yugoslav robot type UMS-3 is shown in Fig. 8.52. The robot stands in front of the press which on its moving part carries the upper part of tool while the lower half of the same tool is on the worktable of the press. Two vacuum grippers on the "arm" tip grasp the rear vacuum up from the end of the inclined feeder, one raw workpiece and the front one in the finished presspiece from between the opened presstool, carrying them out of the working space of the press and let them fall down on to an inclined guide into the transport cart, while from the rear gripper the rawling is dropped into the lower half of the presstool. While the press is working automatically the robot also works when the tool halves are open for the described operations. The whole installation is under the control of a microprocessor-based control system 10, which also supervises the data of the press, such as oil temperature, oil and air pressure, etc. In this way the human worker has been removed from this monotonous and dangerous workplace.

Fig. 8.52. Yugoslav UMS-3 robot working on excenter press

Fig. 8.53. PUMA robot on workplace (UNIMATION – USA)

Figure 8.53 illustrates serving a special grinder with a Puma robot, while Fig. 8.54 demonstrates a universal lathe installation being served by a Milacron T^3 robot, which is equipped with a double gripper: one serves for taking the processed workpiece and the other for bringing a new raw one and putting it into the machine.

Figure 8.55 gives a schematic presentation of the arrangement of working machine tools of a production cell served by a T^3 robot with six degrees of freedom. Robot 1, controlled by microprocessor-based control system 1A, first

Fig. 8.54. Serving a turning lathe by robot (Cincinnati–Milacron–USA)

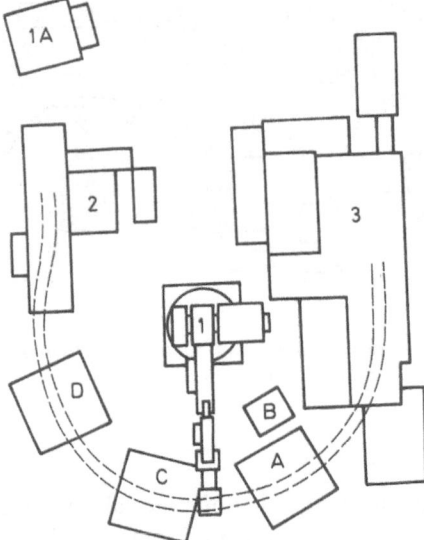

Fig. 8.55. Robotized production cell

serves the hobber 2 and universal lathe 3, via the input magazine A with 30 positions for workpieces, the tool changing device B, intermagazine C and output station D also with 30 positions.

Figure 8.56 shows the arrangement of three UNIMATE 2000 robots on a production line of parts by machining. Designations of the individual positions are given in the drawing.

Figure 8.57 illustrates one more example of using three robots in this case on a gearwheel production line.

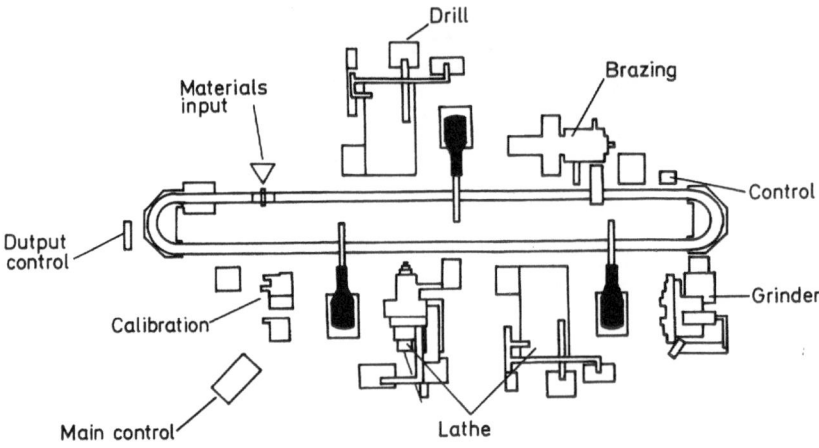

Fig. 8.56. Three robots on a production line

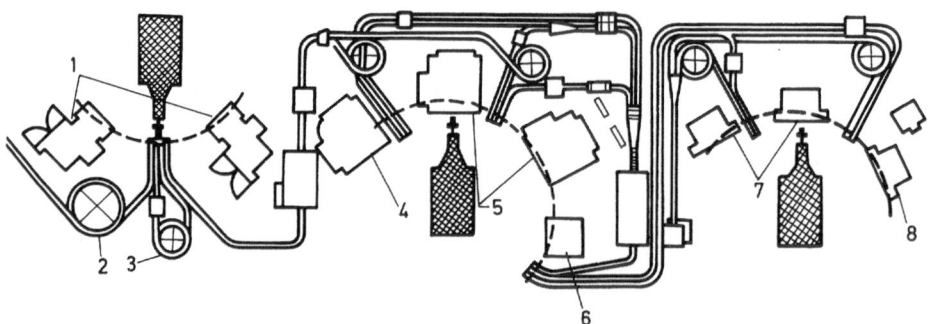

Fig. 8.57. Robots on gear production line. *1* input storage; *2* conveyor; *3* spare loop; *4* milling mac; *5* grinder; *6* thermic treatment; *7* lapping; *8* output control station.

8.3.6   Materials Handling and Palletization

Industrial robots are used more frequently for palletizing of materials on pallets, desks and in magazines, in boxes, etc, as well as taking them from these positions and transferring them from one to another in the same or different arrangement. Such tasks are performed more quickly and efficiently than by men, especially if heavy objects are involved. Figure 8.58 illustrates an ASEA 60 robot transferring and palletizing bricks, taking five pieces each time by means of a special-purpose vacuum gripper.

8.3.7 Processing Castings and Stampings

Taking off the sharp edges or casting tool marks from raw castings, cutting the oncasts off and levelling of functional surfaces. All these are fatiguing, health hazardous and dangerous jobs for the human worker. With robots these tasks are performed quickly, uniformly and efficiently.

Figure 8.59 shows the operation of removing the tool joining marks from a casting being held by the robot in a special gripper performing all the necessary motions of this work process apart from transferring the workpiece from the input to the output position.

Fig. 8.58. Palletizing by robot

Fig. 8.59. Deburring of casting by robot

Figure 8.60 presents the installation for cutting off the oncasts by a grinding plate which is turning on a fixed electric motor, while the robot takes the castings from a pallet one by one, brings it to the grinder and cuts the needed part off then returns it to the old or to a new pallet, as organized.

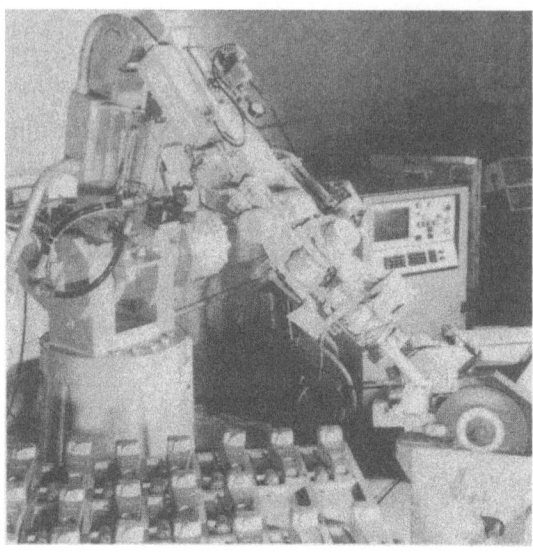

Fig. 8.60. Cutting the caston by robot (Cincinnati–Milacron – USA)

Figure 8.61 shows removing sharp edges of an opening after final processing. The robot brings the workpiece to a specially fast hobber with an oblique workhead and performs the necessary movements aimed at removing the sharp edges from the object.

All these examples are dangerous and unhealthy work for the worker who can also be fatigued and inattentive.

Fig. 8.62. Applying a glue bead by robot

8.3.8   Applying Glue and Leakproofing

In the car industry, which has learnt from experience gained from the airplane industry, more and more glueing is used for joining adjacent components (window glass to its frame, sonorization and heat insulation, carpets, etc.) and also for applying a plastic mass aimed at preventing the ingress of dust and moisture between two adjacent parts connected in a non-leakproof manner (e.g. spot-welding). These work operations demand constant attention on the part of the worker during glue application or the sealing mass (compound) in order to achieve a continuous bead along the prescribed trajectory. Figure 8.62 shows a robot applying glue to a car motor hood, while Fig. 8.63 shows applying the sealant onto the connections of the metal sheets in the car cabin. It can be seen that the robot "enters" into the cabin through the window of the closed car door in order to perform its task demonstrating its reachability and orientability.

Fig. 8.63. Applying of sealing compound by robot (Cincinnati–Milacron – USA)

8.3.9 Assembly by Robots

This work process is one of the most complex in the realm of flexible techno-logical systems implementation using the so-called "intelligent" robots (robots with artificial intelligence elements). The solution can be mainly found in two forms:

1. assembly without visual feedback,
2. assembly with visual feedback.

After several years of research and development experiments, it can be stated that in 93% of the cases of industrial robot applications, the first form that is applied is when a 2D (flat) TV image is used to help in orientation of planar objects. Visual systems which would enable assembly of complex products from arbitrarily piled components are still to be encountered in some laboratories. Moreover, for assembly of coaxial products electric motors, alternators, pneu-matic and hydraulic pumps, etc., assembly without visual feedback is performed.

Figure 8.64 shows the assembly installation of a car generator by a Bendix robot, realized in the Charles Stark Draper Laboratory, MIT, Cambridge, USA. This is a typical example of the assembly of coaxial products because all component parts are stacked one on top of another along the main longitudinal axis of the product: the casings (front and rear), ball bearing, rotor, fan, beltwheel, etc.

Figure 8.65 illustrates the part assembly of car motorheads by three Puma robots; mainly the assembly of inlet and outlet valves, mounting the springs and

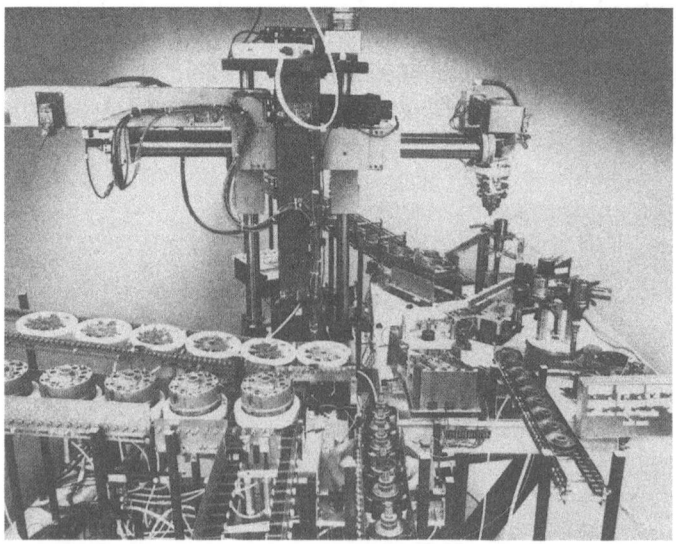

Fig. 8.64. Assembly of car generators by robot (CS DRAPER Lab. – USA)

Fig. 8.65. Assembly of car engines by robots (UNIMATION – USA)

securing them. Similar lines of car engine assembly have been implemented at Fiat, General Motors, Ford, Nissan and some other important car producers.

Figure 8.66 is a schematic view of a complete assembly desk equipped with a Scara robot from Hirata (Japan), used for assembling some smaller sub-assemblies at workplaces along the assembly line. The parts to be assembled are brought in specially planned pallets which are then placed in automatic chargers on the left-hand side of the desk. These chargers automatically dispense to each

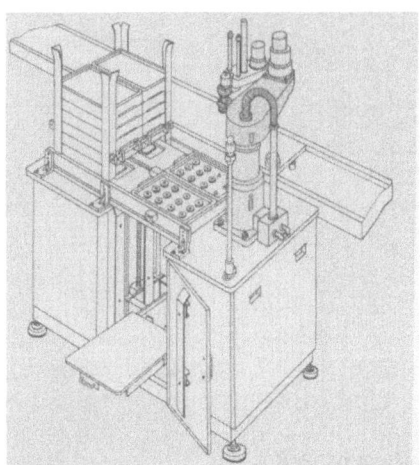

Fig. 8.66. Complete assembly table with robot (HIRATA – Japan)

workplace two pallets each, from which the robot picks out the parts and performs the assembly. When the pallets are emptied, they are taken away and are replaced by two new ones full with parts.

8.3.10 Special Applications of Robots

Finally we come to unconventional robot applications. The very high positional accuracy and notably the precision of the proper position sensors of the robot arms (encoders, resolvers, etc.) have enabled robots to be used instead of the so-called "measuring machines", which are expensive motorized Cartesian measuring systems of appreciable dimensions as far as bigger products are concerned (flying, water and land vehicles, etc.). Figure 8.67 shows an installation for the verification of the airplane body is shown, which in this case is being "touched" by two robots at its characteristic points. The measured data are then recalculated into the workpiece dimensions and memorized to be edited in hardcopy form after the process.

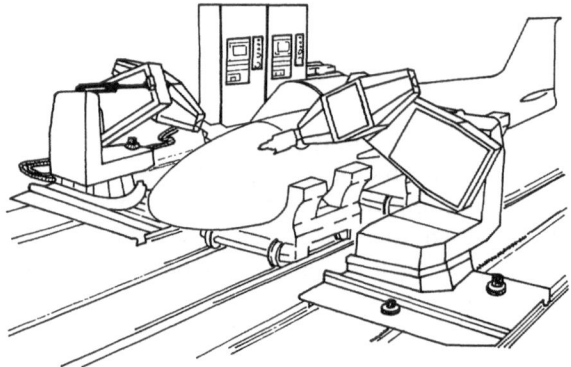

Fig. 8.67. Robots instead of measuring machine

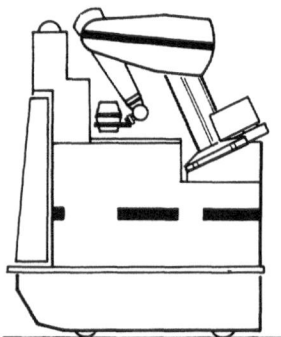

Fig. 8.68. Movable robot ("ROBOCAR") with manipulator on itself

Quite a different application of industrial robots is illustrated in Fig. 8.68, where on a locomotive robot ("Robocar") an industrial Puma robot is mounted to serve in several workplaces in the so-called "clean" rooms of the electronics and some other industries. Presentation of totally specific robot applications in the nuclear environment, in space, at great sea depths, etc. can be found in specific texts, and are not covered in this book.

References

[1] Engelberger, F.J., Robotics in Practice, Kogan Page, London, 1981.
[2] Nof, S.Y., Handbook of Industrial Robotics, John Wiley, New York, 1985.
[3] Bonney, M.C., Young, Y.F., Robot Safety, Springer-Verlag, Berlin, 1985.
[4] Volmer, J., Industrieroboter, Veb Verlag Technik, Berlin, 1981.
[5] Jurevič, E.I. i dr., Ustrojstvo promišljenih robotov, "Mašinostroenie", Lenjingrad, 1980.
[6] Beljanin, P.N., Promišljenie roboti, "Mašinostroenie", Moskva, 1975.
[7] Spinu, G.A., Promišljenie roboti, Konstrirovanie i primenenie, "Viša škola", Kijev, 1985.
[8] Paton, B.E., Promišljenie roboti dlja svarki, "Naukova Dumka", Kijev, 1977.
[9] Warnecke, H.S., Schraft, R.D., Industrie-Roboter, Otto-Krauskopf Verlag, Mainz, 1980.
[10] Lundstrom, G., Gripper Review, IFS, Bedford, 1977.

Chapter 9

Robotics and Flexible Automation Systems

9.1 Introduction

In previous chapters various aspects of robotics have been studied in detail. Some problems, specific for the application of robotic systems in complex flexible manufacturing, not only in a stand-alone work cell, are also outlined. Awareness that robot ought to be considered as an element of complex automation has been recognized throughout this book.

Dynamics of constrained motion under the external reaction forces are very important in many robot applications (polishing, deburring, assembly, etc.). This has been analysed in Chap. 3. Hierarchical control of robotic systems with strategical, tactical, and execution level is defined in Chap. 4. The importance of well designed and versatile IO modules and intelligent sensors controllers has been mentioned in Chap. 5. Also, the need for "good" serial communication channels to the other level computer(s) is recognized (Chaps. 5, 6).

Requirements for an advanced robot programming system and the problems of robot system synchronization with its environment, other machines and devices in complex manufacturing systems, as well as some software design guidelines are given in Chap. 6. Special attention has been paid to sensors in robotics (Chap. 7).

Many examples of robot applications are included in Ch. 8. A flexible manufacturing cell has been introduced (Fig. 8.55), then two examples of flexible manufacturing lines (Figs. 8.56, 8.57), as well as numerous examples of machine tool loading, unloading, material handling, palletizing, finishing operations (deburring, polishing), and assembly tasks as the most challenging in overall flexible automation.

In the following sections we will try to unify in a systematic way major problems (and possible solutions) encountered with factory automation by integrating one or more robots and several machine tools into work cell, flexible manufacturing line, assembly system.

Considerable research and development work has been performed on flexible automation systems [1]. They are usually related to the metal cutting industry, and because of the intensive human labour involved in assembly operations, special interest has been paid to the design of flexible assembly lines. However, the principles are applicable to most manufacturing processes.

To achieve flexible integrated manufacturing, two fundamental requirements can be identified [2]: the need for flexible intelligent modules, and the need for integrated, and at the same time distributed software systems. This chapter will consider some aspects of those requirements.

9.2   Functional Structure of Flexible Automation Systems

A hierarchical structure of computer control of manufacturing plant is illustrated in Fig. 9.1 [3]. Industrial robots, CNC machine tools, and other stand-alone computer controlled manufacturing processes are at item lowest level. Several CNC machine tools under the control of a single computer comprise a direct numerical control system (DNC). A system which produces a particular part or few parts with similar geometry, configured around several CNC machine tools and a robot, is usually recognized as the manufacturing cell. They are classified as a station or a cell level.

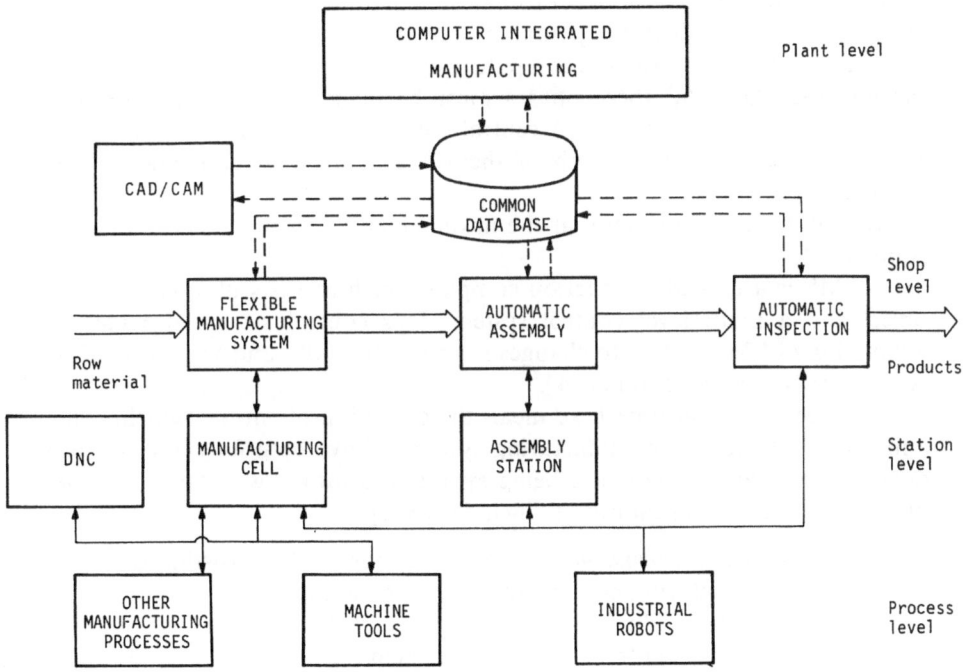

Fig. 9.1. Hierarchical structure of manufacturing systems

The third hierarchical level in computer control of manufacturing is denoted as a flexible manufacturing system. The FMS produces parts which have to be assembled into a final product. In each assembly station, a robot is supposed to assemble parts into a subassembly or into a final product. The final product should be tested by an automatic inspection system.

As far as robotic systems application in flexible manufacturing is concerned, they are primarily used in FMS as material handling devices, for machine tool loading/unloading, for palletizing and for some parts finishing operations. However, in assembly stations and overall assembly lines, in the future, robots will be a major flexible machinery. Already, PC boards assembly lines are fully robotized.

Therefore, we will distinguish between flexible manufacturing systems, based on CNC machine tools and machining centres with robots as auxiliary devices and flexible automation systems based on intelligent robots – a notable example being robotized welding lines and assembly systems.

Computer-aided design and computer-aided manufacturing (CAD/CAM) systems are used in FMS and assembly systems to integrate the design and manufacturing of parts in order to minimize the production time. The design procedure should end up with the product suitable for robot assembly.

At the highest hierarchical level (plant) there are computer-integrated manufacturing systems (CIM). They coordinate all phases of a manufacturing process: the design of the product, the planning of its manufacturing, the production of parts, assembly, testing and the computer-controlled flow of materials and parts through the plant. All these phases must be integrated into one computer network, supervised by the CIM central computer which monitors the interrelated tasks and controls each of them based on an overall management strategy [3].

We will have a closer look at the flexible manufacturing systems and their components.

A FMS is a unified production complex which consists of a technological object and a multi level control system. This control system provides the adaptation of FMS system to changes in production jobs and situations within economically justifiable limits [4].

By a technological object we mean the collection of entities which, before appearance of "flexible manufacturing systems", have been fully and independently developed, and now are being synthesized into new automation complexes. The "technological object" includes [3, 4]:

1. *A flexible manufacturing module* (a production unit) consisting of one or several manufacturing equipments; the module has a local control system and can operate off-line, as a stand-alone unit. It can be a CNC machine tool, a machining centre, or a robotic system. In this sense, even a robotic work cell can be viewed as a flexible manufacturing module.
2. *A computerized storage module* with a local control system, which can operate off-line or under commands from the upper control level.

3. *A flexible measuring and inspection module* used to update the technological process parameters or operation program according to inspection and quality data.
4. *A computerized transport module* with a local control system, which can operate off-line or under the command from an upper control level. It is designed for transporting finished parts, tools and attachments.

A common schematic diagram of FMS in the metal cutting industry is shown in Fig. 9.2. In the following paragraphs we will briefly discuss some of the above-mentioned components of FMS. Robots and computer robot control have been extensively studied in previous chapters.

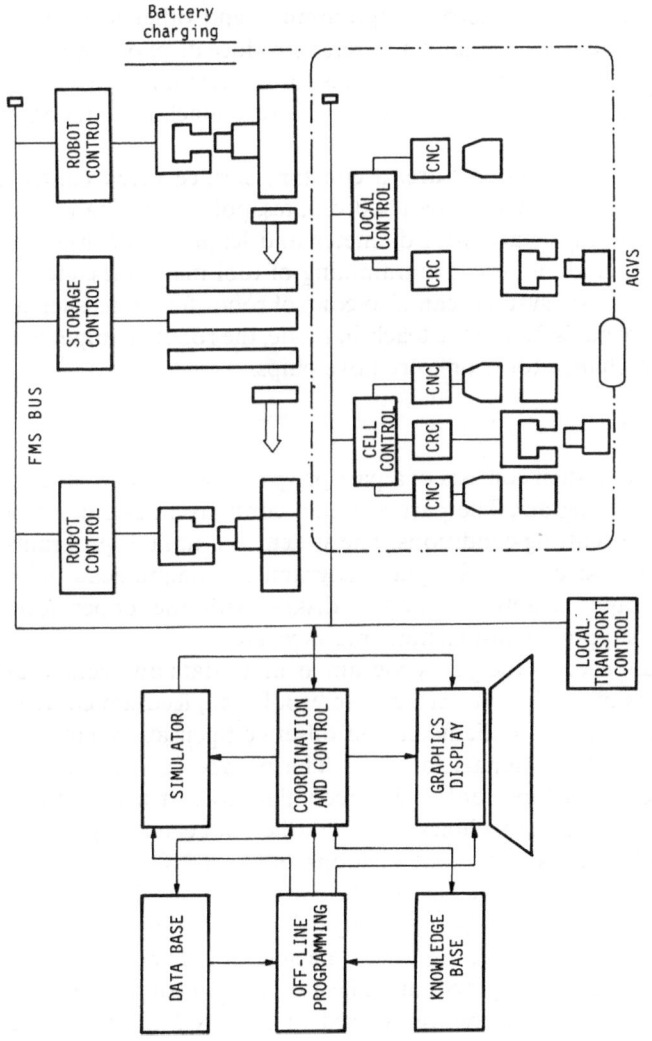

Fig. 9.2. Example of a FMS

9.2.1 Computerized Numerical Control

MIT in the early 1950s developed a control which could be applicable to a wide variety of machine tools [5]. The system allowed design changes without modifications to tooling and fixing. It drove a lead screw through an interface, under the instructions by the output of a computer. The new technology was named Numerical Control (NC). Since then, it has evolved through technological generations. Rapid development of computed hardware led to the appearance of the Computer Numerical Control (CNC) which is nowdays introduced to practically every manufacturing process. Around 1970, drilling, milling, turning were performed on "machining centres" and "turning centres".

Along with the many functional options, programmable interfaces, so called "canned" cycles, subroutines, parametric programming, and other features of advanced software techniques, CNC manufacturers introduce displays for visual editing of part programs in memory, graphics aids for programming, checking and simulation of program execution. Monitoring of every machine function is performed.

Standard features of CNC systems are: a constant surface speed control, linear and circular interpolation in cartesian coordinates, polar coordinates and helical interpolation with automatic tool diameter and length compensation; automatic tool changing and variable programming of tool magazines, etc.

New microcomputer CNC systems can also control robot functions such as loading and unloading of parts. Using the teach-in mode, the robot-manipulator can be programmed to change tools or to remove chips.

9.2.1.1 CNC System

The software of the CNC system consists of a part program, system and utilities programs, and a control program. The part program contains a description of the part geometry and the cutting conditions. The system and utilities programs are used to edit, verify and correct the part program, to communicate with peripherals (punched tape, cassette or floppy disks), with the upper level computer (in FMS) to support man-machine interface, etc.

The control program accepts the part program as input data and generates signals to drive the axes of motion. It performs interpolation, feedrate control, acceleration and deceleration, tool diameter and offset compensation, etc.

It also calculates the drive signals for axes. The number of axes varies, depending on the type of machine tool and is reconfigurable. It can be three linear axes for milling and boring machines, two for lathes, but can also go up to six linear and/or revolute axes. The control loops of the CNC systems are designed to control the position and velocity of the axes. The algorithm depends on the type of drive system. Each axis is separately driven: open-loop or closed-loop. Open-loop controls use stepping motors as the actuator. They are used in simple point-to-point, or positioning systems. The closed-loop configuration is more accurate and reliable and therefore is used in most CNC contouring systems. Although many systems still use hydraulic motors, the DC servodrives practically dominate the field. The position feedback element can be incremental encoder, resolver, or inductosyn. For velocity feedback the tachometer is used. The typical CNC servo-loop is configured as in Fig. 9.3.

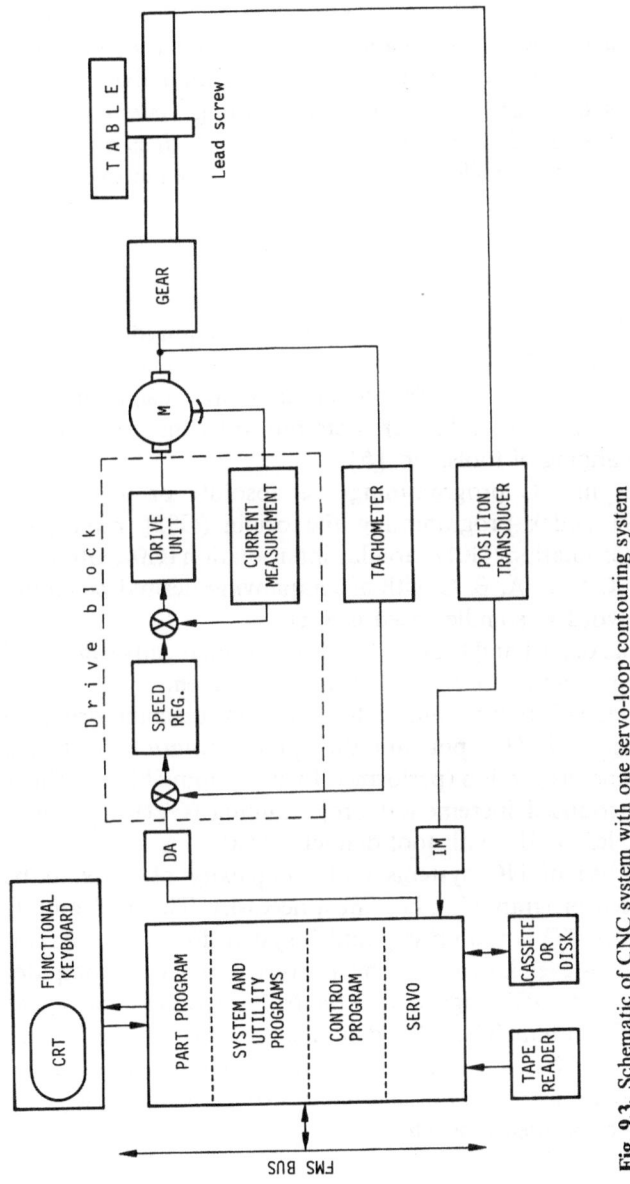

Fig. 9.3. Schematic of CNC system with one servo-loop contouring system

The hardware of the CNC system is often configured as a multi-micro-computer system with common memory. Usually, one microcomputer is dedicated for system and utilities programs and it is responsible for overall functioning of the CNC controller; it also contains and controls the execution of the part program. The second microcomputer is responsible for control functions (part program sequencing, interpolation, feedrate control, tool compensation, etc.), and also for servo-loops, i.e. for driving the axes. The algorithm, used in control loops configured as in Fig. 9.3, are usually of PID type with possibly some kind of velocity feedforward command. The third microcomputer is actually a programmable controller (PC) used for programmable machine interface (interlocks, tool changers, tool magazines, cooling system, etc.). In modern feed drives systems, the drive block (shown in dashed lines in Fig. 9.3) can also be realized as a microcontroller or microcomputer, one per axis or one for more (up to six) axes.

9.2.1.2   NC Programming

NC part programming comprises the collection of all data required to produce the part. The geometry is taken from the engineering drawings. The tool path is calculated based on this information. The other data are: machining parameters–feedrates, spindle speed, etc.; data are determined by the programmer, e.g. cutting direction, changing of tools, etc. [5].

The basic principles in NC programming are absolute or incremental programming (G90/G91 code); programming the origin (G92); rapid positioning (G00); linear interpolation (G01); circular interpolation (G02, G03), etc. The axes addresses are X, Y, Z, A, B, C with accompanying desired positions; feedrate has a reserved word F, spindle speed is S, etc.

In Fig. 9.4, an example of part and tool center path (automatically calculated by the CNC system) and a sample of the NC program is given.

The outlined sequence will rapidly move the tool from machine origin to point A. The spindle speed is 900 rpm and the spindle rotation is started clockwise by M03 machine instruction (performed by programmable interface). In this example, we introduced incremental programming (G91) and cutter diameter compensation, left (G41), with tool diameter D30.

With the increased use of NC systems and complexity of parts to be machined, computer-aided programming became a necessity. The APT and its variations (UNIAPT, EXAPT) is the most popular system for part programming. It is a textual high-level language. The postprocessor is a computer program which translates the APT program into the NC program for particular CNC/machine tool configuration. The APT contains definitions for geometry forms (POINT, LINE, PLANE, CIRCLE); motion statements for point-to-point (e.g. GOTO/symbol for a defined point) or contouring (GOLFT, GORGT etc.), tolerance and cutter specifications, etc.

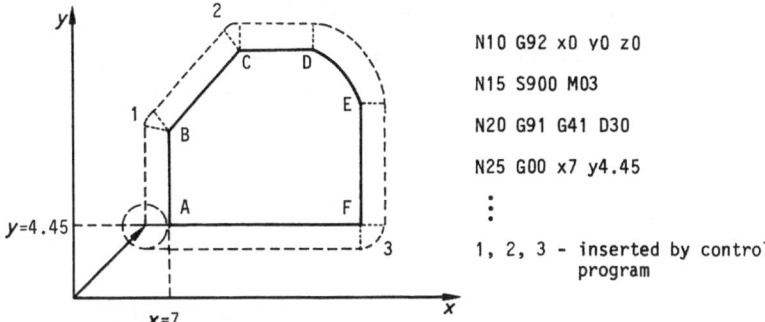

N10 G92 x0 y0 z0

N15 S900 M03

N20 G91 G41 D30

N25 G00 x7 y4.45

: :

1, 2, 3 - inserted by control
          program

Fig. 9.4. Example of part program

In CIM systems, the output of the CAD system (drawing of the part) is interrelated with some computer-aided programming system (e.g. APT). This chain should end with an automatically generated NC part program. This part program will, by a FMS control system, when required, be downloaded into a scheduled CNC/machine tool configuration for part machining. Actual machining will be activated on command from the coordination and synchronization program, responsible for the overall operation of the FMS at the cell or even shop level.

9.2.1.3 *Adaptive Control*

The adaptive control of metal cutting processes is a natural extension of the CNC system [3]. These systems are based on automatic control of operating parameters with reference to measurements of the machining process variables. We basically distinguish between adaptive control with optimization and adaptive control with constraints. The example of the first scheme is given in [3]. The sensors measure the cutting torque, tool temperature, and machine vibration. These measurements are used by the adaptive controller to obtain the optimal feedrate and spindle speed. The second approach does not utilize a performance index. It is based on maximizing the operating parameters subject to process and machine constraints. For example, the operating parameter is a feedrate and constraints are an allowable cutting force on the tool, or maximum power of the machine.

Although there has been a lot of work in the field of adaptive control of machining, quite a few adaptive systems are used in practice. There is a lack of suitable sensors and it is difficult to formulate and to formalize realistic performance indices. But a fully automated "unmanned factory of the future" cannot be designed without adaptive control as termed above. Even in FMS, the provision must be made for trivial tasks such as tool wear monitoring, tool breakage detection, sophisticated cooling systems etc.

9.2.2   Computerized Transport Systems and Mobile Robots

Material transfer, material handling and storage take 30–40% of all operations in manufacturing processes and therefore in flexible automation systems too. Various types of transfer have been utilized in FMS and robot automation systéms.

Conveyor belt and roller systems are among the simplest and the least expensive. They are suitable for flexible manufacturing lines and are used for so-called internal transport – following the sequential and linear flow of materials and parts through production operations.

For transport of tools within FMS, of parts between manufacturing cells and machining centres, of sub-assemblies between and within assembly systems, and for final product to the storage two types of transport means generally are used: the overhead transfer conveyers and floor transport. The latter is the most important one. The oldest is the rail transport, but because of evident drawbacks and unflexibility, new automation systems do not use it.

Automatically guided vehicle systems and mobile robots are the most important transport systems for the FMS and robot automation systems [6, 7]. Their advantages over other transport schemes are: flexibility and reliability, high positioning accuracy, good loading charácteristic, ease of path modifications, manoeuverability and no external power supply needed, etc.

In order to control the AGV or mobile robot reliably, one of the following approaches can be adopted:

(a)   fixing the paths of the vehicle and off-boarding the vehicle;
(b)   a fixed frame of reference is provided for the vehicle; the vehicle refers to this frame by using on-board sensors;
(c)   a "fixed map" of the surroundings is stored on-board the vehicle. Using on-board sensors, the vehicle produces the "current map" which is compared with the fixed map to determine the current vehicle position.

Wire-guided AGVs use the first approach. At present, this is the most widely used technique for industrial applications. Guide wire is laid in the floor, constituting the transport route, arranged in the form of various paths and closed loops. Each loop carries a different frequency A.C. signal. Magnetic plates are fixed at each junction in the floor network, and enable the AGVs to detect junctions during motions. There are also communication points, which provide for the link to the transport system controller. An AGV has its local controller (usually a programmable controller with appropriate modifications due to specific "remote" control problems). AGV identifies itself at a communication point and reports its status to the transport supervisory computer which is often also a programmable controller. It tracks the progress of each AGV in the network, detects and avoids possible collisions, and communicate with the FMS computer (Fig. 9.2).

The major drawback of wire-guided (or analogously painted-line guided) AGVs is that the floor network is not so easily alterable. The other approaches are vehicles guided by on-board, software programmable paths, or so-called "free ranging" robots, which is the right solution for the FMS and generally for flexible automation systems.

Various guidance techniques are used for "free-ranging" vehicles:

1. calculations of vehicle expected position, knowing its starting point and precisely measuring the rotation of drive wheels;
2. positioning the reference beacons at appropriate locations in the space; the vehicle controller knows these positions; with on-board sensors the exact distance and direction from any one beacon is measured and the vehicle position in the surroundings is calculated.
3. Inertial navigation with the axis of the gyroscope set up parallel to the direction of motion; any deviation from vehicle path is detected by the gyroscope, and this is used by a servo to correct the path; the path is programmed by teach-in technique, similarly as for fixed-based robots.
4. Optical or ultrasonic imaging of the surroundings; as explained above, a "fixed map" is stored on-board the vehicle; using video camera or ultrasonic transducers, the vehicle controller periodically generates a "current map"; by the recognition of various objects in this map, the estimates for the

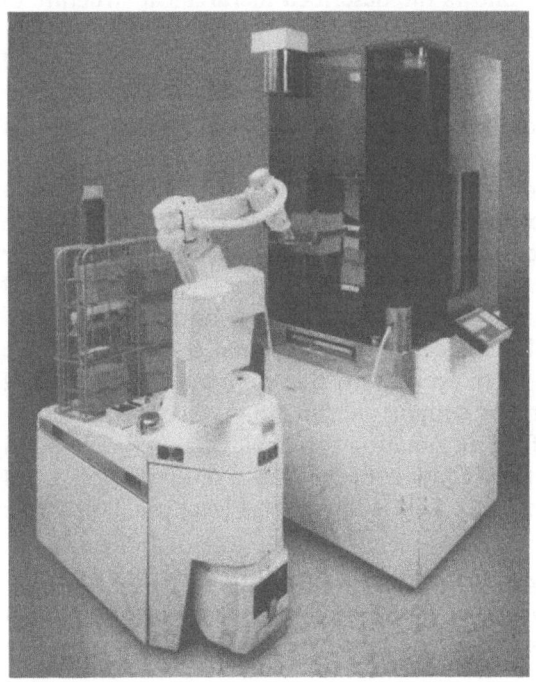

Fig. 9.5. FMS autonomous transport vehicle (FMS INC., USA)

position of the vehicle are calculated and processed to obtain an overall estimate for its position.

The first and the third method are the easiest to implement, but their position measuring method is very prone to errors such as: wheel slippage and gyroscope drift with time respectively. Surrounding imaging methods provide for more intelligent control of mobile robots which can dynamically change their paths due to presence of obstacles, but at present these techniques are still in the research stage.

An example of commercially available mobile robot is shown in Fig. 9.5. It is a product of Flexible Manufacturing Systems, Inc., USA. The autonomous transport vehicle uses the inertial navigation system which permits complete freedom of motion. The local controller is linked to the transport system by IR communication. Vehicle tasks are initiated using a teach pendant; arm movements are taught at each stopping point. Each route can be identified and tailored to specific manufacturing sequences for future recall. Any route or task is stored in the data base of the transport system controller and can be downloaded to any vehicle at any time.

9.3   Control of Flexible Automation Systems

Generally, the control structure follows the described hierarchical structure of manufacturing systems (Fig. 9.1).

The lowest level resolves the task of controlling the single or associated group of operations. Examples are: support of all cutting processes and other machining functions; execution of the programmed robot task; supervision and synchronization of the transport system; monitoring and low level synchronization of AGVs and mobile robots; control of inspection station, etc. As a general solution, this level is implemented by local computerized control. Those local controllers support the "elemental" technologies – components of FMS or robot assembly lines. In Fig. 9.2, they are designated as CNC, CRC,, Local Transport Controller, etc. They can be a cell or a station controller.

The second, or sub-system, level is considered as actual control of flexible automation. It can correspond to the cell or to the assembly station level if respective "local" controllers cannot support all functions of cell (i.e. station). Depending on the complexity of the automation system, it can also correspond to the overall flexible system control and for coordination and supervision of cell and station controllers (the shop level). This is the control level which provides for:

(a) sequencing and monitoring of technological operations;
(b) synchronization and supervision of all system components operations (by a component, in the above-mentioned sense, we mean even a cell or a station);

(c) dynamic scheduling of manufacturing and information processing resources;

(d) coordination of cooperative work of robots in assembly stations, etc.

The third level resolves the CIM tasks of overall plant management with production control, CAD/CAM support, supervision of maintenance, stock control, supervision of inspection and quality systems, computer-aided production planning, etc.

Those levels and their respective responsibilities overlap in hardware as well as in software organization in various flexible automation approaches.

Let us look at the proposed hierarchical control of industrial robots (Chap. 4, Fig. 4.1, p. 121) with respect to flexible system control decomposition. If the "local" (local in the sense of previous arguments) robot control system is intelligent enough and responsible for all sub-levels, strategic, tactical and execution, within the scope of control responsibilities described in Sect. 4.1, then, obviously some of important functions of the second level (sub-system level in the sense of the above introduced control decomposition) are already taken care of. The main differences are the means of data entry and the synchronization with the rest of the automation system. Instead that an operator, through a manual interface, defines globally the robot task (strategic level, Fig. 4.1), in FAS, provision must be made that the task can be downloaded or selected from the repertoire of available programs in the robot controller, from the upper level when required by that upper level. Also synchronization, some supervision of the task execution as well as some on-line replanning, must be left to the upper level (sub-system) control computer, in order that the overall system operates correctly and fully automated (unmanned in second and third shift, for example).

In Chaps. 5 and 6 an intelligent robot controller has been proposed and described functionally. Also, some software requirements have been outlined and the need for advanced techniques to be incorporated in robot controller have been recognized. Such a controller would be capable of controlling a flexible manufacturing cell and it probably would suffice if the cell is an "island of automation" [8]. But its operation has still to be coordinated and synchronized with the other islands of automation at the factory (shop) level.

As we can see, it is not easy to distinguish between shop and cell or station level. However, the proposed decomposition does make a strict difference between CIM (plant) level as the one which exclusively deals with the information on manufacturing process flow and the global production control, and the lower two levels (or three functionally interrelated sub-levels), which deal either with information on material flow and processing, and control of manufacturing, or by actual material handling, part machining and assembly.

In the following paragraphs, we will consider some approaches and software requirements for the subsystem level (shop and/or cell level) and the communication with the lowest (process) level. The CIM level is outside the scope of this book.

For further analyses we will look at the problem of flexible automation system control from two functionally separable aspects.

1. Coordination and control of flexible manufacturing systems, consisting of machining stations, robots primarily as auxiliary devices, transport and storage system, inspection systems, etc.
2. Coordination and control of robot automation systems consisting of assembly stations – complex robotic systems where the cooperative operation of number of robots is of primary interest.

9.3.1   Control of Flexible Manufacturing System

The control system has to resolve at least two kinds of tasks:

1. Synchronization of the manufacturing operations sequencing – the basic control; referred to later as *control*.
2. Combinatorial control of manufacturing operations, related to system state analysis, conditions testing and the choice of the "optimal" action; referred to as *coordination*.

In order to develop control algorithms for those classes of control tasks, it is necessary to have a precise framework for system analysis and control design [9].

The classes of systems mostly studied by control engineers deal with continuous variables and have dynamics which can be described by either differential or difference equations. This is, with manufacturing systems, valid for most of the process level controls (machine tool, robot, motion of AGVs control, etc.). However, higher levels (cell and above) are governed mainly not by differential equations but by the complex interactions of discrete events. Such systems are known as discrete event dynamical systems (DEDS). Besides flexible automation systems, examples of such systems are computer/communication networks, traffic systems, etc.

There is a lack of good dynamically oriented models, analysis and control techniques which can be readily applied to DEDS. Finite-state Markov chains, Petri nets and queueing networks have been used to model some classes. However, neither of those approaches is well suited to flexible manufacturing systems particularly for the purpose of control.

Mathematical models of manufacturing processes must ensure description of both the material and the control flow. Also, problems should be solvable both analytically and by means of simulation. Existing simulation languages of DEDS (e.g. GPSS, SIMSCRIPT) are not suitable for analytical solution of problems. That is why attention of the researchers is attracted to Petri nets. However, a PN is primarily a model of the control algorithm and not the manufacturing process itself [10]. While the modification of Petri nets to coloured Petri net and the first-in-first-out (FIFO) nets are free of this major drawback, they still cannot be used as a coherent framework for flexible automation system analysis and design. Future research should include a concentrated investigation of the modelling problem. For example, recently posed models involving *generalized semi-Markov processes* could prove very useful. We here make an elementary introduction to coloured Petri nets in order to emphasize the importance of the modelling problem in the design stage of flexible automation (Appendix A.9).

It is intuitively clear that control tasks, as briefly outlined above, contain a substantial amount of heuristic logic. Control and coordination of simple manufacturing cell consisting of two machinery components, one robot as a machine loading/unloading device, can be implemented with a programmable logic controller. The number of logic rules is limited and can be programmed by traditional programming tools for PCs (ladder diagrams, block structured languages, BASIC-like industrial languages, etc.). The majority of existing manufacturing cells are actually controlled in such a way. However, more powerful systems require more powerful and systematic methods.

Expert system methodologies provide a systematic approach for dealing with heuristic logic. Expert systems try to model the knowledge and procedures used by a human expert in solving problems within a well-defined domain [11, 12]. Knowledge representation is a key issue in expert systems. Many different approaches have been proposed: first order predicate logic, procedural representations, semantic networks, frames and rule-based expert systems. The last approach is the most important for the problems with which we are concerned. We have implicitly introduced this approach early in this discussion by placing the *knowledge base* box in an FMS example in Fig. 9.2.

9.3.1.1   Control Synthesis

Loosely speaking, the coloured Petri net (Appendix A.9) contains elements of the type set of positions (states), set of transitions, set of tokens. When synthesizing the control, to each position we assign a related technological operation. The sequence of the technological operations (flow of materials and parts through a manufacturing process) is defined by the structure of the Petri net. The process of the synchronized, concurrent execution of the operations is modelled by the tokens passing through the net. Those token movements are mechanized into a flow control program.

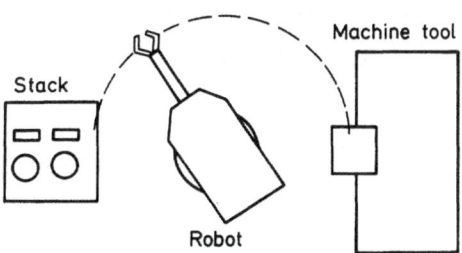

Stack

Machine tool

Robot

Fig. 9.6. An example of flexible manufacturing cell (FMC)

Let us look at one simple example. A flexible manufacturing cell consists of a machine tool, robot and a storage unit – a stack (Fig. 9.6, taken from Ref. [1], Appendix A.9). The stack keeps components c_1 and c_2 and machined parts c_3 and c_4. There are n_1, n_2, n_3 and n_4 of each, respectively. Let us assume that at the beginning $n_1 = n_2 = 2$, $n_3 = n_4 = 0$.

The sequence of operations is as follows: a robot picks a component c_1 or c_2 from the stack, transfers it to the machining location and places it there. The machining is done according to a prescribed program, producing a part c_3 out of component c_1, and part c_4 out of component c_2. The robot then removes the part from the machine tool and returns it to the stacker. The model should take care of the following very obvious constraints: only one component is in the work table at the time; a robot cannot remove a component from a machine tool, until machining is done; a robot cannot remove the next component from the stacker until a previous component is completely machined and returned to the stacker.

The non-primitive coloured net which corresponds to the above example is shown in Fig. 9.7(a). The net is interpreted as follows: position p_1 corresponds to a stack, position p_2 to a machine tool, transition t_1 and t_2 represent transfer of the parts from a stack to a machine tool by a robot; t_3 and t_4 are machining of each component respectively; t_5 and t_6 represent transfer of parts from machine tool to a stack. Figure 9.7(b) presents the normalized coloured Petri net which corresponds to the net shown in Fig. 9.7(a). Positions $d_1, \ldots, d_6$ of the normalized net correspond to the same operations as transitions $t_1, \ldots, t_6$ of the non-primitive net, and transitions $t'_1, \ldots, t'_6$ and $t''_1, \ldots, t''_6$ correspond to the beginning and the end of those operations.

The example is included here only for illustrative purposes. To be more realistic, the model should provide for correctness and should consider the imposed physical constraints. However, even at this level it should help the reader to get some idea how the proposed formal modelling technique may be used in designing the (logical) control system for manufacturing equipment.

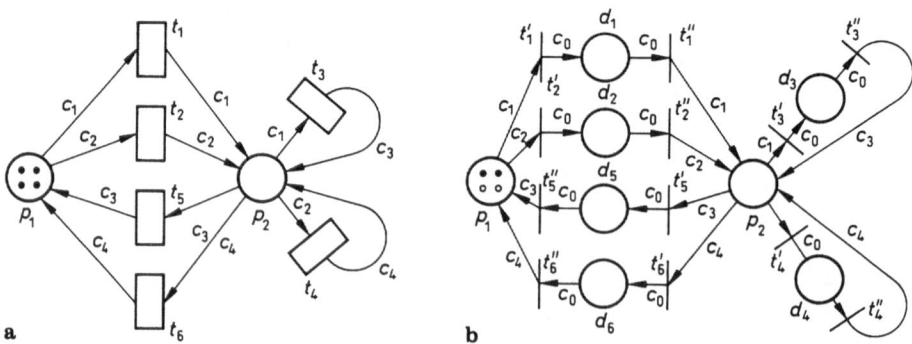

Fig. 9.7 a, b. A model of flexible manufacturing cell from Fig. 9.6

9.3.1.2   Coordination Synthesis

As we have stated in the introduction to this section, the rule-based systems seem to be a most appropriate approach for knowledge representation in flexible automation systems. A typical rule-based expert system has four principal components (Fig. 9.8).

The data base contains facts, evidence hypotheses and goals. Facts are static data (e.g. machining tolerance, working zones, etc.), while the evidence is dynamic data from the process representing the current status of the machine, robot, overall transport system, etc. Hypotheses in an expert system are analogous to the hypotheses a human operator would have observing incomplete and possibly contradictory process data, sensor measurements, etc. The hypotheses are generated and stored in the data base in order to cope with limitations in known facts or measured evidence and are derived based on evidence (data) and other assumptions (facts). Goals can be also static and dynamic. Static goals include various system performance objectives as "find optimal trajectory for the robot n", "find operating point for the manufacturing process x". Dynamic goals include the same type of objectives but established on-line, either by external command or from the control program itself.

Rulebase contains the production rules which are typically described as "if ⟨situation⟩ then ⟨action⟩". The ⟨situation⟩ represents facts, evidences, hypotheses and goals from the data base. The ⟨action⟩ can be to add a new entry to the data base or to modify other entries, but also the ⟨action⟩ is to start a particular operation in the manufacturing process, to activate a controller, to start the current position estimation algorithm for a mobile robot, etc.

We actually define the states of the elements of the system (facts, evidence, hypotheses, goals) as a set of conditions $f_i \in F$, the current state of the overall system $s \in S$ as a set of conditions f_i in a current time instant. Also, we define the actions $a_j \in A$. The rules may be viewed as functions operating on the state. The control strategy consists of the rule r_k of type $r_k: \{ f_{i1}, f_{i2}, \ldots, f_{in} \} \to a_j$, defined for each state of the system $s \in S$. This rule maps the system conditions into a set of actions. The purpose of the *inference engine* is to decide from the context

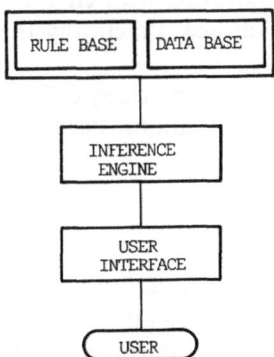

Fig. 9.8. General organization of a rulebased expert system

(current data base of facts, evidence, hypotheses and goals) which rules to select next.

The user interface can be divided into two parts. The first is the development tools (rule editor and rule examiner for development of the system knowledge base). The other part is the runtime user interface. This contains questioning facilities (why a certain operation is executing) and tracing capabilities (how some rule has been executed).

Conditions and actions may be expressions of the type "station $\langle z \rangle$ is free", "process $\langle 1 \rangle$ executes on station $\langle z \rangle$". The example of the rule could be: "*if* station $\langle z \rangle$ is free *then* process $\langle 1 \rangle$ executes on station $\langle z \rangle$". We can find few types of symbols as system states (conditions): symbolic identifiers of technological operations; variables ($x, y, z, \ldots$); procedure names (e.g. procedure for choosing the operation from the set X).

9.3.2   Control of Multirobot Systems

In flexible manufacturing systems, and particularly in assembly lines, the task of cooperative operation of two or more robots is very important. Quite obvious examples are robot handing the part to the other robot, common operation of two robots on the same part (in some manufacturing processes but very often in assembly). The problem is analogous to the (static) obstacles avoidance for a single robot motion.

General control strategies in cooperative operation of two or more robots sharing the same working zone can be viewed as: (1) common zone is excluded from the working envelope of each robot; (2) they share the common zone on the time coordination base; (3) they share the common zone, but their trajectories must be planned to avoid mutual collisions; (4) they share the common region, but the cooperative operation requires their mutual contact. The first two approaches are trivial to implement, but they are usually either technologically unacceptable or too time consuming. The other two approaches are complex and involve extensive computations.

It has been explained in Sect. 4.1 that at the strategic control level trajectory planning can be, in tasks well defined and known in advance, with environmental and operational conditions also known, done off-line. Practically the same reasoning is valid for multirobot tasks.

However, in most assembly tasks the trajectories cannot be precisely defined in advance, and the real time corrections and path modifications, based on sensors information, are necessary.

9.3.2.1   *Off-line Multirobot Trajectory Planning*

Off-line control strategy is generated in two steps: description of the overlapping work zones in the state space; global trajectory planning for all robots.

Firstly, as stated above, the geometric modelling of the working zones of cooperative robots is performed in the space of robots' internal coordinates. The elements of the state space vector are members of each robots' internal co-ordinates vector. To each configuration of multirobot system corresponds a point in the m dimensional state space, $m = \Sigma n_i$, n_i dimension of the i-th robot internal coordinate vector. By the aid of a computer, it is possible to calculate the map in the m dimensional space for the "forbidden" robot configurations.

Secondly, global trajectory planning is performed. It is practically always formulated as the optimization problem of the motion in the space with obstacles (e.g. to find a shortest path, given mutual constraints upon internal coordinates of all robots, derived at the first step).

9.3.2.2   On-line Multirobot Trajectory Corrections

The on-line multirobot trajectory control is far more difficult than the previous one because of quite obvious reasons. The multirobot task changes dynamically,

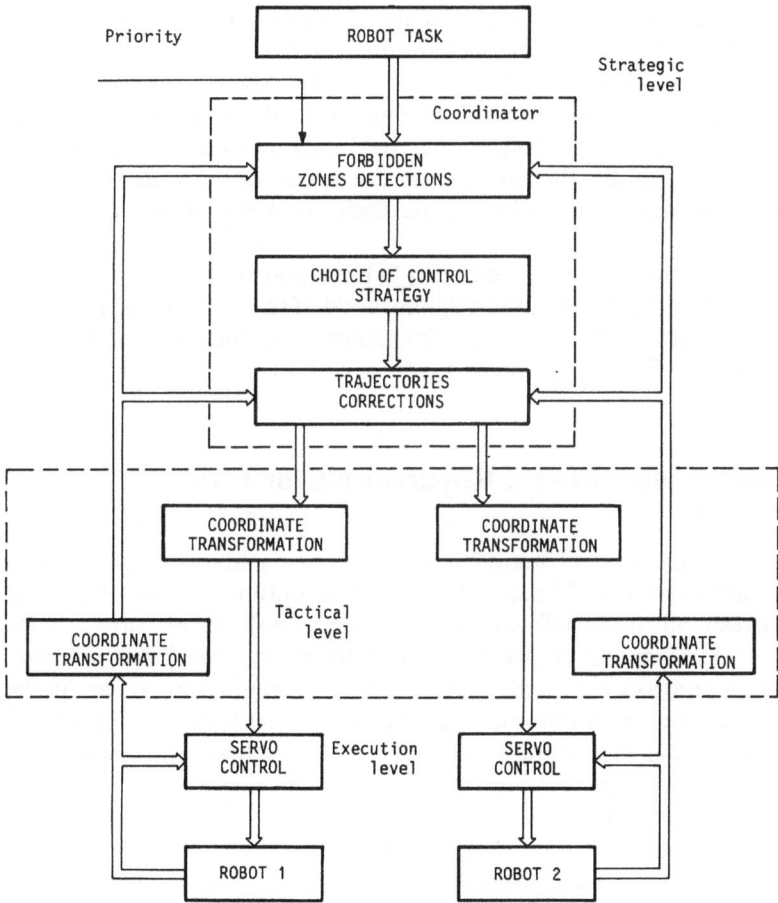

Fig. 9.9. Multirobot hierarchical control structure

that is, the motions are in the presence of unknown obstacles, with sensors inaccuracies and robots' mutual actions not known in advance, etc. The other problem is the restricted time available to detect the obstacles and mutual relationships, to resolve possible conflict situations, to calculate the compensations of the robots' trajectories or replan the whole task, and, in parallel, execute previously prepared segments of the trajectories.

Several techniques are posed in order to resolve these problems.

One is based on the hierarchical control structure proposed in Fig. 4.1 (p. 121) and references [13] and [14]. The analogous control scheme for the control of two robots is presented in Fig. 9.9. The main element of the control structure is the "coordinator" belonging to the strategic level. Let us suppose that one robot has a priority over the other one. The "coordinator" calculates the trajectory of the "slave" robot not to be in the way of the "master" robot. If there was no predicted collision between two robots, they perform the desired, previously calculated motion. If there was the possibility of collision, the correction of the robot trajectories is performed (at the "coordinator").

The corrections may be calculated based on the expert system approach. There is the stored map of the "forbidden" zones, formalized and stored in data base as facts (Sect. 9.2.1). Conditions are derived based on facts and on evidence. The inference engine selects the rule r_k from the rule base, knowing conditions (facts and evidence) and, by the selected rule, decide on the ⟨action⟩–correction of the trajectories.

There are the other possible methods [14]: method of the gradient of the "force" field potential function in the neighborhood of the "forbidden" zone and the geometric definition of the overlapping zones using hidden lines removal algorithms from the field of computer graphics, and the others.

9.4   Hardware and Software Requirements for FAS

The hierarchical structure of manufacturing systems and respective control components are shown in Figs. 9.1 and 9.2. The outlined levels of control generally correspond to the software organization as well. As we have already pointed out, local computer controllers exist at the process level, but generally those computer controls are configured to provide a manual interface. At the sub-system level (cell or shop) very few software packages are capable of coherently handling all control tasks available. At the highest level (plant) there exist many of the management and system design functions (CAPP systems, CAD/CAM tools, etc.), but they have not been developed in the way appropriate for easy integration into the overall design and plant management system.

Generally, the design and the implementation of flexible automation systems worldwide has produced some specific, customized applications, but no unified and standardized solutions of integration of computer hardware and software elements exist.

A number of observations can be made with respect to the design of CIM systems where essentially the problem is to integrate computer-based plant functions by designing some form of computer network [2].

The functions must inevitably be distributed in the sense that they run on different computer hardware located at different sites within a plant. The functions are performed concurrently with digital information being exchanged between computers. The architecture of such a distributed computer system can be designed to be more reliable than the centralized processing. The expandability, flexibility and modularity are the main features of distributed systems. Also, such architecture can simplify the problems of using existing software products which are often hardware dependent. However, well defined communication protocols must be adopted to accomplish effective and reliable information transfer between computers.

In order to generate standard and portable solutions, it is necessary to define communication protocols, at all levels of organization and control.

Research and examples from practice have shown that for the plant (CIM) and possibly shop level of flexible automation system, computer local area networks (LANs) can be used for the data transfer rate of 10^6-10^{10} b/s and distances of 1–5 km. The requirements for this level are flexible and expandable structure; possibility to connect different computer hardware and different terminals; possibility to communicate with other local or public computer networks; reliability; relatively low cost.

For lower control levels, the following parameters are important: a large number of the elemental controlled components, distributed in functional groups; a data transfer rate of 10^2-10^3 b/s; an interval of control action of 10–1000 ms; a distance between controlled components of 20–200 m.

Before LANs start to gain popularity, in the field of programmable automation standard point-to-point communication was used. The asynchronous and synchronous serial protocols (RS 232C, RS 422) were utilized. Their basic drawbacks are the low transfer rate and small reachable distances.

Various LANs topologies are nowadays considered: bus or ring, with different access (direct, random) and different transmission lines. There is no unique answer which topology, access type or transmission line to use in higher control levels of flexible automation. For hudge CAD/CAM packages, the amount of data exchange is significant, and the information is diverse, the frequent access to the centralized data base and program libraries is required, and data block to be transferred can be quite long. In such organizations the bus topology may have some advantage over ring.

Systems with emphasis on computer planning and production management usually do not require the long data blocks for transfer between component computers. The data processing task can be remarkable, but the data "origins" are limited. For such control structures, ring topology may prove more adequate than the others. In those cases, the data base is also distributed.

At lower control level (CNC machine tools, control of fixed base and mobile robots and other computerized manufacturing units) it is necessary to apply so-called small LANs of simple bus topology. They basically have the same characteristics as ordinary LANs, but are less expensive, simpler to implement and can use traditional asynchronous channels. The data transfer rate is of the order $\langle 100\,000$ b/s, but the distances are relatively small (≈ 100 m).

Such progress and breakthrough of integrated but distributed hardware and software, as well as widespread use of LANs, have led to the definition by the international standards organization (ISO) of a "seven layer model for open systems interconnection". General Motors has also announced the manufacturing automation protocols (MAP) which attempt to specify the industrial working protocol for all seven layers of the ISO model.

9.4.1 An Example of the Information Flow in FAS

Let us consider the example of flexible manufacturing system in Fig. 9.2. We can assume that the bus topology communication network between "local" controllers and the supervisory computer is utilized.

To illustrate the requirements and principles involved, we will consider data transfer between supervisory computer and the robots. The type of information transmitted to a robot controller of the intelligence proposed in Chaps. 4, 5 and 6 are: download of robot task program; the choice of the mode of operation; emergency commands; position offsets in world coordinates for on-line modification of robot path, e.g. in response to the measurement of a part dimension by a vision system (not shown in Fig. 9.2); robot status request, etc.

The type of information flow transmitted from a robot controller to the supervisory computer is: robot task program upload (after teaching-in and manual editing at the process level); error report; robot status report; robot position (joint, world, tool coordinates); sensor image; IO image, etc.

The other example of distributed processing is a data exchange between supervisory computer and the local transport control. Types of data transferred from the supervisor to the local controller are: downloading of AGVs route programs; request for the AGVs sensor image; request for transport system status, etc. Data transferred from the local controller to the supervisory computer are: notification of error; sensor images; status report, etc. We should have in mind that local transport controller communicates with the on-board AGVs controllers remotely, by quite different data exchange protocols.

References

[1] Ranky, P.G., The Design and Operation of FMS, IFS, Bedford, UK, 1983.

[2] Weston, R.H., Sumpter, C.M., Gasloigne, J.D., "Distributed Manufacturing Systems", Robotica, Vol. 4, pp. 15–26, 1986.

[3] Koren, Y., Computer Control of Manufacturing Systems, McGraw-Hill, 1983.

[4] Proceedings V IFAC Symp. on Robotics and Flexible Manufacturing Systems, Susdal, USSR, 1986.

[5] Pusztai, J., Sava, M., Computer Numerical Control, Prentice-Hall, 1983.

[6] Miller, A., Automatically Guided Vehicle Systems, IFS, Bedford, UK, 1983.

[7] Premi, S.K., Besant, C.B., A review of various vehicle guidance techniques that can be used by mobile robots or AGVs; Imperial College Internal Report, London, UK, 1984.

[8] Morris, H., "Profitable Robotic Work Cell Result From Interconnecting the Islands of Automation", Control Eng., May 1985.

[9] Challenges to Control – A Collective View, Report of the Workshop, Santa Clara University, Sept. 18–19, 1986.

[10] Budnikov, Y.V., Rudnev, V.V., Tal', A.A., "FIFO Nets and Modelling of Manufacturing Processes in FMS", Proc. V IFAC Symp. on Robotics and Flex. Man. Systems, Susdal, USSR, 1986.

[11] Buchanan, B.G., Shortliffe, E.H., Rule-based Expert Systems, Addison-Wesley, 1985.

[12] Astrom, K.J., Anton, J.J., Arzen, K.E., "Expert Control", Automatica, Vol. 22, No. 3, pp. 277–286, 1986.

[13] Vukobratović, M., Stokić, D., "A Model-based Expert System for Strategical Control Level of Manipulation Robots", Proc. of the 5th ROMANSY, Cracow, 1986.

[14] Anon, "Control of Robot Complexes and Flexible Automated Manufacturing", in Robots and Flexible Automation Systems, edited by Makarov, I.M., Book 3, Moscow, 1986 (in Russian).

Appendix A.9

Coloured Petri Nets

A.9.1 General Notions on Petri Nets

The *structure of a Petri net* is defined by a four-tuple [1]

$$C = (P, T, I, O) \tag{A.9.1.1}$$

where $P = \{p_i, i = 1, n\}$, $T = \{t_j, j = 1, m\}$, $P \cap T = \Phi$, are finite sets of positions and transitions respectively. Input $I(t_j)$ and output $O(t_j)$ are functions of transitions $t_j \in T$ and they are sets of input, i.e. output position of transition.

Graph of Petri net is a bipartite oriented graph $G = \{V, A\}$ where $V = \{v_1, v_2, \ldots\}$ is a set of nodes and $A = \{a_1, a_2,\}$ is a set of arcs $(a_j = (v_j, v_k))$, v_j, $v_k \in V$. Set V can be split into two disjunctive sets, P and T, such that $V = P \cup T$, and $P \cap T = \Phi$.

Marking μ of Petri net C is a function over the set of positions P, $\mu: P \to N$. In other words, the marking μ is the assignment of *tokens* to the positions of Petri net. A number and a position of token can be changed during the *execution* of a net.

A.9.2 Execution Rules of Petri Nets

An execution of a Petri net is controlled by the number and distribution of tokens. The net is executed by firing the transitions.

A transition is fired (if enabled) by removal of input positions and by distribution of those tokens to the output positions of the transitions.

A transition is *enabled* if each of its input positions has at least as many tokens as there are arcs from the position to the transition. In other words, a transition $t_j \in T$ in a marked Petri net C, with a marking μ, is enabled if for each $p_i \in P$

$$\mu(p_i) \geqslant \#(p_i, I(t_j)) \tag{A.9.2.1}$$

An activation of a transition can be carried on as long as there is at least one enabled transition. When there are no such transitions, the execution of the Petri net is aborted.

During execution of Petri net, two sequences result:

1. a sequence of marking $(\mu^0, \mu^1, \ldots)$,
2. a sequence of activated transitions $(t_{j0}, t_{j1}, \ldots)$. These two sequences are related by

$$\delta(\mu^k, t_{jk}) = \mu^{k+1} \ , \quad \text{for} \quad k = 0, 1, \ldots \tag{A.9.2.2}$$

A.9.3   Non-primitive Coloured Petri Net

A formal definition of non-primitive coloured Petri net utilizes terms from the theory of bags [1, 2].

Bag K in the domain A is a set consisting of elements a^γ, where $a \in A, \gamma \geqslant 0$. The function $\gamma = \#(a, K)$ prescribes the number of copies of each element $a \in A$ in the bag K. The set of all such bags for which $a \in K \rightarrow a \in A; \forall a \, \#(a, K) \leqslant m$ is denoted through A^m, and the set of all possible bags in the domain A is denoted through A^∞.

The graph structure of a non-primitive coloured net is defined by a five-tuple:

$$CN = (P, T, C, I, O) \tag{A.9.3.1}$$

where $P = \{p_i, i = 1, n\}$, $T = \{t_j, j = 1, m\}$, $C = \{C_h, h = 1, l\}$ are finite sets of positions, transitions and colours (attributes) respectively. Is and Os are input and output which establish a one-to-one correspondence between each transition of the net $t_j \in T$ and a bag in the domain $P \times C$ (i.e. a bag composed by the pairs $p_i c_h$):

$$\begin{aligned} I &: T \rightarrow (P \times C)^\infty \\ O &: T \rightarrow (P \times C)^\infty \end{aligned} \tag{A.9.3.2}$$

An example of a non-primitive coloured Petri net is shown in Fig. A.9.1. A graph of the non-primitive coloured Petri net is a bipartite oriented graph in

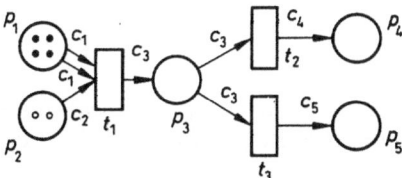

Fig. A.9.1. A non-primitive coloured Petri net

which the positions $p_i \in P$ are shown by circles, non-primitive transitions $t_j \in T$ presented by squares, the arcs go from the positions to transitions and vice versa, and they are marked by the colours $c_h \in C$. Therefore, if bag $I(t_j)$, i.e. $O(t_j)$, incorporates the pair $p_i c_h$ repeated γ times, then the γ arcs, marked with the colour c_h have to be drawn from the position p_i to the transition t_j, i.e. from t_j to p_i.

The graph in Fig. A.9.1 corresponds to the above defined non-primitive coloured Petri net:

$$P = \{p_1, p_2, p_3, p_4, p_5\} , \qquad T = \{t_1, t_2, t_3\}$$

$$C = \{c_1, c_2, c_3, c_4, c_5\} ,$$

$$I(t_1) = \{p_1 c_1, p_1 c_1, p_2 c_2\} , \qquad I(t_2) = \{p_3 c_3\} , \qquad I(t_3) = \{p_3 c_3\}$$

$$O(t_1) = \{p_3 c_3\} , \qquad O(t_2) = \{p_4 c_4\} , \qquad O(t_3) = \{p_5 c_5\}$$

The net positions includes the tokens marked by colours $c_i \in C$ and there may be tokens of different colours in one position, or several copies of tokens of the same colour. The totality of all tokens in all net positions is designated as marking and it is described by the bag M in the domain $P \times C$.

Any transition t_j of the non-primitive coloured Petri net may be fired if all its input positions contain tokens of the same colour by which the arcs going from those positions to the transitions are marked, and the number of those tokens is not less than the multiplicity of the arcs drawn from those positions to transition t_j.

A.9.4   Normalized Coloured Petri Nets

Let $CN = \{P, T, C, I, O\}$ be a above-defined non-primitive coloured net. The net transition $t_j \in T$ is replaced by a pair of transitions t_j' and t_j'' in such a way that all arcs previously entering t_j are brought to the transition t_j', and all the arcs previously drawn from transition t_j leave the transition t_j''.

An additional position d_j (called d-position) is introduced. The unique arc should be drawn into it from the transition t_j', and from which the unique arc should be drawn to the transition t_j''. Both arcs ought to be marked with an additional colour c_0.

Analogous procedure has to be fulfilled for all transitions of the net CN. As a result, the normalized coloured Petri net is obtained [2]:

$$\overline{CN} = (P', T', C', I', O') \tag{A.9.4.1}$$

where $P' = P \cup D$, $D = \{d_j; \ j = 1, \ m\}$, $T' = \{t_j', \ j = 1, \ m\} \cup \{t_j'', \ j = 1, \ m\}$, $\forall t_j \in T \ \exists \ t_j t_j'' \in T'$, $T \cap T' = \Phi$; $C' = C \cup \{c_0\}$; $I'(t_j') = I(t_j)$, $I'(t_j'') = d_j c_0$; $O'(t_j') = d_j c_0$, $O'(t_j'') = O(t_j)$.

The normalized net $\overline{CN}$, which corresponds to the non-primitive coloured Petri net CN (Fig. A.9.1), is shown in Fig. A.9.2. The above procedure of normalization corresponds to the operation of substituting a transition for a sequential block.

The normalized coloured net operates in the following way: its transitions fire instantaneously and they relate to the beginning and the end of an event, modelled by a non-primitive transition. The sequencing of this event is represented by the presence of a token in a d-position.

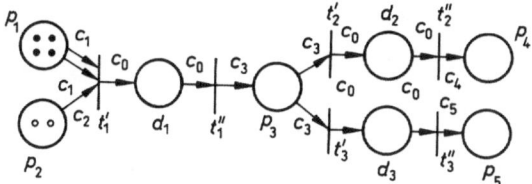

Fig. A.9.2. A normalized coloured Petri net

References

[1] Peterson, J.L., Petri Net Theory – the Modelling of Systems, Prentice Hall, Englewood Cliffs, New Jersey, 1981.
[2] Yuditskiy, S.A., Belousov, O.O., Zaytsev, S. Yu., "A Model of Robotic and Flexible Manufacturing Systems Based on Coloured Petri Nets With Interactions of Transactions", Proc. V IFAC Symp. on Robotics and Flexible Manufacturing Systems, Suzdal, USSR, 1986.

Subject Index